Gynecological Cytology

Hans Friedrich Nauth, MD, FIAC

Head of Cytology Laboratory
Stuttgart, Germany

780 illustrations
18 tables

Thieme
Stuttgart · New York

Library of Congress Cataloging-in-Publication Data
is available from the publisher.

This book is an authorized and revised translation of the German edition published and copyrighted 2002 by Georg Thieme Verlag, Stuttgart, Germany. Title of the German edition: Gynäkologische Zytodiagnostik.

Illustrator: Karin Baum, Mannheim, Germany

Translator: Ursula Vielkind, PhD, CTran, Dundas, Canada

Important note: Medicine is an ever-changing science undergoing continual development. Research and clinical experience are continually expanding our knowledge, in particular our knowledge of proper treatment and drug therapy. Insofar as this book mentions any dosage or application, readers may rest assured that the authors, editors, and publishers have made every effort to ensure that such references are in accordance with **the state of knowledge at the time of production of the book.**

Nevertheless, this does not involve, imply, or express any guarantee or responsibility on the part of the publishers in respect to any dosage instructions and forms of applications stated in the book. **Every user is requested to examine carefully** the manufacturers' leaflets accompanying each drug and to check, if necessary in consultation with a physician or specialist, whether the dosage schedules mentioned therein or the contraindications stated by the manufacturers differ from the statements made in the present book. Such examination is particularly important with drugs that are either rarely used or have been newly released on the market. Every dosage schedule or every form of application used is entirely at the user's own risk and responsibility. The authors and publishers request every user to report to the publishers any discrepancies or inaccuracies noticed. If errors in this work are found after publication, errata will be posted at www.thieme.com on the product description page.

© 2007 Georg Thieme Verlag,
Rüdigerstrasse 14, 70469 Stuttgart, Germany
http://www.thieme.de

Thieme New York, 333 Seventh Avenue,
New York, NY 10001, USA
http://www.thieme.com

Typesetting by primustype hurler GmbH, Notzingen
Printed in Germany by Grammlich, Pliezhausen

ISBN 978-3-13-136081-6 (TPS, Rest of World)
ISBN 978-1-58890-254-2 (TPN, The Americas)

1 2 3 4 5 6

It is important that we recognize what we see.
Far more important is that we understand what we recognize.
But the most important thing is to put into effect what we understand.
The Author

Foreword

With this textbook and atlas, Professor H.F. Nauth has presented an excellent monograph of gynecological cytology. Current knowledge on cytomorphology of the female genital tract is discussed in great detail and critically reviewed. Sound diagrams and over 700 superb color photographs illustrate the text.

Cytological knowledge of pure historic interest has been omitted. Instead, important new information on modern diagnostic procedures has been included. This actualization has created a textbook that excels by its logic, clarity, and relevance for the practice. The text is unambiguously organ-specific, straight-forward, and comprehensible. The work has the further advantage that the didactic presentation of morphological connections is surprisingly easy to understand.

The synoptic approach to cytology and histology is of special importance for the understanding of physiological and, in particular, pathological connections. The concept of including a detailed presentation of supplementary diagnostic methods—such as phase contrast cytology, endoscopy, ultrasonography, DNA cytophotometry, thin-layer cytology, HPV analysis, and automation—is quite new for a textbook of cytology. Furthermore, the clinical/colposcopic correlations are of utmost importance for the gynecologist, in the same way as the cytological/histological comparisons are important for the pathologist. These correlations and comparisons lead to the essential final decision to opt either for a less invasive therapy or for a more radical therapy of female genital carcinomas and their precursors. This approach reflects the author's long-term experience as a practicing gynecologist and clinical cytologist.

Problems of differential diagnostics have been presented and critically analyzed in a special chapter. A special section is dedicated to quality assurance. The nomenclature corresponds to the Munich as well as the Bethesda classification in its most recent modification of 1997. The reference section lists publications relevant for each topic.

With this book, both author and publisher scored a big hit—last but not least due to the high standard of graphic design. *Gynecological Cytology* by H.F. Nauth promises to become an important *up-to-date standard textbook.* We highly recommend this book to cytological laboratory assistants, clinical gynecologists, and pathologists—from the beginner to the advanced.

Prof. M. Hilgarth
Prof. G.L. Wied

Preface

Several considerations gave rise to the idea of introducing a new textbook of gynecological cytology, although numerous compendia and atlases already exist in this specialty. Part of the concept for a new textbook was the idea of a book that was more relevant for medical practice, with due regard to clinical parameters, with a clear subdivision according to the different organs and an adequate presentation of different cytological problems. Nevertheless, this book addresses primarily the practicing cytologist and the cytological laboratory assistant.

I would like to thank Georg Thieme Verlag, Stuttgart, Germany, for supporting this project and Dr. M. Becker, in particular, for suggesting that I write this book. Special thanks are due to Prof. G.L. Wied, Chicago, IL; Prof. M. Hilgarth, Freiburg, Germany; Prof. U. Schenck, Munich, Germany; and Dr. W. Sipli, Stuttgart, Germany, for critical evaluation of the text.

Realizing the idea to make the book more relevant to medical practice was only made possible by using illustrations kindly made available by various colleagues.

The colposcopic images were provided by Prof. M. Hilgarth, Freiburg, Germany (Nos. 2.**2**, 2.**37**, 2.**38**, 2.**40**, 2.**41**, 3.**76**, 3.**96b**, 3.**120**, 3.**140**, 3.**141**, 3.**142**, 3.**165**, 4.**13**, 4.**14**, 4.**16a**, 4.**17**, 4.**18**, 4.**20**, and 4.**73a**) and by Dr. S. Seidl, Hamburg, Germany (Nos. 2.**6**, 2.**35**, 3.**88**, 3.**91**, 3.**96a**, 4.**16b**, and 4.**71a, b**).

Dr. H. Stehle, Stuttgart, Germany, provided the endoscopic images (Nos. 2.B, 2.**64**, 3.**172**, 3.**178**, 4.**84**, and 4.**99**); Dr. W. Dürr, Nürtingen, Germany, provided the sonographic images (Nos. 2.**52**, 2.**53**, 2.**65**, 3.**173**, 3.**179**, 4.**85**, and 4.**100**); and Prof. H. Breinl, Rüsselsheim, Germany, provided the cytological images (Nos. 2.**32**, 2.**33**, 2.**34**, 3.**70**, 3.**77**, 3.**78**, 3.**83**, 3.**85**, and 3.**98**).

The following colleagues kindly made available several histological sections: Dr. P. Fritz, Stuttgart, Germany (Nos. 2.**8**, 2.54, 3.**97**, 3.**107**, 3.**121**, 3.**123**, 3.**124**, 3.**125**, 3.**166**, 3.**168**, 3.**170**, and 3.**175**); Prof. D. Schmidt, Mannheim, Germany (Nos. 3.**124**, 3.**136**, 3.**144**, 3.**146**–3.**148**, 3.**198**, and 4.**24**); and Dr. G. Schumann, Stuttgart, Germany (Nos. 4.**19**, 4.**73b**, and 4.**86**). Two additional cytological images and one histological one were provided by Prof. M. Hilgarth, Freiburg, Germany (No. 3.**37**) and Dr. V. Schneider, Freiburg, Germany (Nos. 2.**63**, 3.**161**). The DNA cytometric histograms (Nos. 7.**4a–c**) were provided by Dr. R. Bollmann, Bonn, Germany.

Dr. G. Schalasta, Stuttgart, Germany, and Mr. P. Schwarzmann, Stuttgart, Germany, assisted me in writing the sections on HPV analysis and automation.

My special thanks go to my coworkers, Mrs. E. Derschka, Mrs. N. Demren, and Mrs. A. Röck, for their considerable help in selecting the photographic material from my collection of cytological micrographs and for word processing, and to Mrs. M. Holzer and Mr. C. Bilharz of Georg Thieme Verlag for editorial and production management.

Hans Friedrich Nauth

Contents

4 Malignant Changes of the Female Genital Tract 165

5 Differential Cytology 255

6 Specimen Collection and Processing 333

7 Analysis and Efficiency 347

Introduction

When it was first introduced, gynecological cytology was limited to the portio vaginalis uteri, the vaginal part of the cervix. Over the years, the neighboring organs have increasingly gained importance, particularly the endocervix, endometrium, ovary, and vulva. The present monograph therefore subdivides the text according to these organs and in the order of their clinical relevance.

There is a close connection between the cytomorphological findings and the results of histopathology, colposcopy, and endoscopy, as well as the clinical and therapeutic consequences. It seemed to make sense, therefore, to discuss the normal anatomy of the organs and their pathological changes from different points of view.

For didactic reasons, a macroscopic or endoscopic description precedes the microscopic description. Each description includes the diagnostic methods used for the particular organ, such as native cytology, colposcopy, hysteroscopy, laparoscopy, or ultrasonography. The histological description and, finally, the cytological description follow. The cytology has occasionally been supplemented by native cytological preparations, because these are indispensable in gynecological practice. To make sure the reader obtains sufficient insight into the clinical consequences of pathological processes, an attempt has also been made to present the particular therapeutic approaches commonly used today.

A short **historical summary** on gynecological cytology precedes the systematic morphological descriptions. This is followed by three chapters on the **normal morphology** as well as **benign** and **malignant changes** of the female genital tract. Another chapter is dedicated to **differential cytology** because this aspect is so important in practice. Finally, there are two organizational chapters on **specimen collection and processing** and on cytological **evaluation and efficiency**.

History of Gynecological Cytology

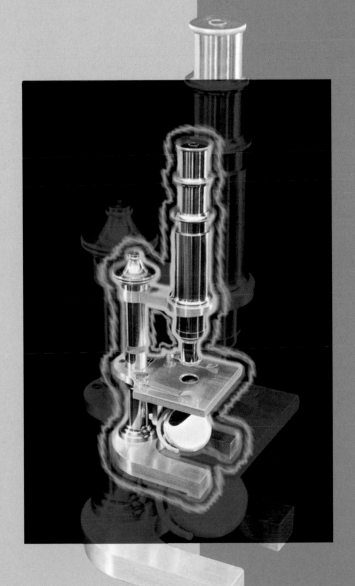

1

1 History of Gynecological Cytology

Founding Period

The microscope was invented by the Dutch lens grinders Hans and Zacharias Janssen in 1600 (158) and made it possible to view objects at a magnification of 60×. This allowed the Dutch natural scientist Jan Swammerdam, some 50 years later, to study insects and erythrocytes (380). The Dutch microscopist Anton Van Leeuwenhoek in Delft achieved a magnification of 275× and, in 1678, was thus able to identify spermatozoa for the first time (196). The Estonian anatomist and embryologist Karl Ernst von Baer, in 1827, was the first to describe the human ovarian follicle (20).

The founder of clinical cytology was Johannes Müller in Berlin, who published a monograph on malignant tumor cells in 1838 (231). At about the same time, Alfred Donné in Paris discovered trichomonads in the vaginal discharge (73). In 1847, Felix Pouchet carried out studies on vaginal cells during the menstrual cycle (289). In the same year, Carl Bruch in Heidelberg published an atlas of tumor cytology (45), and Lionel Beale in London published his atlas of sputum cytology in 1860

(25). In 1855, Rudolf Virchow of Berlin presented his textbook *Zellularpathologie* (Cellular Pathology) (395).

Following these early works, a large number of publications appeared throughout Europe and in the United States, predominantly in the fields of urinary cytology (61), effusion cytology (295), and cytology of the bronchial and gastrointestinal tracts (126, 141). At the same time, the cytology of needle puncture and aspiration (209) as well as hematological cytology developed (86). Apart from isolated observations during the founding period of cytology, gynecological cytology in the narrow sense started with George Papanicolaou in 1943.

Pioneer Period

George N. Papanicolaou (Fig. 1.1) was born in 1883, in Kymi on the Greek island of Euboea, as the son of a general practitioner. He completed his medical studies in Athens in 1904, and received his PhD in Munich in 1910. During his military service he made friends with Greek-

Fig. 1.1 George N. Papanicolaou, 1883–1962.

Fig. 1.2 A. Babes, 1880–1962.

American soldiers, and, in 1913 he decided to emigrate to the United States. He began his anatomical education at Cornell University, in New York State, and studied the behavior of the vaginal epithelium during the hormonal cycle—initially in animals and later in humans (273). For this purpose, smear samples from neighboring clinics were made available to him. In smears of the uterine cervix, he repeatedly observed cancer cells more or less by chance. In order to better recognize and describe these cells, he developed the alcoholic fixation of vaginal smears and also a method of differential staining.

He published his results for the first time in 1928 in a paper entitled "New cancer diagnosis" (272). Strangely enough, the Romanian pathologist Babes (Fig. 1.**2**) published the same observation in the same year in an article entitled "Diagnostic du cancer du col uterin par les frottis," but his methods of specimen collection, processing, and staining were different (17). Initially, both publications received hardly any attention, and the two scientists obviously did not know of each other for a long time. Many years later, in 1943, Papanicolaou, together with the gynecologist Herbert Traut, published the world-famous monograph *Diagnosis of Uterine Cancer by the Vaginal Smear*. This represented the final international breakthrough of gynecological cytology (279).

In the following years, the new method spread exponentially thanks to the initiative of numerous scientists. Papanicolaou retired in 1951, but he continued working in his department for 10 more years. In 1961, he became the director of the Papanicolaou Cancer Research Institute in Miami, Florida, established especially for him and his scientific work. Three months after starting to work there, he received a fatal visit from a former coworker at Cornell, who wanted to show him Babes' 1928 publication (251). Three days after this event, in 1962, Papanicolaou died of a heart attack at the age of 79. Strangely enough, Babes died in the same year at the age of 82. Although Papanicolaou was nominated repeatedly for the Nobel Prize in medicine, he never received it—possibly because he never referred to Babes' publication as he did not know about it.

Consolidation Period

Once the "Pap test" had gained recognition, which took only a few years, the following decades were marked by refinements of the method and by its continuing spread.

The significance of the new method was known in the United States by 1943. Because of the turmoil of the postwar years, the Pap test was introduced in Germany only in 1947, i.e., 4 years after its introduction in the United States. Initially it met with stark rejection by German pathologists (155). Within a very short time, though, the cervical smear gained acceptance there as well—if only in gynecology, for the time being. The method finally fell on fertile ground, especially in Germany, because there was a great impetus to make up for time lost during the war. A number of famous individuals among Papanicolaou's students were of German descent and tried to renew their connections with their home country. After all, Papanicolaou himself had received his PhD in Munich.

This is why, during the founding assembly of the International Academy of Cytology (IAC) in Brussels in 1957, the new association elected a German scientist as the first president. This was H.K. Zinser, who had introduced phase contrast cytology as early as 1949 (423, 425) and was able to report the early results of mass screenings, as did P. Stoll and G.L. Wied (424, 376, 406).

The following years were marked by difficulties in finding the right terminology, especially because the causal relationships between cytological and histological findings were still very much questioned by pathologists. The term **dyskaryosis**, already coined by Papanicolaou and still in use today, should be mentioned here, though its direct correlation with the histological term **dysplasia** (J.W. Reagan) was not yet commonly understood (299, 300). Furthermore, monomorphic and polymorphic cell types were distinguished from each other to facilitate the recognition of a potentially invasive carcinoma (307).

The phenomenon of koilocytosis was first described by Ayre in 1949 (15, 16) and then again by Koss in 1956 (181), although its viral genesis was not demonstrated by electron microscopy until 1987 (179). The quantitative measurement of cells (cytometry) developed during the 1960s and 1970s, thanks to S.F. Patten (282) and others. In Germany, the first chair in clinical cytology was established in 1971, and it was occupied by H.J. Soost.

Gynecological (and also nongynecological) cytology spread mainly through the founding of cytological associations. The German Society of Cytology (DGZ) was founded in Munich in 1960, and the European Federation of Cytological Societies (EFCS) was founded in Prague in 1970 (34, 397). Regular meetings were initially strictly scientific in nature, but increasingly developed into a forum for practical training—in particular, after the IAC (under the guidance of G.L. Wied) successfully introduced tutorials in the United States and Europe (412). The scientific journal *Acta Cytologica*, established in Chicago in 1957, has become the most important journal in the field. Numerous monographs and textbooks have appeared since then; they supplement the practical and theoretical knowledge in cytology.

The Pap test was introduced in Germany in 1971 as a test for the early recognition of cancer; ever since then, it has been free of charge for the patient.

Normal Morphology of the Female Genital Tract

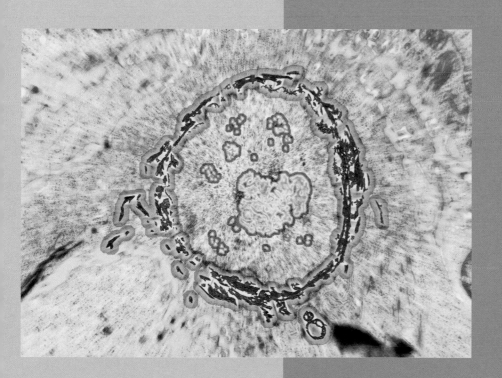

2

2 Normal Morphology of the Female Genital Tract

Portio and Vagina

Macroscopy

The entire vaginal epithelium is visualized macroscopically by spreading the elastic vaginal walls with a speculum.

Types of epithelia. The vaginal epithelium is free of glands and pigment, so it appears pale pink in color. Unlike vulvar epidermis, the epithelium of the vagina does not keratinize. The vaginal epithelium extends from the inner thirds of the labia minora to the portio vaginalis. (The abbreviated terms for parts of the uterus are "portio" or "portio vaginalis" for portio vaginalis uteri and "cervix" for cervix uteri.) In the area of the external cervical os (ostium of uterus), the vaginal epithelium turns into the glandular mucosa of the cervical canal (Fig. 2.**1**). Whereas the transition from the vulvar epidermis to the vaginal epithelium is gradual, the transition from the vaginal epi-

thelium to the cervical mucosa is abrupt. This abrupt transition is characterized by a change in color from pink to red and by a change in surface structure from smooth to papillary. The boundary between the squamous epithelium of the vagina and the columnar epithelium of the endocervix is called the **squamocolumnar junction** (SCJ) and may be located on the ectocervix or inside the endocervix.

Effect of sex hormones. If the effect of sex hormones is prominent—as it is during sexual maturity and particularly during pregnancy—the junction is located on the ectocervix and therefore easy to see. The intensely reddened mucosa is called **ectopia** or **ectropium** (Fig. 2.**2**). It is sensitive to touch and may easily bleed on contact. After menopause, the boundary between vaginal epithelium and mucosa is usually no longer visible because it has withdrawn into the interior part of the cervical canal under the decreasing influence of the sex hormones.

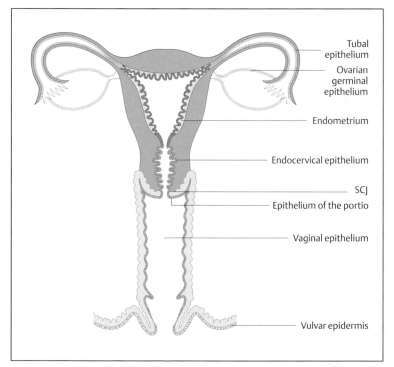

Fig. 2.**1** **Types of epithelia** of the external and internal female genitals. SCJ, squamocolumnar junction.

Tubal epithelium

Ovarian germinal epithelium

Endometrium

Endocervical epithelium

SCJ

Epithelium of the portio

Vaginal epithelium

Vulvar epidermis

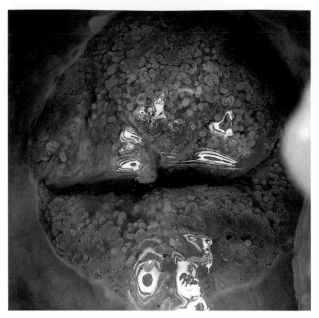

Fig. 2.**2** **Ectopia of the portio vaginalis.** The peripheral smooth, pink squamous epithelium turns abruptly into the red columnar epithelium, which exhibits a villiform surface.

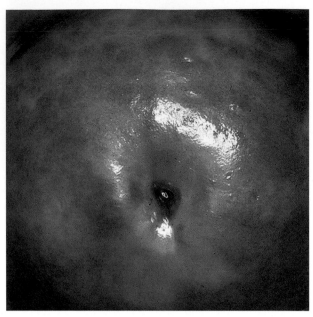

Fig. 2.**3** **Normal portio vaginalis.** The squamous epithelium is smooth and pink. The boundary to the columnar epithelium lies inside the cervical canal and cannot be seen here.

In nulliparous women, the external cervical os has the form of a pit (Fig. 2.**3**), but in parous women it is a transverse slit, thus forming the anterior and posterior **lips of the ostium of uterus** (Fig. 2.**2**).

Structure and function. The portio vaginalis is the cone-shaped lower part of the uterus that extends into the vaginal tube. It serves as an occlusion area for the uterine cavity and is subjected to a complicated opening mechanism during childbirth. The reflections of the anterior and posterior lips of the ostium of uterus toward the vaginal wall are called the anterior and posterior parts of the **vaginal fornix**. The **anterior** vaginal fornix lies posterior to the floor of the urinary bladder, and the **posterior** vaginal fornix lies anterior to the rectum.

Blood flow. The vaginal mucosa consists of an epithelial part and a connective tissue part. The connective tissue is called **stroma** or **dermis**; it supplies the epithelium with blood and nutrients (Fig. 2.**4**). Countless vascularized dermal papillae, a few millimeters apart, extend like cones into the epithelium. The epithelium is thinner over the tips of the papillae than it is between them, and the vascularized papillae appear as a fine reddish punctation when viewed from above at low magnification (Fig. 2.**5**). Under certain conditions (e. g., acute inflammation), there may be an increase in blood flow through the papillae, thus intensifying the pattern of punctation (see Fig. 3.**96 b**, p. 107).

Atrophy of the portio vaginalis. An atrophic portio exhibits signs of regression due to an increase in connective tissue, a decrease in blood flow, and a thinning of the epithelial layer (54). In old age, the size of the portio may be reduced to such an extent that it comes to lie at the same level as the cervical mucosa. The vaginal epithelium is pale pink in color and less reflective than during sexual maturity (Fig. 2.**6**). Contact often causes petechial bleeding, since there is no protective effect of the once highly developed epithelium.

Histology

The nonkeratinized squamous epithelium consists of four layers:
- ► Basal cell layer (stratum basale)
- ► Parabasal cell layer (stratum parabasale)
- ► Intermediate cell layer (stratum spinosum)
- ► Superficial cell layer (stratum superficiale)

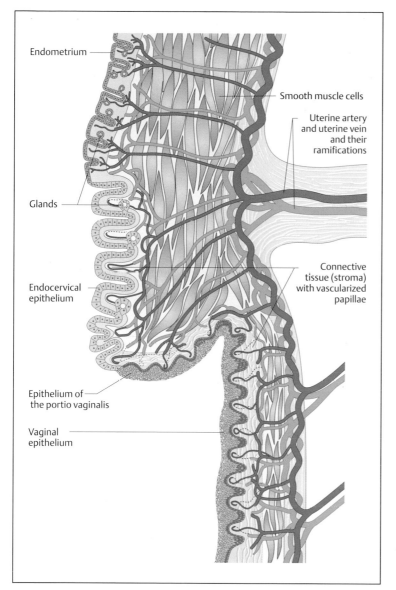

Endometrium

Smooth muscle cells

Uterine artery
and uterine vein
and their
ramifications

Glands

Connective
tissue (stroma)
with vascularized
papillae

Endocervical
epithelium

Epithelium of
the portio vaginalis

Vaginal
epithelium

Fig. 2.**4 The walls of vagina, cervix, and uterus.**
A diagrammatic cross-section showing the vascular
supply to the walls. In the area of the portio and
vagina, the surface of the epithelium is smooth.
Numerous dermal papillae bring blood vessels close
to the epithelium. In the area of the endocervix, the
epithelial surface is enlarged due to the formation
of villi.

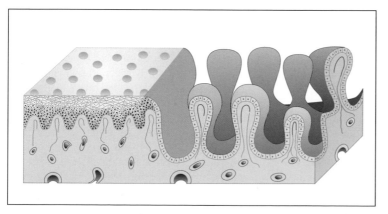

Fig. 2.**5 Epithelia of portio and endocervix.** Dia
gram showing the histological structure of the two
different epithelial surfaces. The squamous epithelium has a smooth surface with a fine, reddish punctation, while the endocervical epithelium has a villiform surface.

Fig. 2.**6** **Atrophy of the portio vaginalis.** Numerous blood vessels shine through the thin epithelial layer. Due to sclerosing processes, the skin is easily injured and tends to hemorrhage (→).

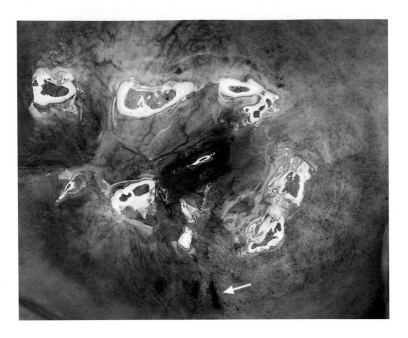

Papillary body. In thin tissue sections, which are normally cut perpendicularly to the surface of the epithelium, the connective tissue part and the epithelial part of the tissue can be clearly differentiated from each other. The blood and lymph vessels run through numerous dermal papillae and bring nutrients to the epithelium. This structure provides a firm anchorage for the epithelium onto the supporting connective tissue. The layer of dermal papillae as a whole is called the **papillary body** and is separated from the epithelium by a basement membrane. This is a semipermeable barrier at which the exchange of metabolites between dermis and epithelium takes place (191). Leukocytes can also penetrate this membrane (345). Immediately on top of the **basement membrane** is the basal cell layer of the epithelium, and the basement membrane is thought to be a product of the basal cells (284). In the case of epithelial atrophy, the papillary body is visibly flattened.

Epithelial structure. The squamous epithelium of the vagina is **multilayered** and—depending on hormonal influences—consists of 5–50 cell layers (390) that do not keratinize at the surface. The vaginal epithelium is therefore a stratified, nonkeratinized squamous epithelium (Fig. 2.**7**).

▶ **Basal cell layer.** The lowest stratum of the epithelium is a single layer of small, cuboidal basal cells, each with a large nucleus that often contains a nucleolus. These cells are able to divide, and they respond to numerous stimuli with increased proliferation (266).

▶ **Parabasal cell layer.** The parabasal cells form about 5–10 layers on top of the basal cell layer; they are slightly larger than basal cells and have a large nucleus that sometimes contains a nucleolus. The cytoplasm stains dark because it is rich in RNA.

Several studies suggest that the actual **renewal of cells** does not originate from the basal cells but from small, deep-seated parabasal cells (333). These are pluripotent cells that produce a stratified squamous epithelium under normal conditions; under certain conditions, however, they may follow other pathways of differentiation. It is thus possible that they develop either into reticular cells, especially histiocytes, or into regenerative epithelial cells, or even into glandular cells (355).

Once these cells have developed into large, more superficial parabasal cells, differentiation into a squamous epithelium is no longer reversible (375, 377). The layers of basal cells and deep-seated parabasal cells represent the germinative center for cell renewal (**stratum germinativum**), which gives rise to all cell layers above. According to their topographic position, these layers are called the parabasal, intermediate, and superficial cell layers.

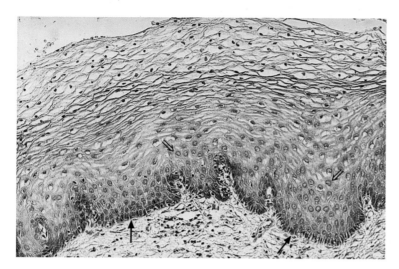

Fig. 2.**7 Normal (highly developed), nonkera-tized stratified squamous epithelium of the vagina.** The basal cell layer, visible as a dark row of nuclei (⟶), rests on the papillary body of the connective tissue. The layers of parabasal cells and intermediate cells that follow are characterized by vesicular nuclei (⟹). The superficial cell layer, which forms the surface of the skin, contains pyknotic nuclei (dark dots). Hematoxylin-eosin (HE) staining, ×100.

Fig. 2.**8 Atrophic vaginal epithelium after the menopause.** The papillary body is markedly flattened, and the epithelium consists of at most 10 cell layers. Above the dark basal cell layer, the epithelium contains almost exclusively parabasal cells. Isolated intermediate cells are visible at the surface (⟶). HE, ×100.

2

▶ **Intermediate cell layer.** With increasing maturity, the cells become larger and produce glycogen, which becomes incorporated into their cytoplasm (85). The intermediate cells formed in this way occupy a space about 15–30 cell layers thick. Because of the increasing pressure, the cells flatten. The cytoplasm is light, and the nucleus is markedly smaller than in the parabasal cell layer.

▶ **Superficial cell layer.** This layer is about 5–10 cell layers thick. Under the influence of estrogen, mucopolysaccharides are formed instead of glycogen (85) and provide the cells with a resistant, keratin-like scaffold. At the same time, the nucleus degenerates and shrinks (pyknosis).

▶ **Process of cell maturation.** The uppermost cell layers may easily peel off (exfoliation) because the intercellular adhesion sites (desmosomes) are progressively reduced during epidermal cell maturation (379). The entire intercellular space contains tissue fluid which enables both the transport of substances and the migration of leukocytes. The superficial and intermediate cells exfoliate into the vaginal cavity. The sugar molecules contained in the glycogen of exfoliated cells are degraded into lactic acid by the Döderlein's bacilli of the vagina (vaginal lactobacilli) (410), and the cells dissolve (**cytolysis**). The entire process of maturation lasts less than a week under normal conditions during sexual maturity.

Fig. 2.**9** **Superficial cells.** Mostly eosinophilic and some basophilic superficial cells. The cytoplasm of these flat, polygonal cells is spread out, the cell margins are well-defined, and the nuclei are pyknotic. ×400.

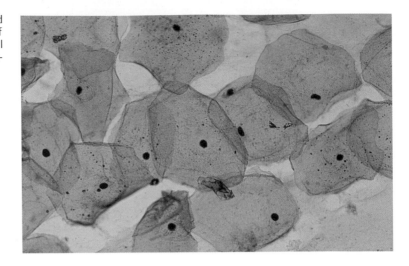

Fig. 2.**10** **Superficial cells.** Wrinkled superficial cells with pyknotic nuclei. The cells are partly eosinophilic, partly basophilic. ×790.

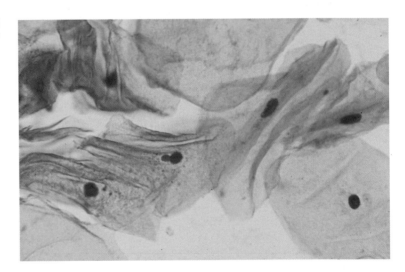

Epithelial atrophy. In the absence of sex hormones after the menopause, the squamous epithelium consists of only a few cell layers (Fig. 2.**8**). A similar situation exists during childhood, during the postpartum period, and in certain endocrinological diseases associated with a greatly reduced production of sex hormones.

Cytology

The gynecological smear shows four types of squamous epithelial cells:
▶ Superficial cells
▶ Intermediate cells
▶ Parabasal cells
▶ Basal cells

In the cytological smear, the cells are no longer associated to form a tissue. Instead, the exfoliated cells lie isolated, flattened, and spread out. Hence, they are viewed from above, which makes them appear larger and more voluminous than when they are viewed in histological sections. Individual cells can be better assessed in this way.

Smears from the female genital tract almost exclusively undergo staining according to the method introduced by Papanicolaou (see p. 339).

Superficial cells. Superficial cells are flat and stain either basophilic or eosinophilic. They are polygonal in shape, with a diameter of 50–60 µm. Their cytoplasm is transparent, homogenous, and has well-defined margins. The cells are more spread out during the first half

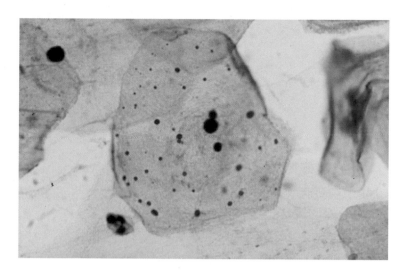

Fig. 2.**11 a–c   Superficial cells.** Each micrograph shows an eosinophilic superficial cell with the flat cytoplasm spread out. ×1000.
a Cell with a pyknotic nucleus and keratohyaline granules embedded in the cytoplasm.

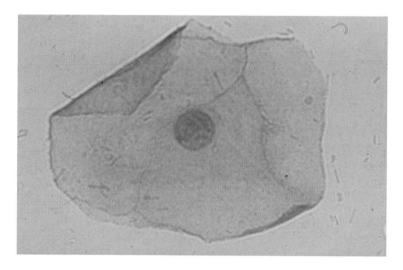

b Cell with a polygonal shape, clearly defined cell margins, and an active, vesicular nucleus.

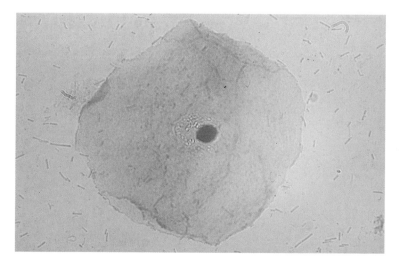

c Cell with a degenerated, pyknotic nucleus.

Fig. 2.**12 Intermediate cells.** Flat, intermediate cells with the cytoplasm spread out and with partly vesicular, partly pyknotic nuclei. ×400.

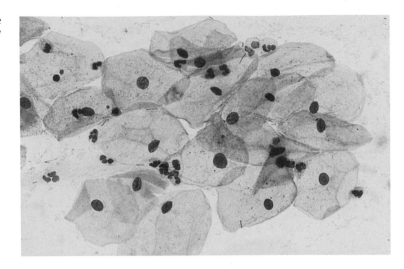

Fig. 2.**13 a, b Intermediate cells.** Each micrograph shows a flat, basophilic intermediate cell with the cytoplasm spread out. ×1000.
a Cell with a polygonal shape, clearly defined cell margins, and an active, vesicular nucleus.

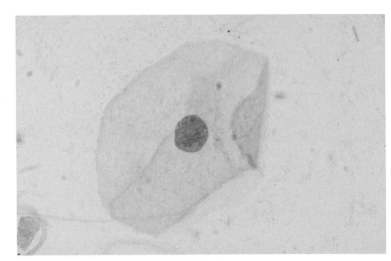

b Cell with a degenerated, pyknotic nucleus.

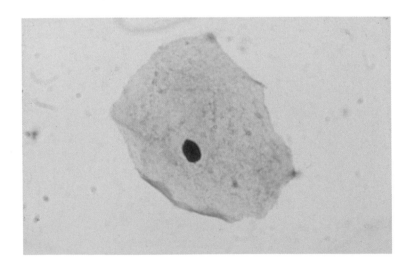

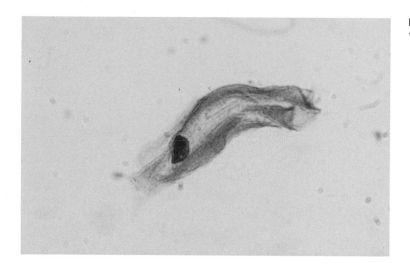

Fig. 2.**14** **Intermediate cell.** This cell has a wrinkled cytoplasm and a pyknotic nucleus. ×1000.

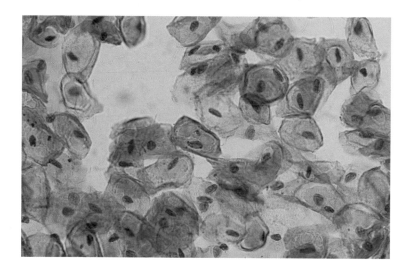

Fig. 2.**15** **Navicular cells.** Boat-shaped intermediate cells with wrinkled cytoplasm and thickened cell margins are called navicular cells. ×400.

of the menstrual cycle and appear more wrinkly during the second half (Figs. 2.**9**, 2.**10**) (88). The cells occur isolated or, especially during the second half of the cycle, in loose clusters where individual cell margins are always well defined. Occasionally, keratohyalin granules are found in the cytoplasm (Fig. 2.**11 a**). The cells have either a large, active, vesicular nucleus of about 7 µm in diameter with easily recognizable, fine, and evenly dispersed chromatin (Fig. 2.**11 b**), or a shrunken pyknotic nucleus with condensed, darkly staining chromatin due to degeneration (Fig. 2.**11 c**). Due to the effect of estrogen, eosinophilic superficial cells with pyknotic nuclei dominate at midcycle, and basophilic superficial cells with active, vesicular nuclei dominate at the beginning and at the end of the cycle (293).

Intermediate cells. These are about 30–50 µm in diameter and have a large nucleus about 8 µm in diameter. They have basophilic staining characteristics and are polygonal or round in shape. Their nuclei are usually active but sometimes also pyknotic (Fig. 2.**12**, 2.**13 a, b**). During the second half of the cycle, the intermediate cells tend to form clusters and have wrinkles like the superficial cells (Fig. 2.**14**). We distinguish between large and small intermediate cells; the large ones already contain glycogen, but the small ones do not. During pregnancy, and under the strong influence of gestagen or androgen, these cells are often boat-shaped, with thickened cell margins. They are therefore called **navicular cells** (271) (Fig. 2.**15**).

Fig. 2.**16a, b   Parabasal cells.** Round cells with basophilic cytoplasm. ×400.
a   Cells with nuclei that are partly active and vesicular, partly degenerated (⟶).

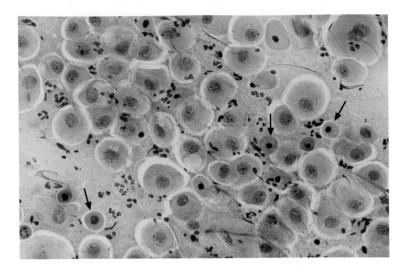

b   Cells with degenerated (condensed) nuclei.

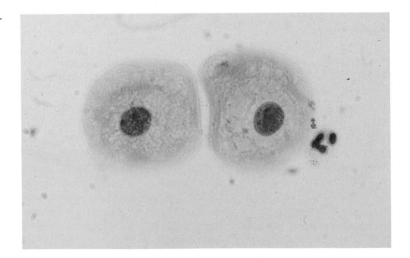

Fig. 2.**17   Two parabasal cells** with rounded cytoplasm and active vesicular nuclei. ×1000.

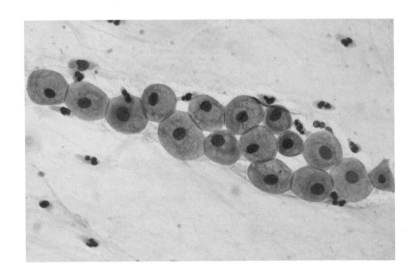

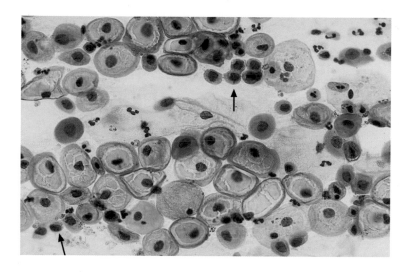

Fig. 2.**18** **Parabasal cells in the postpartum period.** They contain glycogen and show a light endoplasm with thickened cell margins. Some basal cells are also present (→). ×400.

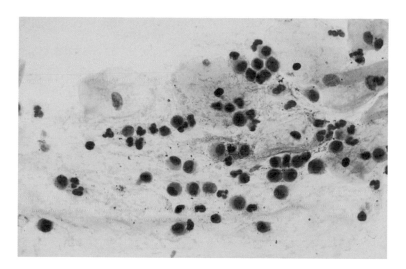

Fig. 2.**19** **Basal cells** with rounded cytoplasm and partly active, partly degenerated nuclei. Some superficial and intermediate cells are also visible. ×400.

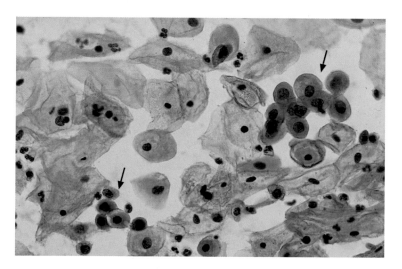

Fig. 2.**20** **Basal cells.** Two clusters of basal cells (→) with rounded cytoplasm and pyknotic nuclei, surrounded by parabasal and intermediate cells. ×400.

2

Fig. 2.**21** **Two basal cells** with sparse, rounded cytoplasm and active, vesicular nuclei that contain nucleoli. One intermediate cell and one superficial cell lie nearby. ×1000.

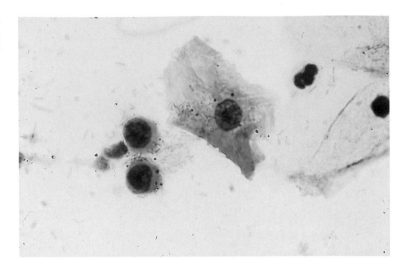

Parabasal cells. Parabasal cells are about 20 µm in diameter and have a nucleus of about 9 µm in diameter. They are strongly basophilic and have an oval to round shape (Fig. 2.**16 a, b**, 2.**17**). There are small and large superficial parabasal cells. After the menopause in childhood, and in the postpartum period, and also in certain endocrine diseases, they are the dominant cell type in the smear. They often show degenerative changes due to their low resistance (81). In the postpartum period, they incorporate glycogen which makes them more resistant; they frequently have thickened, well-defined cell margins and a nucleus that sometimes appears displaced toward the edge of the cell. These cells are called **postpartum cells** (Fig. 2.**18**).

Basal cells. These cells have a diameter of 12–14 µm and a nuclear diameter of 8–10 µm, i. e., the nucleocytoplasmic ratio is clearly shifted toward the nucleus (Fig. 2.**19**). The cells are round and show strong basophilic staining, and their nuclei are often degenerate (Fig. 2.**20**). If the chromatin structure is still intact, a nucleolus may be visible (Fig. 2.**21**). In severe epithelial atrophy, the basal cells exfoliate into the vaginal cavity. They are also regularly found during the reparative phase of inflammatory processes; because the basal cell hyperplasia accompanying the inflammation causes a thickening of the basal epithelial layer, the cells come to the surface (81).

Hormonal Cytology

The nonkeratinized squamous epithelium is primarily the target organ of sex hormones, but other hormones and hormone-like substances also influence the epithelium:

- ▶ Estrogens (estradiol, estriol, estrone), which are follicular hormones
- ▶ Gestagens (progesterone), which are corpus luteum hormones
- ▶ Other hormones (androgens, corticoids, gonadotropins, etc.)

The sex hormones have a decisive effect on the structure and thickness of the vaginal squamous epithelium. The composition and the activity levels of these hormones are determined by the woman's age, possible hormonal therapies or replacement, and also by pregnancy and lactation (385).

Cell indices and **Schmitt staging** are used to determine the degree of cellular proliferation.

Mechanism of action of sex hormones. The stimulating effect of sex hormones on the proliferation of the vaginal squamous epithelium is triggered either by endogenous production or by exogenous supply (Fig. 2.**22**). The type of supply is not important in achieving an effect. Application of hormones may be oral, parenteral, transcutaneous, or topical, although topical intravaginal application has the fastest and strongest effect (210).

The activity of sex hormones is based on their direct binding to a receptor molecule on the cell membrane. The activated receptor is then transported to the nucleus, where the hormone induces DNA synthesis within a few minutes (162, 163). If the hormonal effect continues, it takes about a week for the epithelium to develop from a state of complete atrophy to a fully differentiated epithelial structure (368). Estrogens are considerably more effective than

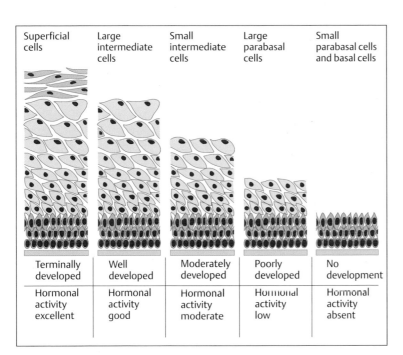

Superficial cells	Large intermediate cells	Small intermediate cells	Large parabasal cells	Small parabasal cells and basal cells
Terminally developed	Well developed	Moderately developed	Poorly developed	No development
Hormonal activity excellent	Hormonal activity good	Hormonal activity moderate	Hormonal activity low	Hormonal activity absent

Fig. 2.**22** **Vaginal epithelium.** Diagram showing the changes in histological structure depending on the activity of sex hormones.

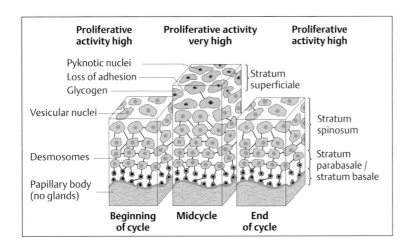

Fig. 2.**23** **Vaginal epithelium.** Diagram showing the changes in histological structure depending on the phase of the menstrual cycle.

gestagens or androgens; the latter induce only a moderate level of proliferation of the squamous epithelium when given on their own (419).

Age-dependent differences. In a sexually mature woman, the vaginal epithelium is normally highly developed, containing not only the basal and parabasal cell layers but also the intermediate and superficial cell layers (400, 401). In childhood, during lactation, and after the menopause, the epithelium develops only as far as producing a parabasal cell layer, since stimulation by sex hormones is absent during these periods of life.

Phases of the menstrual cycle. During the normal cycle, the thickness of the epithelium differs depending on the hormonal situation (Fig. 2.**23**). At the beginning and end of the cycle, epithelial development proceeds only to the upper intermediate cell layer, whereas at midcycle, when estrogen activity is at its peak, the epithelium is fully developed and includes a superficial cell layer. The phases of the ovarian cycle (follicular phase, luteal phase, and menstrual phase) are reflected by the cells present in the vaginal smear (297).

▶ During the **follicular phase**, which lasts from about day 5 to day 15 of the cycle, the cells are predominantly isolated, flat, and

Fig. 2.**24 Follicular phase.** Flat, spread-out basophilic and eosinophilic superficial cells with partly vesicular, partly pyknotic nuclei. ×400.

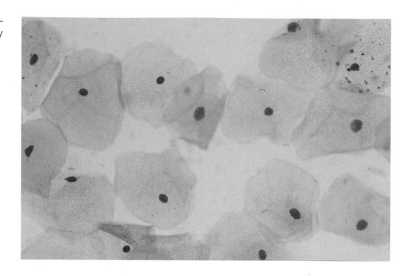

Fig. 2.**25 Ovulation phase.** Flat, spread-out eosinophilic superficial cells with pyknotic nuclei. ×400.

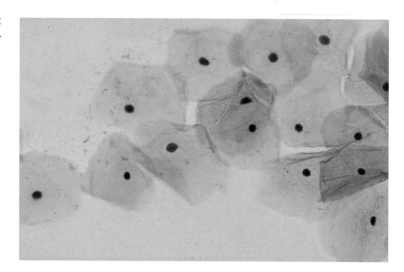

Fig. 2.**26 Fern-leaf phenomenon.** Crystallization pattern of the cervical mucus during the ovulation phase. ×125.

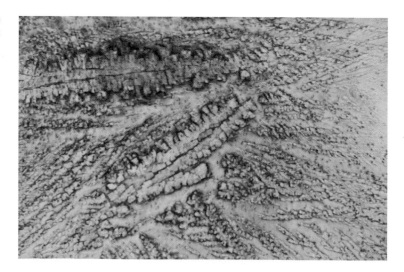

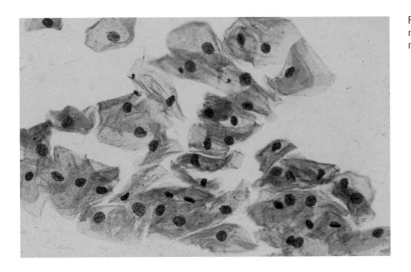

Fig. 2.**27** **Luteal phase.** Clusters of wrinkled intermediate cells with partly vesicular, partly pyknotic nuclei. ×400.

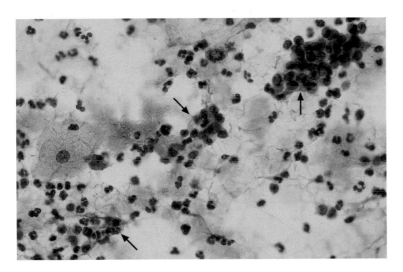

Fig. 2.**28** **Menstrual phase.** The background of the preparation suggests inflammation. Endometrial cells (→) and leukocytes. ×400.

spread out (Fig. 2.**24**). The number of eosinophilic superficial cells with nuclear pyknosis peaks on the day of ovulation, and cytolysis stops (Fig. 2.**25**). At this point, the accompanying cell image is usually very clean and free of leukocytes (291). At the same time, the cervical mucus liquefies; clinically, it shows an increase in threadability and a special crystallization pattern, called fern-leaf artifact, in native preparations and occasionally also in stained preparations (Fig. 2.**26**) (274).

▶ During the **luteal phase** (days 15–28), when the cells are under the influence of progesterone, they often lie in clusters, and wrinkles or rolled-in margins can be seen in the cytoplasm (Fig. 2.**27**) (298). At the same time, the number of cyanophilic intermediate cells with active, vesicular nuclei rises

again, indicating cytolysis. The accompanying cells include increasing numbers of leukocytes and histiocytes.

▶ Beginning with the **menstrual phase** (days 1–5), huge numbers of erythrocytes as well as endometrial and stromal cells are also present. Hence, the background of the preparation appears dirty and inflammatory (Fig. 2.**28**).

Cell indices. To determine the effects of estrogen and gestagen on the squamous epithelium, cell indices are used to report the percentage of mature or wrinkled superficial cells (320).

▶ The karyopyknosis index gives the percentage of pyknotic nuclei of squamous epithelial cells.

▶ The eosinophilia index gives the percentage of superficial cells that are stained red (215).

▶ The wrinkle index gives the percentage of wrinkly superficial and intermediate cells (408).

The eosinophilia index is about 20% at the beginning and the end of the cycle and reaches 50% at midcycle. The karyopyknosis index is always about 10% higher than the eosinophilia index, and follows a similar cyclic course. The curves of both indices correspond to estradiol concentrations in the serum—the only difference being that the hormone peaks 1–2 days before the indices reach their maxima (260). The squamous epithelium requires this latency period in order to respond to the hormonal changes.

The influence of ovulation inhibitors. The spontaneous course of the cycle may show variations, depending on the length of the cycle and whether or not ovulation has occurred. The course is also altered by intake of ovulation inhibitors, since the production of sex hormones in the ovary is artificially suppressed (199, 280). By means of a feedback mechanism, the administration of hormones leads to the inhibition of gonadotropin release from the hypophysis, since the production of follicle-stimulating hormone (FSH) and luteinizing hormone (LH) is slowed down. Both these hormones are responsible for the normal spontaneous course of the cycle (370). The dosage and type of application of ovulation inhibitors differ from preparation to preparation:

▶ For **sequential preparations**, the normal cycle is imitated by administering the hormones at different times. However, the peak values of cell indices lie clearly below those of the spontaneous cycle (279).
▶ For **combination preparations**, the same dose of hormones is administered throughout the cycle, and cell indices remain at a constant low level of 20–30% (202).
▶ For continuous administration of low doses of gestagens, in the form of the **minipill**, the course of the cycle remains largely undisturbed because ovulation is not inhibited (228).

Hormone replacement therapy. Nowadays sex hormones are prescribed more and more frequently, not only as ovulation inhibitors during sexual maturity but increasingly also after the menopause because hormone replacement therapy has many advantages. Bone density, skin turgor, lipid metabolism, vaginal flora, and menopausal deficiency symptoms, such as hot flushes and migraines, are favorably influenced by the replacement of sex hormones (194). Replacement of hormones influences vaginal cell proliferation in the same way as the administration of ovulation inhibitors. The levels of the cell indices are 20–30% (133).

Practical significance of the cell indices. Measuring the level of sex hormone activity in the serum, as it is commonly done today, is far more exact than the calculation of cell indices. For this reason, these indices are now rarely calculated. Cell indices provide little information about the absolute hormone activity; rather, they indicate the relative ratio of one sex hormone to another (259). Furthermore, more reliable clinical methods are now available for the timing of ovulation, particularly ultrasonography (123). However, a rough estimate of the degree of epithelial development by means of Schmitt's classification is still used today in routine diagnostics. It is a simple and inexpensive way of providing the clinician with information about the approximate levels of sex hormones (340).

Schmitt's original classification. In this classification Arabic numerals are assigned to the various stages of maturation of the squamous epithelium. The various stages correspond to the following cell types:
▶ Stage 4: superficial cells
▶ Stage 3: large intermediate cells
▶ Stage 2: small intermediate cells
▶ Stage 1: parabasal cells

The quantitative presence of the four cell types is estimated subjectively and recorded as a maturation score. Depending on which cell types predominate, this yields a score of 4–3 or 3–4 for a **high degree of development**, 3–2 or 2–3 for a **moderate degree**, and 1–2 or 2–1 for a **low degree**.

Modified Schmitt classification. One disadvantage of the original classification is the fact that it does not take into account eosinophilic maturation of superficial cells and the detection of basal cells. Furthermore, the visual assessment of large and small intermediate cells seems to be far more difficult than the distinction between eosinophilic and basophilic superficial cells. The modified classification attempts to address these problems (Fig. 2.**29**). Thus, the various stages now correspond to the following cell types:

- ▶ Stage 4: eosinophilic superficial cells
- ▶ Stage 3: basophilic superficial cells
- ▶ Stage 2: intermediate cells
- ▶ Stage 1: parabasal cells
- ▶ Stage 0: basal cells

High degree of epitheliae development. If the Schmitt score is 4–3, one may assume that the activity of sex hormones is adequate for a sexually mature woman. Such a high degree of development is also detected during the first week after birth, as a result of the residual effects of maternal hormones (361).

During pregnancy, on the other hand, the increased level of progesterone does not permit complete maturation of the epithelium, despite the high level of estrogen production at the same time. As a result, intermediate cells—namely, **navicular cells**—predominate, thus corresponding to stage 2 according to Schmitt (156). In this situation, even artificially supplied estrogens cannot achieve complete maturation into superficial cells (292). If the pregnancy is disturbed, the progesterone level falls and the effect of estrogen predominates; as a result, superficial cells may appear in the smear (30).

Low degree of epithelial development. If the fetus dies, the mother's levels of sex hormones drop, and parabasal cells predominate in the smear (319).

In childhood and during postpartum lactation, hormonal deficiency results in the appearance of **atrophic cells**, corresponding to Schmitt stages 0–1. Upon intake of certain ovu-

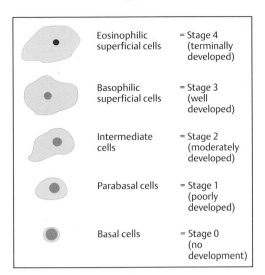

Fig. 2.**29** **Vaginal epithelium**. Diagram showing the changes in cytological structure of the vaginal epithelium depending on the effect of sex hormones (modified Schmitt classification).

lation inhibitors at low dosage (e. g., the micropill), but more often because of inflammatory processes, atrophic cells are observed temporarily (Fig. 2.**30**) (346). Under physiological conditions, however, the typical image of atrophic cells is seen only when ovarian function ceases at menopause. If the ovarian function is eliminated by sterilization, the same effect is reached prematurely. If ovarian function is artificially maintained by means of estrogen replacement, this effect can be reversed.

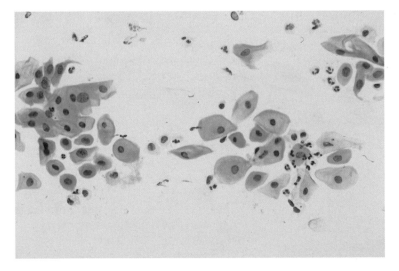

Fig. 2.**30** **Atrophic cells.** Effect of taking low doses of ovulation inhibitors. Parabasal cells and basal cells predominate. ×250.

Fig. 2.**31** **Androgen-induced cell image.** Parabasal cells and intermediate cells predominate. ×400.

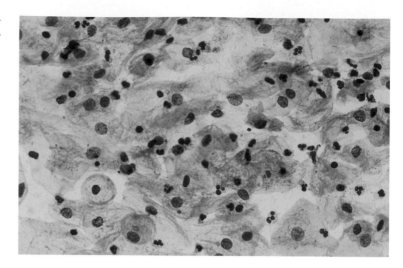

Mixed degrees of epithelial development. In about 10% of all postmenopausal women who do not receive estrogen replacement, mixed cell images (Schmitt stages 0–4, or even cells with a high degree of development) are observed despite the absence of ovarian function (214). This phenomenon is explained by the presence of other endogenous, nonovarian sources of estrogen, particularly the adrenal cortex and fatty tissue (19).

Other influences. A moderate degree of epithelial development, corresponding to Schmitt stage 2, is attained in many endocrinological disorders and also under the influence of various exogenous hormones, such as antiestrogens, gestagens, androgens (Fig. 2.**31**), or gonadotropins (133). In each individual case, the degree of development depends on the interaction of endogenous hormones with the substances supplied exogenously. Certain steroid hormones, digitalis preparations, vitamins, histamines, and other chemicals may also cause a moderate degree of development of the vaginal epithelium (249).

Relevance for cancer prevention. Determining the degree of development of the vaginal squamous epithelium is very important for early cytological detection of cancer, since atrophy makes it difficult to detect carcinomas and their precancerous stages (368).

Phase Contrast Cytology

Phase contrast microscopy is not important for early detection of cancer. Nevertheless, it is a valuable quick method in gynecological practice, where it serves the following purposes:
▶ Detection of pathogens
▶ Estimation of the degree of epithelial development
▶ Identification of spermatozoa

Principle. Phase contrast microscopy is used in gynecological practice to assess the unstained smear in the native state (native cytology). The phase contrast microscope differs from the more commonly used bright-field microscope by the insertion of an annular aperture and a phase plate into the projection of the optic beam, thus causing a phase shift of the light (161). In this way, transparent objects show up dark on a bright background, surrounded by reflective halos of light (378).

Collection of cells. The cells are collected by taking discharge from the vaginal region, dissolving it in physiological saline on a microscope slide, and mounting it with a coverslip. This rapid procedure permits a quick survey of the level of proliferation of the squamous epithelium, and its microbiology, without having to wait for the evaluation of the stained smear (40).

Morphology. For the detection of fungi, trichomonads, and spermatozoa, phase contrast microscopy is superior to viewing stained smears because these microorganisms show a

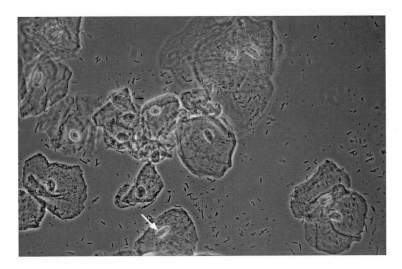

Fig. 2.**32 Superficial cells of the vaginal squamous epithelium.** Phase contrast image showing a halo of light around pyknotic nuclei (→). Döderlein bacilli are visible in the background. ×400

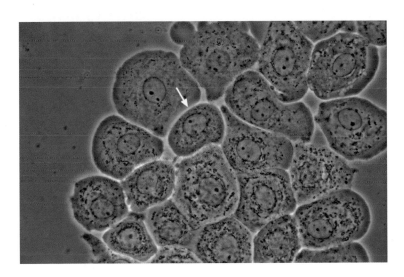

Fig. 2.**33 Parabasal cells of the vaginal squamous epithelium.** Phase contrast image showing halos of light around the cells (→). Chromocenters are clearly visible in the nuclei. ×400

greater contrast in native smears (335). The dark cells of squamous and glandular epithelia are surrounded by bright halos of light, which usually appear the brighter the smaller the cells are. The pyknotic nuclei of superficial cells are also surrounded by halos of light, whereas active nuclei appear uniformly light (Fig. 2.**32**, 2.**33**).

Relevance. The method is not suitable for early detection of cancer because abnormal nuclei show few characteristic features. Except in stained preparations the chromatin appears light and dispersed, and the boundary between nucleus and cytoplasm is often blurred (Fig. 2.**34**) (426). The nucleocytoplasmic ratio

is therefore difficult to assess. Another disadvantage of the method is that the smears cannot be preserved and archived because they dry out within minutes and are then useless.

Phase contrast microscopy is therefore only good for immediate diagnosis, though this is of tremendous help in gynecological practice as it allows one to diagnose and treat acute infections (335). Since the preparations cannot be preserved, documentation of the findings is limited and is only possible through microphotography. Sending native smears to a cytological laboratory is not feasible, because of their transient nature, so the method is left to the practicing gynecologist.

2

Fig. 2.**34 Polymorphic atypical squamous epithelial cells.** Phase contrast image of a smear from a patient with invasive cervical carcinoma. Assessment by native cytology is unsuitable for diagnosis because the cellular and nuclear structures have no characteristic features. ×400

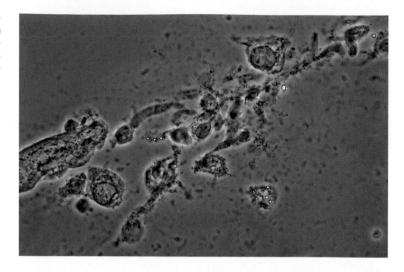

Fig. 2.**35 Erythroplakia of the portio.** Native inspection reveals an undefined redness with blurred peripheral margins.

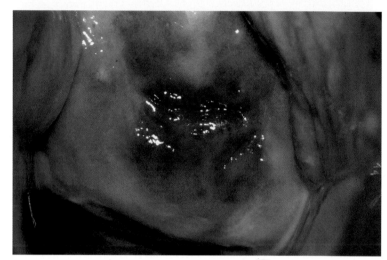

Colposcopy

Colposcopy may be regarded as a supplementary method for the cytological detection of cancer. It serves the following purposes:

▶ Accurate localization of pathological changes and accurate determination of their expansion

▶ Detection of changes that were not detected in the cytological smear

Principle. In gynecological practice, colposcopy allows one to evaluate the surface of the portio as well as the surfaces of the vagina and vulva at low magnification (4–40×). The colposcope consists of a binocular **optical tube** with a central light source positioned in the light path. Because of the focal length of the colposcope, it is possible to view objects from a distance of about 30 cm. Hence, the device does not need to be introduced into the vagina

(148). It is usually attached laterally to the gynecological chair and can be rotated toward the middle. A green filter may be inserted to increase the color contrast.

Inspection of native epithelium. Colposcopic magnification permits a better evaluation of the epithelial surface—particularly of its **color, structure**, and **capillary image**—than is possible using the naked eye. The characteristic pale pink color of the vaginal epithelium results from the quantitative ratio of epithelium and connective tissue and also from the degree of vascularization and blood flow to the stroma (Fig. 2.**35**). The epidermis of the vulva exhibits a characteristic brownish color resulting from the presence of a superficial keratin layer and pigment in the basal cell layer. Numerous factors may lead to changes in the color of the epithelium, resulting in a more reddish, whitish, or brownish tone. The surface is normally flat in the region of the portio, whereas the vagina and vulva show a rippled surface. Many pathological influences alter the height of the epithelium, having either a thickening or a thinning effect. The structural arrangement of such changes in height allows one to draw conclusions regarding the nature of the underlying disease (404). Assessment of the native epithelium is supplemented by two additional examination steps, the acetic acid test and the iodine test.

Acetic acid test. The **obligatory acetic acid test** exploits the property of proteins to coagulate in an acidic environment. In the case of acetic acid, however, this is not coagulation in the narrow sense but rather a kind of swelling, with the exact chemical process being largely unknown. As well as cytoplasmic proteins, it is particularly the nuclear protein of the squamous epithelium that swells temporarily within 30–60 seconds after 2–5% acetic acid is added to the vaginal epithelium. This creates a reversible loss of cellular transparency (149). The affected epithelial area acts as a filter on top of the stroma; while the stroma normally shines through in red because of its vascularization, the epithelium now shows a whitish color (Fig. 2.**36a, b**). The higher the density of nuclei in the squamous epithelium, the more proteins are present and the more distinct is the whitish color. Normal epithelium shows only some whitening, whereas atypical epithelium leads to intense whitening (acetowhite epithelium) because it is rich in nuclear proteins (Fig. 2.**37**).

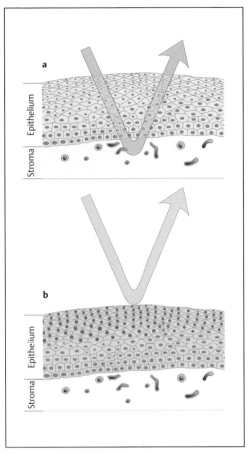

Fig. 2.**36a, b Light path through the epithelium.** During native inspection **(a)**, the light is reflected by the stroma, thus letting the skin appear **reddish.** After adding acetic acid **(b)**, the superficial epithelial cells swell, and the light is reflected by the epithelial surface, thus letting the skin appear **whitish.**

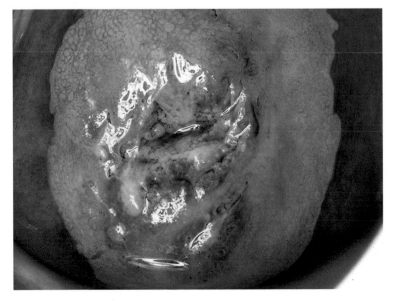

Fig. 2.**37 Acetowhite epithelium.** Demarcated area, in part with mosaic structure, corresponding to a CIN II lesion (the same case as in Fig. 2.**35**).

Fig. 2.**38 Iodine-negative epithelium.** Demarcated area, in part with mosaic structure, corresponding to a CIN II lesion (the same case as in Fig. 2.**35**).

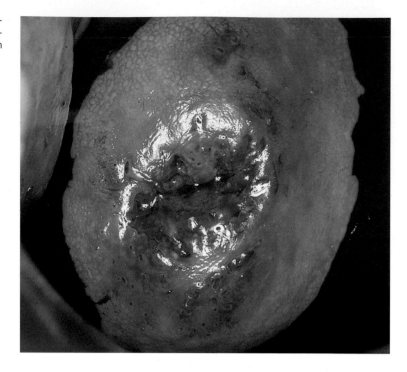

Fig. 2.**39 Variability of the squamocolumnar junction** in the area of the external cervical os. Depending on age and hormones, the boundary between squamous epithelium and columnar epithelium lies inside the cervical canal in the endocervical region (normal finding), or outside in the ectocervical region (ectopia).

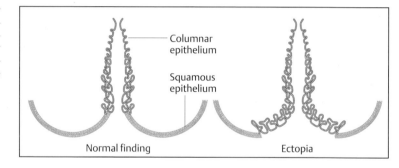

Columnar epithelium

Squamous epithelium

Normal finding   Ectopia

Iodine test. This test, also known as **Schiller's test**, exploits another chemical reaction—the browning of glycogen by iodine. Since the mature layers of the squamous epithelium contain plenty of glycogen, the vaginal skin turns evenly brown if there is adequate estrogen activity (337). An abnormal epithelium contains hardly any glycogen and remains light in color when tested with iodine; it is therefore said to be **iodine-negative** (Fig. 2.**38**). If the portio is atrophic, the iodine test also fails to cause browning because there is no glycogen (223).

The iodine test is especially suited for demonstrating the topography of a lesion and its boundaries with the healthy epithelium. This procedure is specially advantageous just before surgery.

Endocervix

Macroscopy and Colposcopy

The visible portion of the cervix of the uterus is called the ectocervix, and the invisible portion is called the endocervix. The endocervical epithelium becomes macroscopically visible as ectopia only if the squamocolumnar junction is localized on the ectocervix.

Endocervical epithelium. This term specifies the endocervical columnar epithelium, which must not be confused with the portion of nonkeratinized squamous epithelium temporarily localized in the endocervix. The endocervical epithelium reaches from the internal cervical os deep in the endocervix to the external cervical os, where it turns into the nonkeratinized

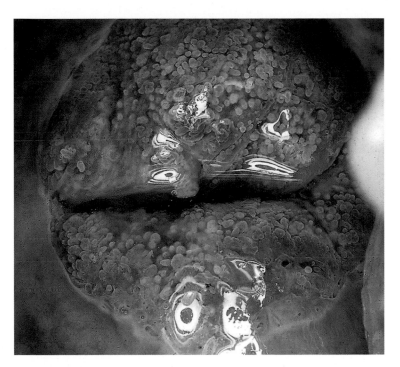

Fig. 2.**40** **Ectopia of the portio.** The peripheral smooth, pink squamous epithelium turns abruptly into the red columnar epithelium, which exhibits a villiform surface. The area shows a grape-like structure.

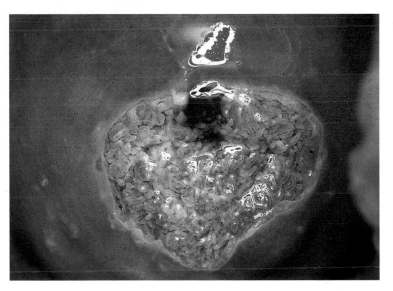

Fig. 2.**41** **Ectopia of the portio after acetic acid test.** The area with grape-like structure shows a slightly whitish reaction of the surface.

squamous epithelium covering the surface of the portio.

Depending on the age of the woman, the boundary between the squamous epithelium and the columnar epithelium may be localized either on the ectocervix or inside the endocervix (Fig. 2.**39**) (264). The ectocervical localization of endocervical epithelium is called **ectopia** or **ectropion** and occurs mainly during sexual maturity (Fig. 2.**40**). The columnar epithelium exhibits a villiform surface structure, presumably to increase the secretory surface as much as possible (264). Unlike the squamous epithelium, the endocervical epithelium consists of a single cell layer, which allows the color of the underlying vascularized connective tissue to show through largely unfiltered. The resulting intensely red color is sharply distinguished from the pink portio. The villi of the endocervical epithelium appear as grape-like structures under the colposcope; they stain whitish after the addition of acetic acid and thus stand out more clearly (Fig. 2.**41**) (223). The iodine test is negative because there is no glycogen in this region.

Fig. 2.**42 Normal squamocolumnar junction of the cervix.** The nonkeratinized stratified squamous epithelium turns abruptly into the simple columnar epithelium, which consists largely of secretory cells. Between the secretory cells, reserve cells are visible as a dark row of basal nuclei (——▶). HE, ×100.

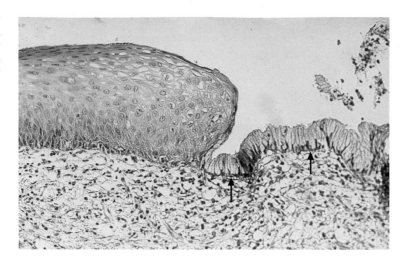

Histology

The endocervical columnar epithelium consists of a single row of glandular cells, the activity of which is hormone-dependent.

Endocervical epithelium. The transition from the squamous epithelium to the endocervical epithelium is abrupt (Fig. 2.**42**). The simple columnar epithelium originates from small **reserve cells**, which lie underneath the columnar cells and are often hardly visible (85). The cohesion between the cells is stronger here than in the upper squamous epithelial layers, and the rate of cell renewal is much slower (75). In contrast to the squamous epithelium, there is therefore very little spontaneous exfoliation of endocervical cells, and there is also no exfoliation during the menstrual phase (261). The endocervical epithelium consists mainly of two types of epithelial cells: mucus-secreting cells and ciliated cells (375). In addition, there are intercalated resting cells (peg cells) and reserve cells.

Cervical mucus. The cervical mucus serves as a barrier against bacteria on the one hand, and as a port of entrance for the spermatozoa into the cervical canal on the other. The activity of the **secretory endocervical cells** is dependent on hormones. Under the influence of progesterone, the secretion is reduced and a viscous mucus obstructs the cervical os (351). Estrogens, on the other hand, cause an increase in mucus production, thus making it more liquid and the cervical canal more penetrable. This process peaks at midcycle, the time of ovulation, so that the spermatozoa can easily

penetrate the mucus. The contraceptive action of the **minipill** is based on the prevention of mucolysis by continuously supplying low doses of gestagens without suppressing ovulation (128). The **ciliated endocervical cells** promote the flow of secretion toward the vagina to combat ascending bacterial infections (143).

The endocervical epithelium not only covers the tissue surface but also forms gland-like cryptae in the cervical wall. The mucus produced here is drained to the outside through pores at the surface.

Cytology

The cytological smear may contain endocervical glandular cells of the following four varieties:
▶ Mucus-secreting cells (secretory cells)
▶ Ciliated cells
▶ Resting cells (peg cells)
▶ Reserve cells

Columnar epithelial cells originate from the endocervix, or—if derived from the ectropion of the portio—from the ectocervix. Unlike squamous epithelial cells, endocervical cells rarely exfoliate spontaneously into the vaginal cavity; they are usually only detached from the tissue during collection of the smear (368). Smears taken from an **ectopia** show an extraordinarily large proportion of endocervical glandular epithelium. If the ectopia is large, the smears may contain only glandular cells and mucus, and the absence of squamous epithelial cells thus makes the smear unsuitable for assessment. In such cases, the clinician should take care to collect also cells from

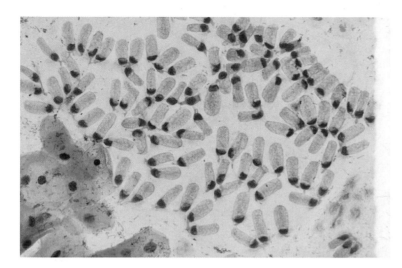

Fig. 2.**43** **Secretory cells from the endocervical columnar epithelium** in diffuse arrangement, with their nuclei located at the cell base. Normal superficial cells are visible in the left lower corner. ×320.

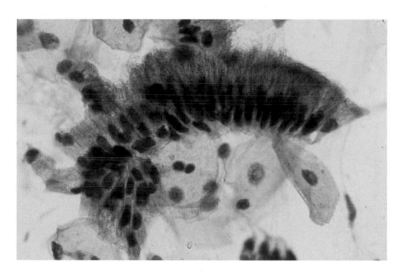

Fig. 2.**44** **Endocervical columnar epithelial cells.** A single-row, palisade-like cluster of cells with basal nuclei. Normal intermediate cells are also present. ×630.

the squamous epithelium located at the periphery of the ectopia. In addition to endocervical cells, the endocervical smear almost always contains granulocytes that serve the local immune defenses.

Mucus-secreting cells. The secretory columnar cells are the predominating cell type. Columnar or peg-shaped when viewed from the side, and cuboidal when viewed from above (143), they are about 16 μm long (Fig. 2.**43**, 2.**44**). The margins of the basophilic cytoplasm are less clearly defined than in the squamous epithelial cells and are very sensitive to environmental influences. As a result, only degenerate endocervical cells may be present. If the smear is properly collected, however, these cells may be well preserved, and their cytoplasm may then exhibit numer-

ous vacuoles. Some cells contain only a single, large secretory vacuole; it occupies most of the cell, with the nucleus being pushed to the base of the cell. These cells are called **goblet cells** (Fig. 2.**45**).

The nuclei are round or oval, sometimes peg-shaped, have a regular chromatin network, and frequently exhibit chromocenters or micronucleoli (360). The position of the nucleus within the cell depends on the phase of the cycle. At the beginning of the cycle, the nucleus is at the base of the cell, at midcycle it lies in the center, and at the end of the cycle it is pushed to the Border by mucus production. When viewed from above, the endocervical cells appear cuboidal with central nuclei, thus giving the cell cluster the appearance of a honeycomb structure (Fig. 2.**46**).

Fig. 2.**45** **Goblet cells.** A single-row, palisade-like cluster of mucus-secreting endocervical columnar epithelial cells, each with a basal nucleus and a single large secretory vacuole. ×630.

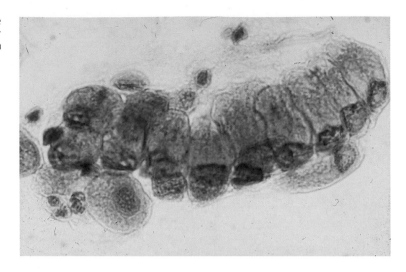

Fig. 2.**46** **Honeycomb structure.** A sheet of mucus-secreting endocervical columnar epithelial cells viewed from above. The arrangement of cells with central nuclei results in a honeycomb structure. ×630.

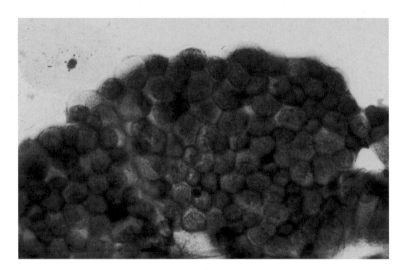

Fig. 2.**47** A cluster of **endocervical columnar epithelial cells.** The cells form an acinus-like structure. ×400.

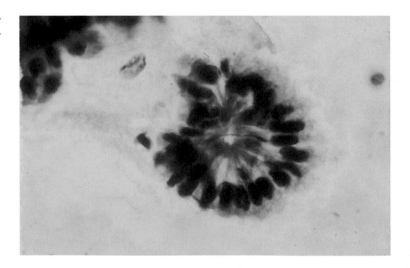

Fig. 2.**48** **Endocervical ciliated cells.** Two clusters of cells in a single-row arrangement, showing active, basal nuclei and distinct borders of eosinophilic cilia at the apex (→). Normal superficial cells lie in the center of the picture. ×630.

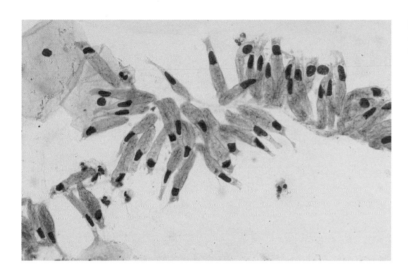

Fig. 2.**49** **Endocervical peg cells.** Loose clusters of cells with cone-shaped cytoplasm and basal nuclei. ×400.

Since the endocervical cells are actively secreting, they often lie within tracts of mucus, usually in loose clusters or palisade-like rows. If exfoliated from endocervical cryptae, the smear may contain an acinus-shaped cell cluster (Fig. 2.**47**).

Ciliated cells. Another type of cells exhibits an abundance of apical **cilia**—hence the term ciliated cells (Fig. 2.**48**). These eosinophilic cilia are anchored in the apical cytoplasm by a dark-staining terminal plate. Occasionally, these cells contain two or even more nuclei, and these may be clearly enlarged.

Peg cells. These cells are rarely detected in the smear (Fig. 2.**49**). They have little cytoplasm and are highly columnar to cone-shaped, with a small, dense nucleus located at the base of the cell. Peg cells evidently represent inactive, resting glandular cells (261).

Reserve cells. These are lined up like pearls at the base of the columnar epithelium and are rarely observed in the smear (375). Sometimes, however, they respond to proliferative stimuli with increased activity and cell division, thus causing **reserve cell hyperplasia**. In this case, they are detected in the cytological smear, but because of their foamy cytoplasm they differ only slightly from small histiocytes and are frequently confused with them (Fig. 2.**50a, b**) (81).

Fig. 2.**50a, b** **Endocervical reserve cells. ×1000.**
a A loose cluster of cells with foamy cytoplasm, undefined cell margins, and active nuclei. Isolated superficial cells lie nearby. ×400.

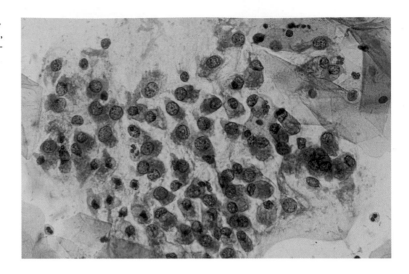

b An elongated cluster of cells with undefined margins of the cytoplasm and partly active, partly degenerate (——➤) nuclei. ×1000.

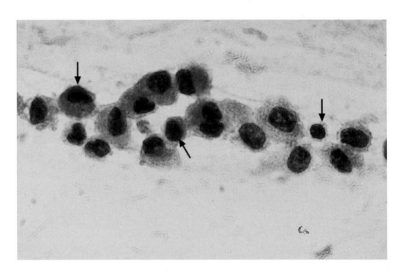

Endometrium

Endoscopy and Ultrasonography

The mucosa of the endometrium can only be visualized by the use of endoscopic methods. Another imaging procedure commonly used today is ultrasonography, particularly vaginal ultrasonography.

Hysteroscopy makes it possible to introduce an optical device into the uterine cavity. The endometrium presents itself as a homogeneously red mucosa, the surface properties and structure of which can now be assessed (Fig. 2.**51**). The endometrium appears uniform in the ultrasonogram as well. It is 5–9 mm thick during the proliferative and secretory phases (Fig. 2.**52a, b**) but only about 1 mm thick when it is atrophic after the menopause (Fig. 2.**53**).

Histology

The endometrium consists of a single row of cuboidal glandular cells; these cover the epithelial surface and also line the walls of the glands. Their growth and secretory activity are hormone-dependent, and so is the metamorphosis of the interstitial connective tissue during pregnancy. We distinguish between:
▶ Proliferative phase
▶ Secretory phase
▶ Menstrual phase
▶ Decidual transformation
▶ Endometrial atrophy

The entire uterine cavity is covered by a simple (single-layered) cuboidal epithelium, which turns into the endocervical columnar epitheli-

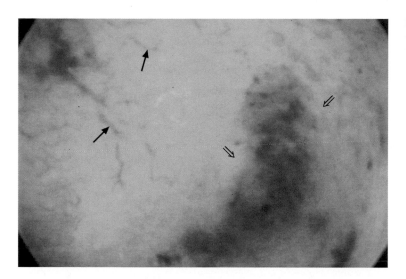

Fig. 2.**51** **Normal hysteroscopic finding.** Smooth endometrium with some delicate vessels (⟶) and circumscribed hemorrhages (⟹).

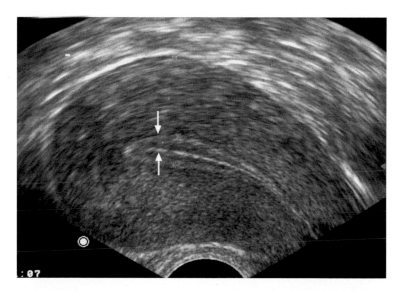

Fig. 2.**52 a, b** **Normal ultrasonographic findings of the uterus.**
a Endometrium in the **proliferative phase** (6.4 mm thick) (⟶).

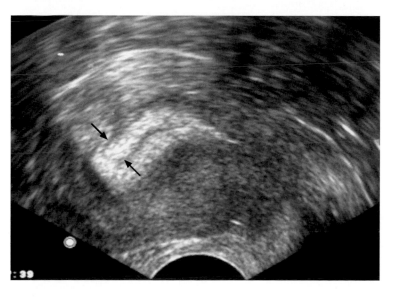

b Endometrium in the **secretory phase** (echo-rich zone of reflection, 8.0 mm thick) (⟶).

2

Fig. 2.**53 Normal ultrasonographic finding of the uterus.** Postmenopausal **atrophic endometrium** (1.0 mm thick) (→).

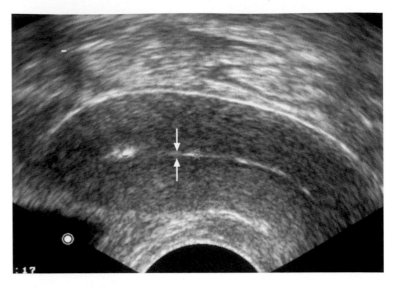

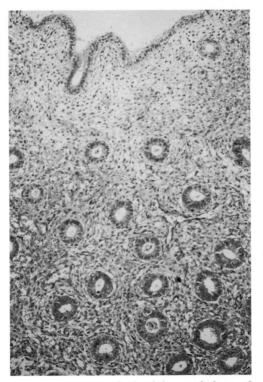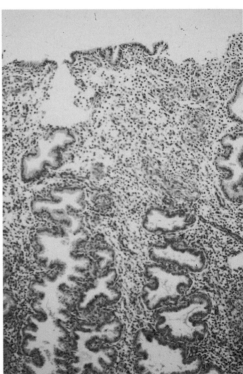

Fig. 2.**54 a, b Normal glandular epithelium of the endometrium**

a Proliferative phase: The glandular lumina are tubular. HE, ×100.

b Secretory phase: The glandular lumina have the characteristic serrated appearance (sawblade pattern). HE, ×100.

um at the internal cervical os. Although ciliated cells may occasionally occur (131), the epithelium consists almost exclusively of secretory cells. It sends many glands deep into the connective tissue of the uterine wall, and the thickness of the epithelium varies with the number and size of these glands, depending on the age of the woman and the phase of the hor-

monal cycle. The gland-containing stroma (consisting of **stratum basalis** and **stratum functionalis**) borders on the myometrium, which is made up of smooth muscle cells.

Proliferative phase. During the first half of the menstrual cycle, there is intense growth of the tubular glands (Fig. 2.**54 a**). The proliferation is caused by estrogens produced in the ovarian follicle.

Secretory phase. Under the influence of progesterone, the corpus luteum—which develops after ovulation—stimulates the secretory activity of the glandular epithelium. As the glands occupy more and more space, their tubular lumina coil intensively and assume a serrated appearance in cross-section (sawblade pattern) (67) (Fig. 2.**54 b**).

Menstrual phase. If conception does not take place, the secretory activity of the glands declines. Leukocytes and erythrocytes penetrate the superficial part of the endometrium, also known as the **pars functionalis**, which is shed at menstruation (**desquamation**).

Decidual transformation. If conception occurs, the fertilized egg implants somewhere in the now well-padded, thickened glandular epithelium. As a result of trophoblast activity, the fertilized egg grows deep into the mucosa and forms the primordial fetus with placenta and embryo. Under the influence of high levels of gestational hormones, stromal cells enlarge and become **decidual cells**; at the same time, there is severe edematous swelling of the interstitial connective tissue.

Endometrial atrophy. After menopause, the absence of sex hormonal stimulation leads to atrophy of the endometrium, which at that time measures less than 1 mm in thickness. However, if hormones are supplied through therapy, the cyclic replacement of hormones by initial intake of estrogen and subsequent intake of gestagen results in regular phases of endometrial proliferation and withdrawal bleeding (135).

Cytology

The cytological smear may contain epithelial endometrial cells, stromal cells from the deeper part of the endometrium, and fetal cells from a pregnancy. We distinguish:
▶ Proliferating endometrial cells
▶ Secretory endometrial cells
▶ Exodus
▶ Stromal cells
▶ Decidual cells
▶ Syncytiotrophoblastic cells
▶ Cytotrophoblastic cells
▶ Cells of the Arias-Stella reaction
▶ Atrophic endometrial cells

The cuboidal endometrial cells are about 10 µm in diameter and thus clearly smaller than the endocervical columnar cells. Their basophilic and chromophobic cytoplasm has undefined margins, which are sensitive to environmental influences; as a result, the smear often contains naked degenerate nuclei (37). Cells taken directly from the endometrium of the uterine cavity are usually better preserved than those following spontaneous exfoliation. Sometimes the cells lie isolated, although most of the time they occur in clusters. Their nuclei are round and show a regular chromatin network.

Proliferating endometrial cells. During the first phase of the cycle, cell clusters are compact and the cells possess relatively little cytoplasm; these are proliferating endometrial cells (Fig. 2.**55 a**). Later in the proliferative phase, the nuclei may show nucleoli (3).

Secretory endometrial cells. During the second phase of the cycle, the cells increase in volume and are more loosely associated; these are then secretory endometrial cells (Fig. 2.**55 b**). Their nuclei may exhibit chromocenters at this time.

During the first half of the cycle, the endometrial cells may still exfoliate physiologically, whereas during the second half of the cycle this usually happens only in the case of a hormonal disorder (178).

Exodus. During the menstrual phase, masses of endometrial cells exfoliate—either as individual cells or as cell clusters—and the simultaneous presence of detritus gives the smear a dirty, inflammatory appearance (see Fig. 2.**28**). Since the occurrence of endometrial cells in the smear is correlated with a rejection reaction within the uterine cavity, the cell clusters are often infiltrated by leukocytes, and other nonepithelial elements may be detected nearby, such as granulocytes, histiocytes, erythrocytes, and protein precipitation (60).

Endometrial cell clusters exfoliating at the end of menstruation or at midcycle are known as **exodus**. They often form large spheres, characterized by a double contour pattern of external glandular epithelium with internal stromal cells, and often show an outer line of retraction (Fig. 2.**56**) (277).

Stromal cells. Occasionally, endometrial stromal cells also exfoliate. The more superficial ones are mostly round, and the deeper ones are

2

Fig. 2.**55 a, b   Endometrial cells.** ×400.
a   Proliferative phase: A loose cluster of cells with sparse cytoplasm and active nuclei.

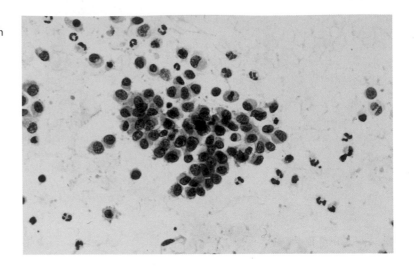

b   Secretory phase: A loose cluster of cells with an abundance of cytoplasm and active nuclei.

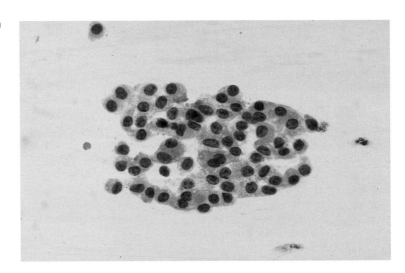

Fig. 2.**56   Exodus pattern.** A large, spherical cluster of endometrial cells exhibiting a dense central core of stromal cells, an external border of epithelial cells, and an outer line of retraction (→). This structure is called "exodus" (mass departure). ×400.

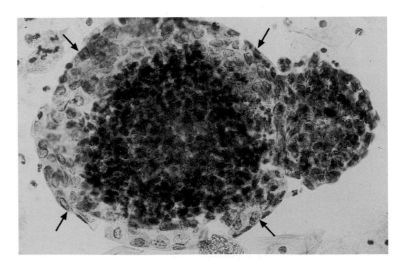

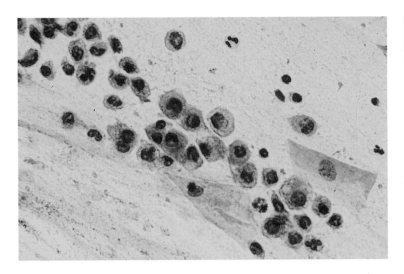

Fig. 2.**57** **Superficial endometrial stromal cells.** A loose cluster of cells with rounded, foamy cytoplasm, undefined cell margins, and active nuclei. ×400.

spindle-shaped (Fig. 2.**57**, 2.**58**, 2.**59**). The cytoplasm has a foamy structure and undefined margins, and the nuclei resemble those of vital endometrial cells or histiocytes (308).

Decidual cells. During pregnancy, the stromal cells change under the influence of hormones and assume a structure reminiscent of squamous epithelium. They are then called decidual cells. Occasionally, such cells show up in the smear and appear as basophilic, loose clusters of cells with vacuolated cytoplasm, well-defined margins, and a large, round nucleus that contains a nucleolus (Fig. 2.**60**) (233). They resemble either endometrial stromal cells or regenerative squamous cells.

Syncytiotrophoblastic cells. At the beginning of a spontaneous abortion, fetal cells may occur in the smear; these are usually portions of the trophoblast (383). Syncytiotrophoblastic cells are multinucleated giant cells with homogeneous cytoplasm (Fig. 2.**61**).

Cytotrophoblastic cells. These cells are usually isolated and exhibit a foamy, basophilic cytoplasm. The nuclei are much enlarged, polymorphic, and often contain macronucleoli. The cells are therefore reminiscent of cancer cells (Fig. 2.**62**).

Arias-Stella reaction. In case of hormonal overstimulation, cell clusters of the so-called Arias-Stella reaction are detected in rare cases. These cells have an abundance of vacuolated cytoplasm and large nuclei with a nucleolus. They originate through budding of the endometrial glands from the intraluminar epithelium, but are often confused with malignant cells (Fig. 2.**63**) (353).

Atrophic endometrial cells. After the menopause, the atrophic endometrial cells are smaller and their cell clusters more compact than during sexual maturity (see Fig. 3.**171**) (67).

These cells are rarely well preserved and appear more often as naked nuclei, thus making it difficult to distinguish them from degenerated endocervical nuclei (see also Fig. 3.**167a, b**).

2

Fig. 2.**58 Deep-seated endometrial stromal cells.** Stromal cells with spindle-shaped cytoplasm (——➤). A cluster of endometrial cells lies nearby (⟹). ×400.

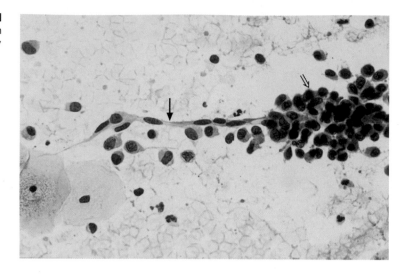

Fig. 2.**59 Deep-seated endometrial stromal cells.** A loose cluster of cells with oval to spindle-shaped (——➤), foamy cytoplasm, undefined cell margins, and active nuclei. ×400.

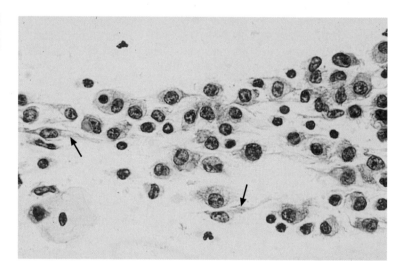

Fig. 2.**60 Decidual cells.** A loose cluster of cells with well-defined margins, foamy cytoplasm, and active nuclei with occasional formation of nucleoli. ×400.

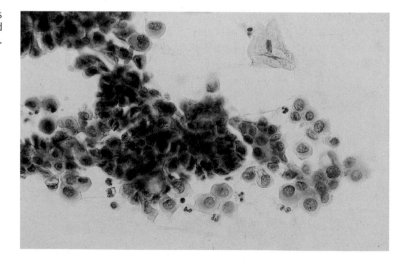

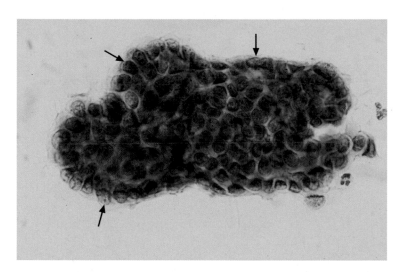

Fig. 2.**61** **Syncytiotrophoblast.** A compact syncytial sheet of cells with uniform central position of the nuclei and with a peripheral border of cells in palisade-like arrangement (——➤). ×630.

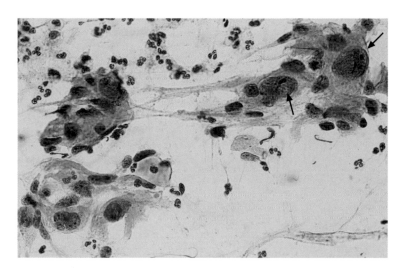

Fig. 2.**62** **Cytothrophoblastic cells.** A loose cluster of cells with voluminous, spindle-shaped, foamy cytoplasm and undefined cell margins; the enlarged, polymorphic nuclei vary in size and contain a macronucleolus (——➤). ×400.

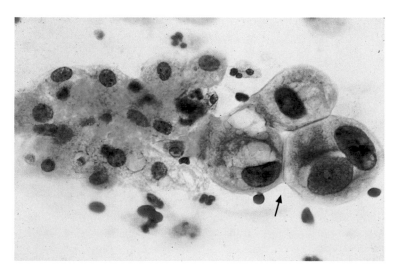

Fig. 2.**63** **Arias-Stella reaction.** A loose cluster of pseudodecidual endometrial cells. The cells at the right show the Arias-Stella reaction (——➤) with enlargement of cells and nuclei, vacuolization of the cytoplasm, and formation of nucleoli. ×400.

2

Fig. 2.**64 Female internal genitals.** Normal laparoscopic finding. The fundus of the uterus (→) is visible on the right, the uterine tube at the top, and the left ovary in the middle of the picture. The fimbriae of the uterine tube are held with pick-up forceps. Part of the colon is visible at the lower left.

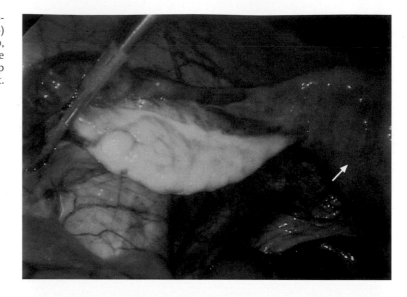

Fig. 2.**65 Left ovary.** Normal ultrasonographic finding. Visualization of several echo-poor follicles (→).

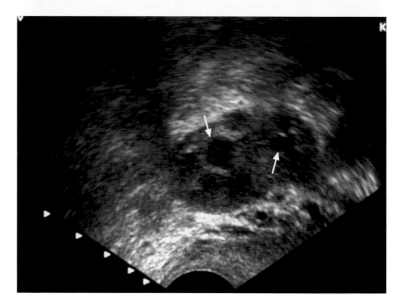

Ovary

Endoscopy and Ultrasonography

The ovaries are assessed by means of endoscopic methods. In addition, ultrasonography is available as an imaging procedure, particularly vaginal ultrasonography.

Laparoscopic examination shows the ovaries as flat, oval bodies the size of a pigeon's egg.

Because of the connective tissue membrane beneath the germinal epithelium, called the **tunica albuginea**, the ovaries exhibit a white reflecting surface (Fig. 2.**64**). Occasionally, the surface shows round protrusions or retracted scars; these are caused by maturing or degenerating ovarian follicles, respectively. The ultrasonogram shows the ovaries as oval structures in which different stages of follicular maturation are distinguished (Fig. 2.**65**).

Histology

The covering epithelium of the ovary, like the epithelium lining the ovarian follicles in the area of the ovarian cortex, is a simple cuboidal epithelium. The ovarian follicles mature under the influence of hypophysial hormones. Upon ovulation, the ruptured follicle transforms into the corpus luteum.

Ovarian cortex. The surface of the ovary is covered by a simple cuboidal epithelium, the **germinal epithelium** (178). Underneath this epithelium lies the extensive cortical stroma; it contains numerous **primary ovarian follicles**, some of which develop into vesicular ovarian follicles during sexual maturity (Fig. 2.**66**). The inner wall of the follicle is lined by estrogen-producing **granulosa cells** (283). The outer follicular wall consists of theca cells derived from the stroma. At the time of ovulation, the follicle is about 2 cm in diameter and easy to identify in the ultrasonogram.

After ovulation, the follicle develops into the progesterone-producing **corpus luteum**, with the theca cells migrating into the follicle and transforming into **granulosa-lutein cells** (18). If conception occurs, the corpus luteum increases in size and forms the corpus luteum of pregnancy, which will produce much larger amounts of hormone. Without conception, the corpus luteum dies after a few days and shrinks. Progesterone production is thus reduced, and the residual follicle is gradually resorbed.

Ovarian medulla. The relatively thick ovarian cortex turns into the ovarian medulla, which consists of cells that produce certain steroid hormones and androgens (207). At the center of the ovary, called the **hilum of ovary**, blood vessels enter the organ.

Cytology

The puncture fluids obtained from follicular cysts and lutein cysts contain granulosa cells and granulosa-lutein cells, respectively. Glandular cells of the ovarian germinal epithelium normally do not appear in the cytological smear. Their detection always raises the suspicion that an ovarian carcinoma is present.

Ovarian cysts. These cysts are often drained by puncture under the guidance of ultrasonography or laparoscopy, and their content is sub-

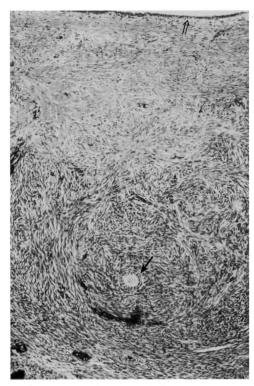

Fig. 2.**66 Cortical region of the normal ovary.** The covering epithelium at the surface consists of cuboidal cells ($\Rightarrow$). A primary follicle ($\rightarrow$) is visible in the rich connective tissue of the stroma in the lower part of the picture. Masson–Goldner stain, ×100.

jected to cytological examination (see also pp. 152 ff.).

▶ Besides protein precipitation and histiocytic foam cells, the puncture fluid from benign follicular cysts often contains an abundance of **granulosa cells** with sparse, foamy cytoplasm, undefined cell margins, and round nuclei (Fig. 2.**67**; see also Fig. 3.**182**). The chromatin structure is regular, though coarse-grained, and may contain a nucleolus (81).

▶ In addition to these, the puncture fluid from lutein cysts usually contains round **granulosa-lutein cells** with an abundance of fine-grained, sharply delimited cytoplasm and relatively small nuclei (Fig. 2.**68**; see also Fig. 3.**183**)

Glandular cells. Glandular cells from the ovary are normally not detected in the cytological smear. However, in the presence of an ovarian carcinoma, exfoliated cancer cells may travel through the fallopian tube and uterus into the vagina and are thus detected in the smear (84).

2

Fig. 2.**67** **Granulosa cells.** Cells from the puncture fluid of a follicular cyst, with loose, foamy cytoplasm, undefined cell margins, sparse cytoplasm, and relatively large, regularly structured, nucleoli-containing nuclei. ×400.

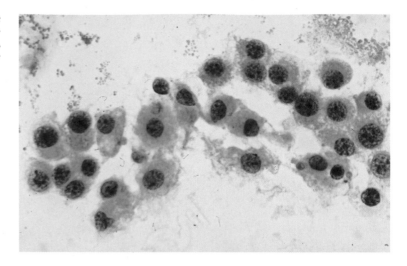

Fig. 2.**68** **Granulosa-lutein cell.** A round voluminous cell from a puncture fluid of a lutein cyst, exhibiting a loose, granular cytoplasm (—▶) and a small nucleus. A cluster of granulosa cells lies nearby. ×630.

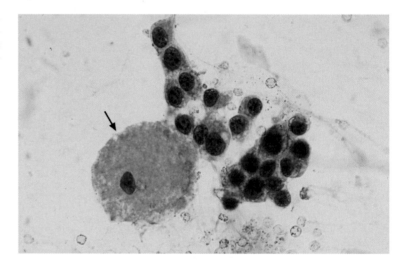

Vulva

Macroscopy

Knowledge of the anatomical structure of the vulva, including the borders of its organs, is an important prerequisite for clinical diagnosis.

Structure. As befits its ectodermal origin, the vulva is covered by keratinized squamous epithelium. This epithelium has all the attributes of the epidermis of normal skin, i. e., it contains sweat glands, sebaceous glands, hair follicles, and pigment inclusions, which give the skin a brownish color (193).

Figure 2.**69** shows a diagram of the anatomical structure of the external female genitals.

▶ Anteriorly, the vulva is delimited by the mons pubis, posteriorly by the anterior perineum, anus, and posterior perineum, laterally by the labiocrural folds, and towards the interior by the vaginal orifice or the edge of the hymen.

▶ Anteriorly to the urethral orifice, the labia minora merge to form the prepuce of the clitoris, which covers the clitoris; posteriorly to the vaginal orifice, they form the navicular fossa.

▶ The labia majora merge in front of the clitoris to form the anterior labial commissure and behind the navicular fossa to form the posterior labial commissure.

▶ The area between labia majora and labia minora is called the interlabial fold.

▶ After menopause, and particularly during old age, there are often physiological signs of involution that are manifested by partial or total atrophy of the labia minora, al-

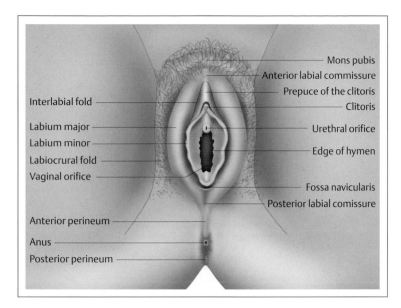

Fig. 2.**69** **External female genitals.** Normal macroscopic anatomy.

Mons pubis
Anterior labial commissure
Prepuce of the clitoris
Clitoris
Urethral orifice
Edge of hymen
Fossa navicularis
Posterior labial comissure

Interlabial fold
Labium major
Labium minor
Labiocrural fold
Vaginal orifice
Anterior perineum
Anus
Posterior perineum

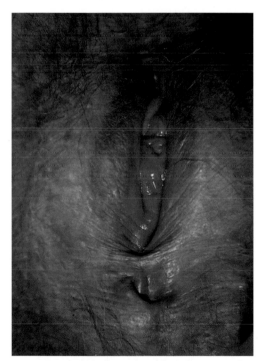

Fig. 2.**70** **Uncomplicated senile vulvar atrophy.** Complete atrophy of the labia minora. No clinical symptoms.

though these signs are not accompanied by subjective symptoms (Fig. 2.**70**).

Toluidine blue test. Since the epithelium of normal skin does not contain glycogen, Schiller's iodine test does not play a role here. Nucleated horn cells, however, stain well with toluidine blue, a test for active cells. The blue-staining parakeratotic areas clearly stand out from the remaining orthokeratotic skin after rinsing with acetic acid (89) (see pp. 242 ff., 346).

Histology

The stratified squamous epithelium of the vulva (vulvar epidermis), unlike that of the vagina, contains pigment and is keratinized on the surface. The normal stratum corneum (horny layer) is devoid of nuclei, a condition that is called **orthokeratosis**. However, the nuclei persist when keratinization accelerates, and this condition is called **parakeratosis.** Both these maturation processes depend on sex hormones.

Epithelial structure. The keratinized squamous epithelium rests on the prominent papillary body of the dermis (Fig. 2.**71**). **Melanin pigment granules** are incorporated in the cells of the basal layer. The epithelial stratification resembles that of the nonkeratinized squamous epithelium of the vagina. The horny layer is noticeably thicker at the labia majora than at the labia minora. In the area of the inner third of the labia minora, the vulvar epidermis turns into the nonkeratinized squamous epithelium of the vagina (176). Like the vaginal epithelium, the vulvar epidermis regularly undergoes cell regeneration. However, the rate of cell regeneration is clearly lower here than in the vaginal epithelium, with regeneration taking about 3–4 weeks (57). The influence of sex hormones manifests itself in the vaginal smear after only 1–2 days, but in the vulva it takes about twice as long.

2

Fig. 2.**71** **Vulvar epidermis.** Diagram of the changes in histological structure depending on the phase of the menstrual cycle.

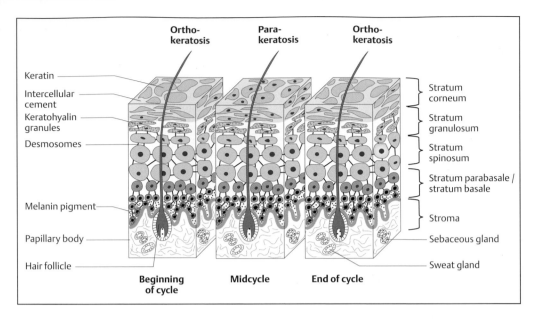

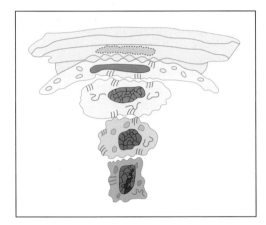

Fig. 2.**72** **Epidermal cell types.** Diagram of a cross-section through the vulvar epidermis (according to E. Christophers, modified).

Process of keratinization. During epithelial differentiation, keratinization of cells occurs relatively suddenly at the transition from the **stratum granulosum** to the **stratum corneum** (Fig. 2.**72**) (93). It causes considerable cellular changes. Keratohyalin granules that are still present in the stratum granulosum are extruded from the cytoplasm into the intercellular space, where they cement the cells to each other and thus contribute to the protection of the skin (107). Exfoliation of cells is therefore low and occurs only in the uppermost cell layers.

The process of keratinization involves dehydration of the cells, a reduction in cellular organelles, and an increase in acid mucopolysaccharides. The latter enhance the resistance of the cytoplasm (324). Simultaneously, the nuclei degenerate and usually disappear, i.e., they undergo karyolysis or are extruded from the cell (178).

The appearance of denucleated cells correlates with normal differentiation and is called **orthokeratosis** (Fig. 2.**73 a**). In about 10 % of cases, however, the nuclei persist. This condition is called **parakeratosis**; it intensifies upon proliferation-activating stimuli, thus leading to the acceleration of epithelial maturation (Fig. 2.**73 b**) (243). This is controlled by mechanical, chemical, inflammatory, and hormonal factors.

Effect of sex hormones. Unlike the vaginal epithelium, the vulvar epidermis is of a constant thickness, independent of the phase of the cycle. At the beginning and the end of the cycle, orthokeratotic keratinization predominates, whereas **parakeratosis** increases at midcycle under the influence of estrogen. In the absence of hormone replacement after the menopause, **orthokeratosis** is found exclusively (248).

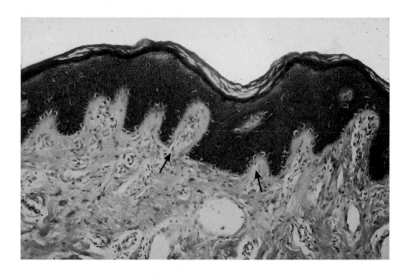

Fig. 2.**73 a** **Orthokeratosis.** Keratinized stratified squamous epithelium of the postmenopausal vulva. The papillary body of the connective tissue is well developed (⟶). The horny layer is free of nuclei, indicating orthokeratosis. HE, ×100.

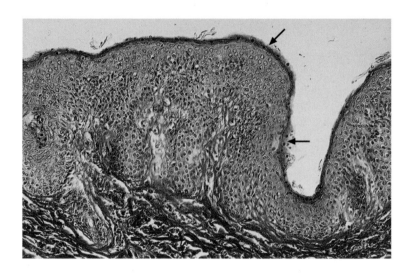

Fig. 2.**73 b** **Parakeratosis.** Keratinized stratified squamous epithelium of the vulva during sexual maturity. The papillary body of the connective tissue is well developed. The horny layer contains nuclei, indicating parakeratosis (⟶). Elastica–van Gieson (EVG) stain, ×100.

Cytology

Normal cells in a vulvar smear include **orthokeratotic cells** and **parakeratotic cells**. The percentage of parakeratotic cells in relation to the total number of exfoliated horn cells is called the **parakeratosis index**.

Non-normotopic cells. As a result of minimal exfoliation, vulvar smears contain fewer cells than vaginal smears. In addition to horn cells typical of the vulva, the smear often contains also non-normotopic, nonkeratinized epithelial cells derived from the vaginal cavity (245). After the menopause, atrophic processes commence in the vaginal epithelium, so less discharge is produced during this phase of life, thus reducing the transfer of vaginal cells to the outside.

Vaginal epithelial cells are not representative for the diagnosis of vulvar diseases and are easily distinguished from horn cells on the basis of their known morphological criteria (Fig. 2.**74**) (71). Under normal conditions of Papanicolaou staining, it is assumed that basophilic and eosinophilic cells are derived from the vagina, particularly when they contain nuclei. The bacterial flora in cytological vulvar smears is heterogeneous. Döderlein's bacilli may be detected because of contamination from the vagina, though the normotopic flora represents a mixed colonization.

Normotopic cells. Denucleated horn cells are called **orthokeratotic cells** and appear as orangophilic or flavophilic cells in the stained smear (Fig. 2.**75 a**). If the nuclei are preserved, the horn cells are called **parakeratotic cells**.

Fig. 2.**74 Comparison of morphological criteria** of a vaginal superficial cell and a vulvar parakeratotic cell.

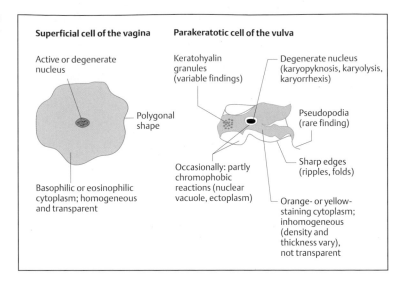

Superficial cell of the vagina

Active or degenerate nucleus

Polygonal shape

Basophilic or eosinophilic cytoplasm; homogeneous and transparent

Parakeratotic cell of the vulva

Keratohyalin granules (variable findings)

Degenerate nucleus (karyopyknosis, karyolysis, karyorrhexis)

Pseudopodia (rare finding)

Sharp edges (ripples, folds)

Occasionally: partly chromophobic reactions (nuclear vacuole, ectoplasm)

Orange- or yellow-staining cytoplasm; inhomogeneous (density and thickness vary), not transparent

Fig. 2.**75 a, b Orthokeratotic cells.** Denucleated horn cells of the vulva.
a Cells with sharp edges and flavophilic cytoplasm; a cluster of vaginal intermediate cells lies nearby. ×500.

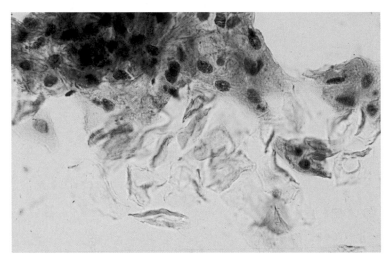

b Cell with orangophilic cytoplasm, unstained ectoplasmatic marginal border, and central nuclear "vacuole" (⟶). ×1000.

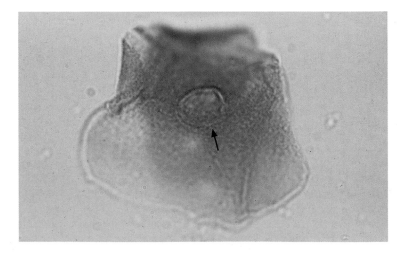

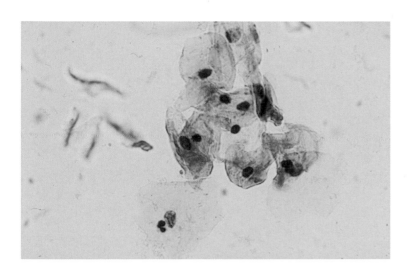

Fig. 2.**76 a, b   Parakeratotic cells.** Nucleated horn cells of the vulva.

a Cells with irregularly structured, flavophilic cytoplasm, and degenerate, condensed nuclei. ×500.

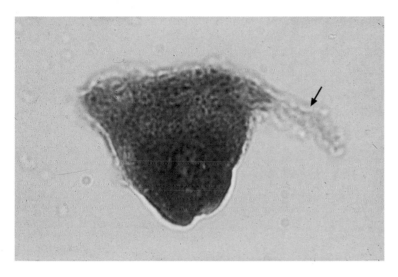

b Cell with irregularly structured, orangophilic cytoplasm, masses of keratohyalin granules and degenerate, karyolytic nucleus. Note the pseudopodium-like cell process (⟶). ×630.

Occasionally, problems may occur when trying to distinguish these cells from eosinophilic vaginal superficial cells with pyknotic nucleus. However, horn cells are usually only half the size and have sharper edges. Their cytoplasm is not transparent but appears optically dense and highly structured. They are rippled or folded, are of variable density and thickness, and occasionally contain keratohyalin granules. In addition, horn cells often exhibit partly chromophobic color reactions, particularly in marginal areas as an **ectoplasmatic marginal border** and, if the nucleus is absent, in the form of a central **nuclear "vacuole"** (Fig. 2.**75 b**). Parakeratotic cells always have a degenerate nucleus without recognizable chromatin structure (Fig. 2.**76 a, b**) (287).

Horn cells more frequently exfoliate as clusters than vaginal cells do. Despite their ir-regular shape and sharp edges, they do not vary significantly in size and shape. Because of their stable keratin scaffold, they are more resistant to environmental factors than vaginal cells. Hence, they do not require immediate alcohol fixation after collection, unlike vaginal cells.

Parakeratosis index. The percentage of nucleated horn cells occurring in the normal vulvar smear in relation to the total number of horn cells present is called the parakeratosis index (Fig. 2.**77**) (248). This index shows individual, age- and hormone-dependent variations; its mean value is 14 % in sexually mature women but only 3 % after the menopause. Application of gestagens and androgens after the menopause does not affect the index, but it may increase to 25 % when estrogen is sup-

Fig. 2.**77 Mean vulvar parakeratotic indices.** Comparison between premenopause and postmenopause under different hormonal starting conditions.

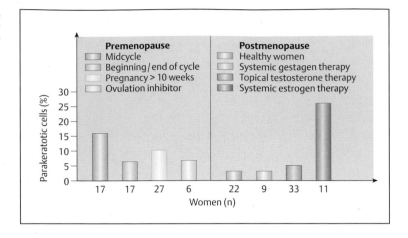

plied. During the menstrual cycle, the index shows periodic variations with a peak at midcycle. It is around 7% at the beginning and the end of the cycle, around 17% at midcycle, and 21% on the day of ovulation.

The course of the vulvar parakeratosis index correlates with the eosinophilia index and karyopyknosis index of vaginal cytology and also with the corresponding serum estradiol concentrations, thus suggesting estrogen dependency (200) (pp. 23, 24). Other morphological signs of cyclic hormonal action—in particular, differences in the color of the cytoplasm or structural changes of the cells as they occur in vaginal cytology—are not detected in the vulva.

Benign Changes of the Female Genital Tract

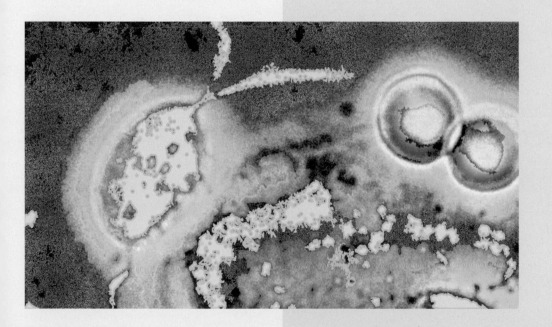

3

3 Benign Changes of the Female Genital Tract

Portio and Vagina

Physiological Changes

Cytolysis

Due to the activity of vaginal lactobacilli (Döderlein bacilli), the cytoplasm of intermediate cells lyses and glycogen is liberated. The sugar molecules derived from the glycogen are then metabolized into lactic acid. This creates an acidic environment and serves as a protective mechanism against potentially pathogenic microbes.

The physiological disintegration of squamous epithelial cells is called cytolysis. As the process requires glycogen-containing cells, it can only take place when the vaginal epithelium has developed at least the **upper intermediate cell layer** (Fig. 3.**1**) (411). No cytolysis is observed in an atrophic smear because parabasal cells do not contain glycogen (410). Superficial cells are protected from cytolysis because of their keratin-like cytoplasmic scaffold; hence, there is no cytolysis even when estrogen activity is high and the eosinophilia index is also high. (409).

Physiological cytolysis affects only the cytoplasm, not the nucleus (379). It is brought about by **Döderlein bacilli**, lactobacilli that exist only in the vagina. These bacilli are immobile rods of various lengths—though they exhibit a relatively uniform length within the same smear (227). Long piliform variants occasionally occur (Fig. 3.**2**). The bacilli completely or partially dissolve the cytoplasm of glycogen-containing intermediate cells, thus creating a fine granular protein precipitate that is sometimes misinterpreted as "coccal flora" by inexperienced examiners (Fig. 3.**3**). The bacilli then ferment the released glycogen to lactic acid. The resulting acidic environment (approximately pH 4) protects the epithelium against infections as it creates unfavorable growth conditions for foreign bacteria (410). Cytolysis therefore largely excludes the presence of pathogens from the vagina, although fungal infections are not affected by colonization of the vagina with lactobacilli.

If cytolysis is pronounced, most of the nuclei exist as **naked nuclei** and often show edematous swelling, although the chromatin structure is still well preserved and the nuclear envelope smooth. The enlarged nuclei pose a problem when cytolysis needs to be differentiated from dysplastic processes (Fig. 3.**4**). Topical application of an antibiotic will prevent cytolysis and may thus be advantageous when establishing the diagnosis (368).

The presence of a lactobacillus flora is not necessarily associated with the image of cytolysis in the vaginal smear. This is particularly true when the cells have been collected with a spatula, since this procedure predominately sets free cells derived from the superficial tissue layer where cytolysis caused by bacteria is not possible. By contrast, collecting the cells with a cotton swab yields cells from the vaginal secretion covering the portio vaginalis, and this essentially consists of defoliated cells that are in the process of undergoing cytolysis (322).

Degeneration

Unlike cytolysis, the degeneration of cells found in the cytological smear affects both cytoplasm and nucleus.

Nuclear changes:
▶ Nuclear swelling
▶ Karyolysis
▶ Karyorrhexis
▶ Karyopyknosis
▶ Hyperchomatic nuclear envelope

Cytoplasmic changes:
▶ Perinuclear halo
▶ Pseudoeosinophilia
▶ Vacuolation
▶ Hyalinization
▶ Structural changes
▶ Heterolysis
▶ Changes in cell shape
▶ Condensation of mucus
▶ Disturbed cell maturation

Fig. 3.**1** **Image of cytolysis.** In addition to well-preserved and partly basophilic, partly eosinophilic superficial cells with pyknotic nuclei, there are large intermediate cells undergoing lysis. Their nuclei are vesicular and often naked (→). There are numerous rod-shaped bacteria, representing vaginal lactobacilli (Döderlein bacilli). ×400.

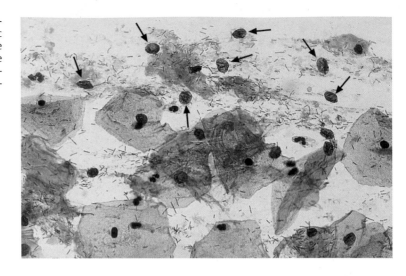

Fig. 3.**2** **Lactobacilli.** A piliform, elongated morphological variant of these bacilli and vesicular cell nuclei are visible between the superficial cells. ×630.

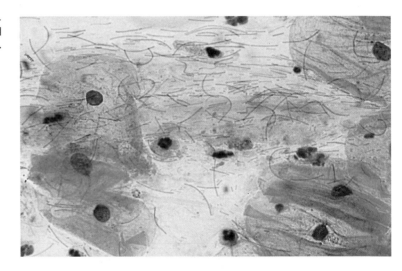

Fig. 3.**3** **Cytolysis by lactobacilli.** Vesicular, naked nuclei (→) and cytoplasmic debris, accompanied by a dust-like protein precipitate. ×630.

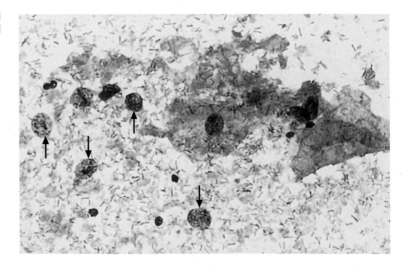

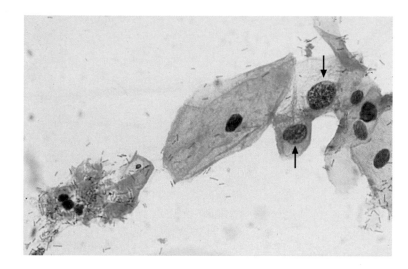

Fig. 3.**4** **Cytolysis by lactobacilli.** The swelling of individual nuclei, yielding nuclei with several times their normal size (→); their chromatin structure remains evenly distributed, and the nuclear envelope is smooth. ×630.

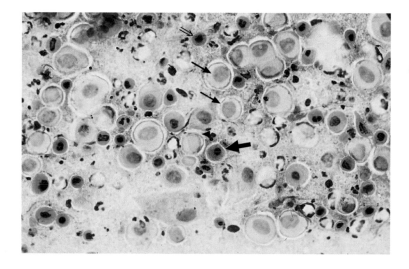

Fig. 3.**5** **Karyopyknosis.** Nuclear degeneration of parabasal cells by both karyolysis and karyopyknosis, resulting in an increased (→) or decreased (⇒) nuclear size. The cytoplasm is often pale and pseudoeosinophilic, or it shows hyaline degeneration (➡). ×400.

Whereas cytolysis by lactobacilli is a sign of vitality of the squamous epithelium with well-preserved nuclei, degeneration is a **process of cell death** (**necrosis**) that is largely induced by the unfavorable environment of exfoliated cells. The absence of lactobacilli causes the pH to rise, thus leading almost instantly to the appearance of foreign bacteria (185). Even without a visible inflammatory reaction, this causes rapid cell death, with the loss of a normal chromatin structure being observed particularly in the less resistant parabasal and endocervical cells, but also in the more highly differentiated squamous epithelial cells and nonepithelial cells (93).

Nuclear Changes

▶ **Nuclear swelling:** Osmotic cell regulation ceases, causing water influx and nuclear swelling (Figs. 3.5–3.7). This may cause problems in distinguishing degeneration from dysplastic processes (160).

▶ **Karyolysis:** The formation of unstructured, hypochromatic, and enlarged nuclei is called karyolysis (Fig. 3.**5**).

▶ **Karyopyknosis:** Loss of water then causes the dead nuclei to shrink and become hyperchromatic; this is called nuclear condensation or karyopyknosis (Fig. 3.**5**) (178, 425).

▶ **Karyorrhexis:** The rupture of the nucleus into fragments is called karyorrhexis. The chromatin structure first becomes very

Fig. 3.**6** **Karyorrhexis.** Nuclear degeneration by karyorrhexis, while the cytoplasm is still intact. The nuclei are hyperchromatic; they show a perforated nuclear envelope and a fragmented chromatin pattern (⟶). At the terminal stage, only some nuclear debris remains (⟹). ×630.

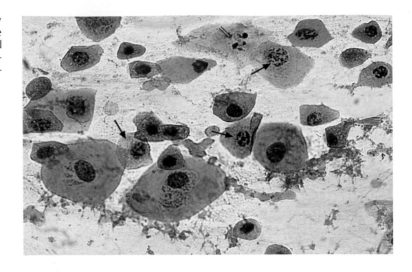

Fig. 3.**7** **Hyperchromatic nuclear envelopes.** The smear shows atrophic vaginitis with partial activation and enlargement of the nuclei, and also formation of nucleoli. Also visible are degenerative changes of nuclei and cytoplasm, such as cytoplasmic signs of cytolysis and pseudoeosinophilia and distinct hyperchromatic nuclear envelopes (⟶). The background of the preparation indicates inflammation. ×400.

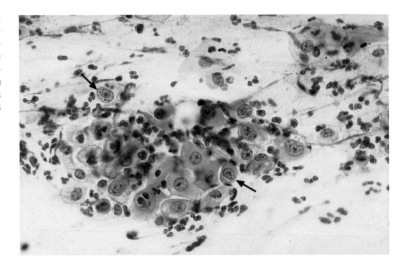

Fig. 3.**8** **Perinuclear halo.** Cytoplasmic degeneration of a parabasal cell (⟶). The nucleus is already condensed, and the cytoplasm is amphophilic and shows a perinuclear halo. A still intact parabasal cell (on the right) and several superficial cells lie nearby. ×630.

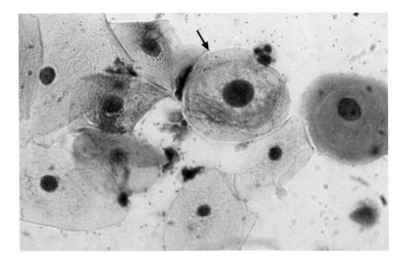

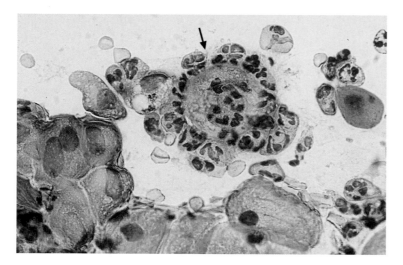

Fig. 3.**9** **Pseudoeosinophilia.** Cytoplasmic degeneration of a parabasal cell after karyorrhexis, with amphophilic and pseudoeosinophilic cytoplasm. Peripheral leukocytes (⟶) begin to phagocytose the dead cell. ×630.

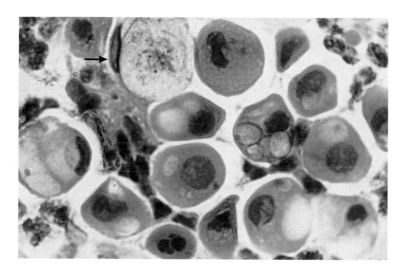

Fig. 3.**10** **Vacuolation.** Cytoplasmic degeneration of parabasal cells by vacuolation, while the nuclei have already degenerated. The vacuoles are of variable size; if very large, they may push the nucleus to the margin of the cell ("signet-ring cell") (⟶). ×1000.

coarse and then disintegrates, while gaps in the nuclear envelope result in a fragmented chromatin pattern (Fig. 3.**6**). The entire chromatin scaffold finally breaks up into many condensed and hyperchromatic fragments.

▶ **Hyperchromatic nuclear envelope:** Chromatin particles adhering to the inner nuclear membrane create the appearance of a hyperchromatic nuclear envelope (Fig. 3.**7**) (93).

Cytoplasmic changes

▶ **Perinuclear halo:** Shrinking of the nuclei leads to the formation of perinuclear halos (Fig. 3.**8**) (160).
▶ **Pseudoeosinophilia:** The degenerative changes eventually affect the cytoplasm as well; it becomes pseudoeosinophilic as a result of denaturation (Fig. 3.**9**) (368).
▶ **Vacuolation:** Water uptake leads to cell enlargement and formation of vacuoles (Fig. 3.**10**).
▶ **Hyalinization:** Condensation of the cytoplasm causes intense red staining, which is called hyalinization (Fig. 3.**11**) (356).
▶ **Structural changes:** Disintegration (or partial lysis) of the cytoplasm is often the result

Fig. 3.**11 Hyalinization.** Cytoplasmic degeneration of parabasal cells by hyaline degeneration following karyolysis and karyopyknosis. The shrunken cytoplasm stains intensely red (pseudokeratinization) (——▶). ×630.

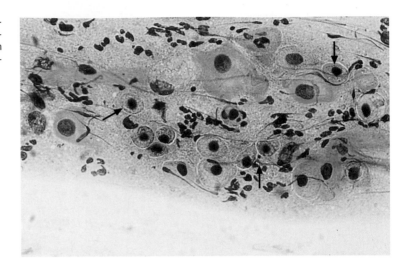

Fig. 3.**12 Polka-dot cells.** Cytoplasmic degeneration of superficial cells with partial disintegration resulting in a dotted pattern. Several normal intermediate cells are also present. ×500.

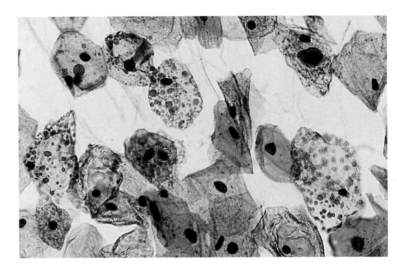

of structural changes. In some areas of the cell, the cytoplasm may exhibit complete chromophobia, although some rudimentary structures are still stained. The stained structures may appear as dots ("polka dot cells," Fig. 3.**12**) or as ridges that form window-like gaps between them ("fenestrated cells," Fig. 3.**13**) (212).

▶ **Heterolysis:** Lysis of the cytoplasm due to enzymes derived from bacteria other than Döderlein bacilli is called heterolysis (Fig. 3.**14**). Especially with atrophic cells, this may lead to confluence of several cells and formation of pseudosyncytial cells, known as **cell cohesion** (Fig. 3.**15**) (358). When the cytoplasm lyses completely, naked nuclei remain and often dominate the entire cell image.

▶ **Changes in shape:** Cell degeneration naturally affects atrophic smears, in particular, because parabasal cells are very sensitive to environmental factors. Elongation of the cytoplasm, possibly due to forces acting on the cell surface, leads to the formation of **spindle-shaped** cells (Fig. 3.**16**).

▶ **Condensation of mucus:** Chemotactic effects cause condensation of the mucus that covers the cells. This is probably responsible for the presence of blue blobs in the smears; these may be easily confused with enlarged naked nuclei with nucleoli (Fig. 3.**17**) (1).

▶ **Disturbed cell maturation:** Severely degenerated smears occasionally exhibit signs of a disturbance of squamous epithelial maturation, showing cells with hugely

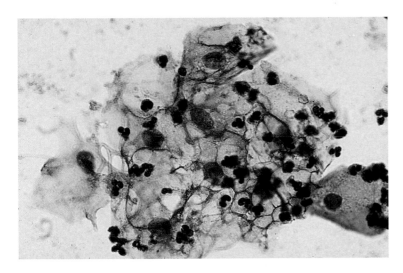

Fig. 3.**13** **Fenestrated cells.** Cytoplasmic degeneration of superficial cells with partial disintegration resulting in ridges. The cells are overlaid by leukocytes. A normal intermediate cell is visible on the lower right. ×630.

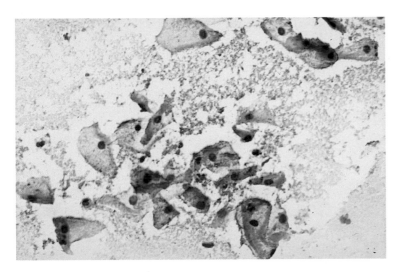

Fig. 3.**14** **Image of heterolysis.** Infection with *Gardnerella* causes the cytoplasm of intermediate cells to disintegrate without simultaneously inducing an inflammatory reaction. ×250.

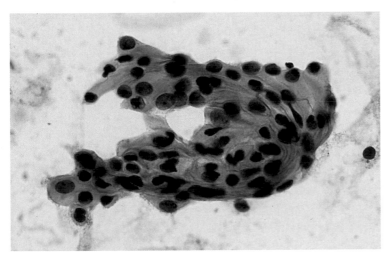

Fig. 3.**15** **Atrophic cell cohesion.** Pseudosyncytial cell sheet formed by parabasal cells. ×630.

Fig. 3.**16 Spindle-shaped cells.** Atrophic cell image with fusiform cytoplasmic elongation of parabasal cells. ×200.

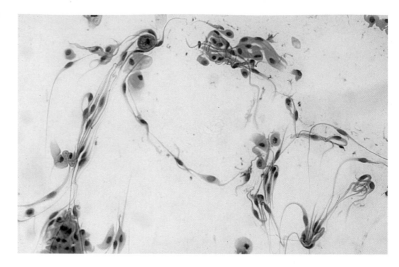

Fig. 3.**17 Blue blobs.** Atrophic cell image with degenerating parabasal cells. Some mucus has condensed on top of individual parabasal cells (⟶), presumably due to chemotactic effects, thus making them look hyperchromatic. They are therefore easily confused with enlarged nuclei of malignant cells. ×400.

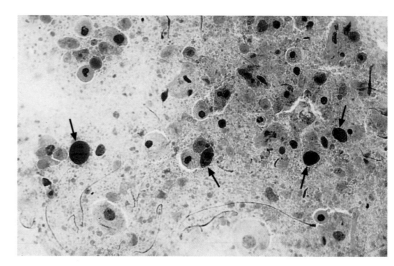

Fig. 3.**18 Polyploidization.** Atrophic smear in old age. Hugely enlarged and hyperchromatic nuclei often lead to confusion with abnormal nuclei (⟶). As they often coexist with degenerative nuclear changes, the chromatin structure cannot be assessed. ×400.

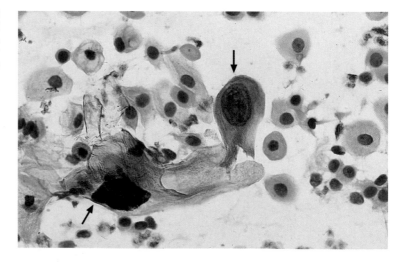

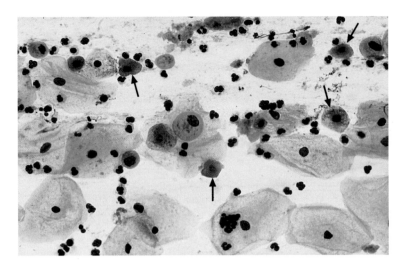

Fig. 3.**19 Miniature superficial cells.** Postmenopausal smear after hormone treatment. In addition to several parabasal cells, there are normally matured intermediate and superficial cells, as well as isolated small, eosinophilic cells that may have matured too suddenly (⟶). ×400.

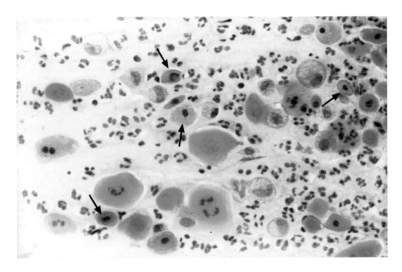

Fig. 3.**20 Atrophic vaginitis.** The degenerated parabasal cells are accompanied by a cell image of chronic inflammation. Granulocytes, lymphocytes, and histiocytes, as well as hyaline, degenerated parabasal cells (⟶) result in a colorful cell image. ×250.

enlarged nuclei (Fig. 3.**18**). These are most likely the result of spontaneous gene mutation in old age, causing **polyploidization** of individual nuclei. Estrogen treatment may hasten the maturation of some parabasal cells, thus creating **miniature superficial cells** (Fig. 3.**19**) (416).

Atrophic vaginitis. The atrophic vaginal epithelium is less resistant to bacterial or fungal superinfections and also to mechanical irritations and injuries. Inflammatory superinfection leads to atrophic vaginitis (Fig. 3.**20**). To prevent this from happening, systemic—or at least topical—estrogen replacement after the menopause is advantageous (194) (see also p. 124).

Fig. 3.**21 A representative cervical smear.** Sufficient numbers of well-preserved squamous epithelial cells and endocervical cells are present. Due to
their separate positions in the smear, these cells
permit to draw conclusions with respect to their
different anatomical origins. ×250.

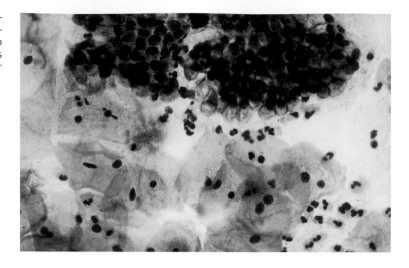

Fig. 3.**22 a, b Squeezed cells.** These are the result
of improper methods of smear collection.
a Cells with longitudinally flattened cytoplasm.
×320.

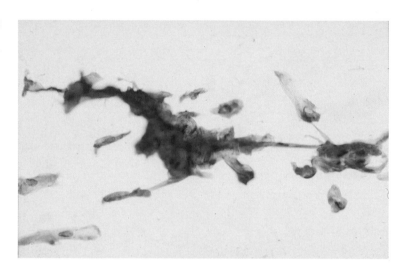

Technical Factors

Cell damage due to technical factors may be
caused by simple errors during collection and
processing of the material, or by complex iatro-
genic cellular reactions such as those following
radiation or chemotherapy.

Sampling Method

Improper collection of samples from the cervix
may be the reason why insufficient numbers of
cells have been harvested, or why the number
of squamous or columnar epithelial cells is too
low for an assessment. The swab technique
therefore greatly influences the assessment of
a smear (Fig. 3.**21**).

Furthermore, only instruments that are suitable for harvesting sufficient amounts of **representative cell material** from the collection
site should be used.

Smear Technique

▶ **Squeezing of cells** caused by a very swift
smear technique may lead to longitudinal
stretching of both cytoplasm and nuclei.
This creates spindle-shaped cells, which are
often confused with abnormal cells
(Fig. 3.**22 a, b**).
▶ **Superposition of cells** may lead to misinterpretations of the correct size of cells and
nuclei (Fig. 3.**23**).

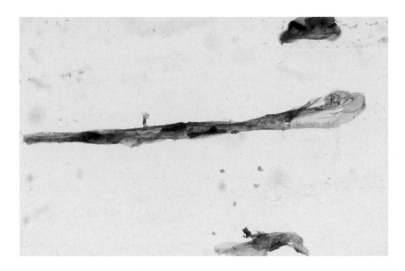

b Cells with longitudinally flattened cytoplasm and nucleus, thus resembling true spindle-shaped cells. ×400.

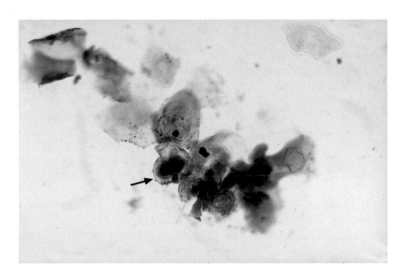

Fig. 3.**23** **Superposition of cells.** When lying on top of each other, cells of variable size appear as pseudohyperchromatic cells due to the additive effects of stained material, thus mimicking nuclear atypia (⟶). ×250.

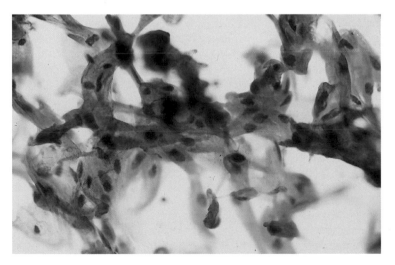

Fig. 3.**24** **Disorganized arrangement of cell material.** This is the result of improper methods of smear collection using circular movements of the cotton swab. ×400.

Fig. 3.**25** **Exodus.** Endometrial cell cluster showing the typical exodus pattern with crowded stromal cell nuclei in the center, epithelial cells at the outside, and an external line of retraction (⟶). ×400.

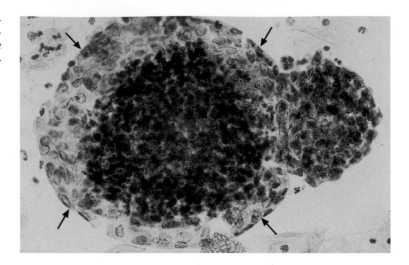

When the material is smeared in an irregular fashion—for example, in circles or in a zigzag pattern—squamous and columnar epithelial cells become mixed up and their ectocervical or endocervical origin is no longer recognizable (Fig. 3.**24**).

Displacement of Cells

Contamination of smears with displaced cells normally occurs when samples are collected during menstruation. Although a lactobacillus flora may be present, the smear has an inflammatory appearance when plenty of leukocytes, histiocytes, endometrial cells, mucus, and blood is present (see Fig. 2.**28**, p. 23). These cells may also occur during midcycle and in the absence of macroscopically relevant breakthrough bleeding—in particular, when the patient is taking one of the low-dose ovulation inhibitors that are preferred today. In these patients, the epithelium often shows a low degree of differentiation or is even atrophic (see Fig. 2.**30**, p. 25) (288). Groups of endometrial cells appearing at the end of menstruation, or even during the first half of the cycle, are called **exodus**. They appear as spherical cell clusters with a stromal center and an epithelial periphery (Fig. 3.**25**) (277).

Anucleated **horn cells** found in the vaginal smear have usually been displaced and are not typical for the collection site; they originate either from the genitals of the sexual partner, or the hands or vulva of the patient (Fig. 3.**26**) (244). In case of a prolapsed uterus, or when the patient uses a vaginal diaphragm, these cells may be the result of epidermization caused by pressure (Fig. 3.**27**) (368).

In rare cases, the staining solution may have been contaminated externally, thus transferring certain fungal, plant, or animal parts to the smear (33):

▶ The fungus **Alternaria** is characterized by segmented conidiospores that resemble snowshoes or beer barrels (Figs. 3.**28**, 3.**29**).
▶ The mold **Aspergillus** is characterized by Y-shaped ramifications and typical conidiophores (Fig. 3.**30 a, b**).
▶ The fungus **Geotricum** forms regular, sharply defined geometrical structures (Figs. 3.**31**, 3.**32**).
▶ **Plant cells** (Figs. 3.**33–35**) and **cerci of beetles** (Fig. 3.**36 a, b**) may be present.
▶ **Oxyurid eggs** (Fig. 3.**37**) and **pubic lice** (Fig. 3.**38**) are found in exceptional cases, due to contamination of the medium when moving the swab across the hair of the vulva.

Some artifacts are not easy to identify (Fig. 3.**39 a, b**). The accidental transfer of cells from one microscope slide to another is very rare, but it may have grave consequences in case of malignant cells (Fig. 3.**40**) (321).

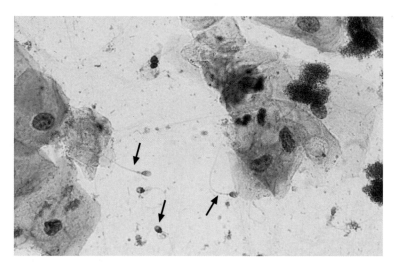

Fig. 3.**26 Orthokeratotic cells.** Anuclear horn cells in the cervical smear. Their presence is frequently due to transmission of cells from the sexual partner, finger, or vulva, rather than to a local disturbance of differentiation. Also visible are normal intermediate cells. In the lower center of the picture, isolated spermatozoa are present (→), suggesting that the horn cells are most likely derived from the partner. ×400.

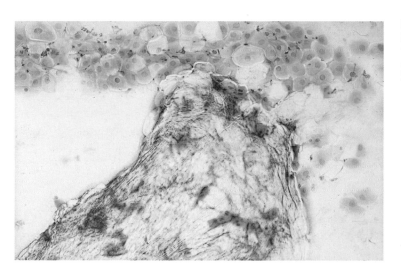

Fig. 3.**27 Orthokeratotic cell sheet.** Here, the horn cells are due to a local disturbance of differentiation leading to epidermization of the portio in response to a vaginal diaphragm. Numerous parabasal cells are visible at the top. ×100.

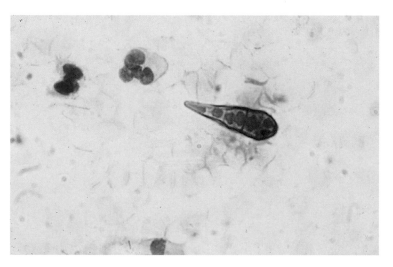

Fig. 3.**28 The fungus Alternaria.** Typical segmentation and snow-shoe appearance. ×1000.

3

Fig. 3.**29** **The fungus Alternaria.** Typical segmentation and beer-barrel appearance. ×1000.

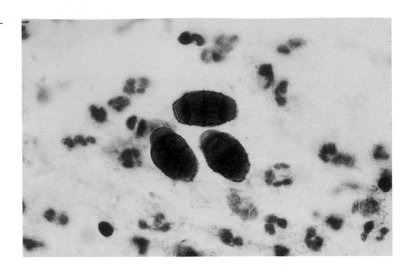

Fig. 3.**30 a, b** **The fungus Aspergillus.**
a Typical Y-shaped ramifications and terminal conidiophores. ×160.

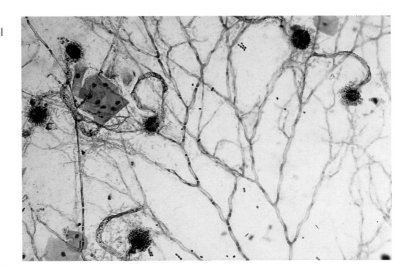

b A typical conidiophore consisting of a dense core and a corona of blastospores. ×500.

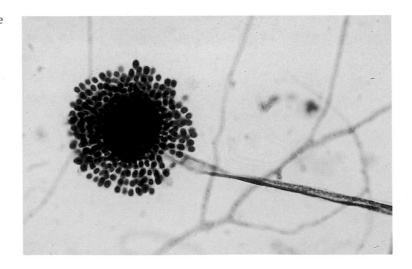

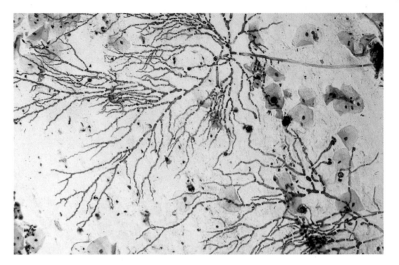

Fig. 3.**31** **The fungus Geotricum.** Regular geometric ramifications. ×120.

Fig. 3.**32** **The fungus Geotricum.** Regular geometric ramifications and sharp marginal contours. ×200.

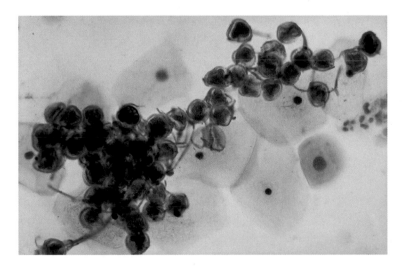

Fig. 3.**33** **Pollen.** A loose aggregation of brown-staining, round structures. Several superficial and intermediate cells are visible in the background. ×400.

3

Fig. 3.**34** **Plant cells.** A group of eosinophilic, small, elliptic cells with a thick, brown cell wall. ×630.

Fig. 3.**35** The diatom Cosmarium. A large segmented imicellular alga with chlorophyll. ×250.

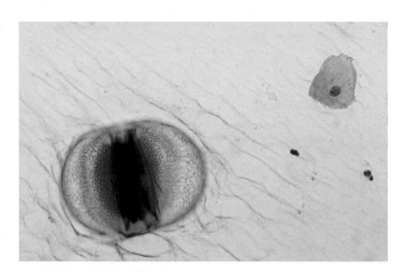

Fig. 3.**36 a**, **b** **Cerci of the museum beetle** (*Anthrenus museorum*).
a ×200.

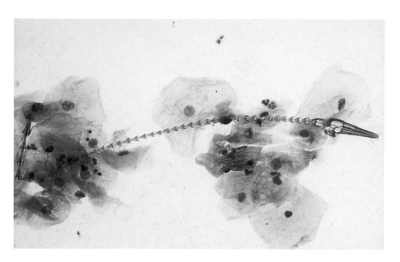

b A single cercus is surrounded by several superficial cells. ×250.

Fig. 3.**37 Oxyurid eggs.** Oval eggs with characteristic birefringence and lateral flattening. Isolated intermediate cells are visible at the top. ×400.

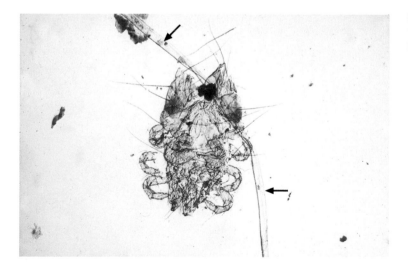

Fig. 3.**38 Pediculosis of the vulva.** Retrograde transmission of a small pubic louse, including the pubic hair (⟶) onto which it clings. ×100.

Fig. 3.39 a, b Unidentified artifacts.

a Artifact consisting of thread-like, vacuolated degenerated structures (muscle cells? fibroblasts?). ×400.

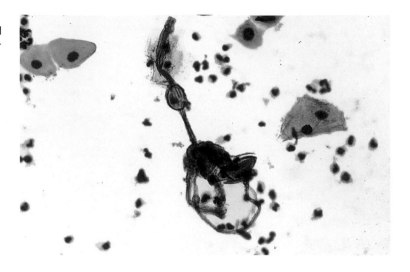

b Artifact consisting of mucous vacuoles and ointment residues (—►). ×630.

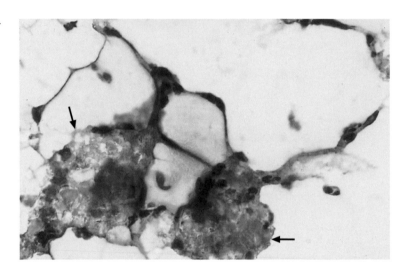

Fig. 3.**40 A cluster of contaminating cancer cells** (right half of image). Cancer cells of a patient with cervical carcinoma have been accidentally transferred from her microscope slide onto that of a young healthy woman (a virgin), presumably during fixation within the same cuvette or during the staining procedure. ×100.

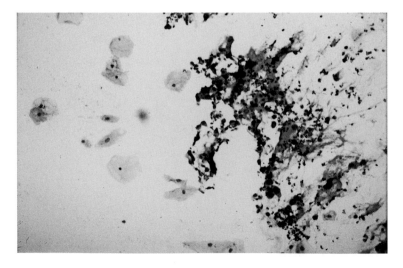

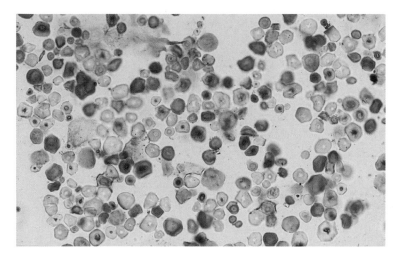

Fig. 3.**41** **Starch granules.** The cervical smear is overlaid with starch granules derived from surgical glove powder. The characteristic central birefringence is caused by light being reflected by the starch granules. ×320.

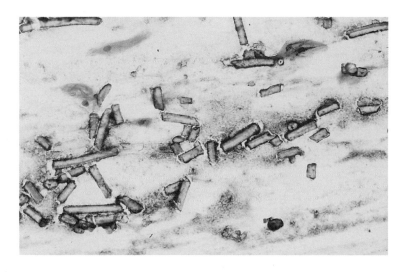

Fig. 3.**42** **Plastic fibers.** The cervical smear is obscured by sharply defined plastic fibers that probably originate from textiles. ×200.

Obscuring of Cells

Assessment of the smear may be severely hampered if the representative cells are obscured by foreign cells, mucus, or artifacts. These may include talcum, starch granules, textile fibers (most often the remains of cotton swabs or tampons), as well as residues of ointments or oils derived from intravaginal application.

▶ **Admixture of mucus:** Mucus often stains cyanophilic or blue-green, less often eosinophilic.
▶ **Starch granules:** These have characteristic structures and birefringent properties (Fig. 3.**41**).
▶ **Plastic fibers:** These are sharply defined and do not disintegrate (Fig. 3.**42**).

▶ **Natural fibers:** These show signs of disintegration and may exhibit leukocytic infiltration and even cell nuclei (Fig. 3.**43**).
▶ **Ointments and oils:** The fat component of these has usually been completely removed by alcohol during the staining procedure, but other components may be detected as blue-green, homogenous artifacts (Fig. 3.**44**) (246). These amorphous substances are sometimes resorbed by phagocytic cells (Fig. 3.**45**).
▶ **Blood:** Admixtures of blood appear pleomorphic. **Erythrocytes** may be still intact, or may have disintegrated to form **lysed blood** (Fig. 3.**46**). When lysed blood contaminates adjacent areas of the smear, it causes **pseudoeosinophilia** at these sites (360). Menstrual blood is subjected to fi-

Fig. 3.**43** **Natural fibers.** The cervical smear is obscured by natural fibers that are beginning to disintegrate, and are probably derived from textiles or from the cotton wool of tampons. Individual nuclei can be identified. ×100.

Fig. 3.**44** **Ointment residues.** After application of intravaginal medication, the ointment remains in the form of blue-green amorphous structures (⟶) surrounded by histiocytic cells. ×100.

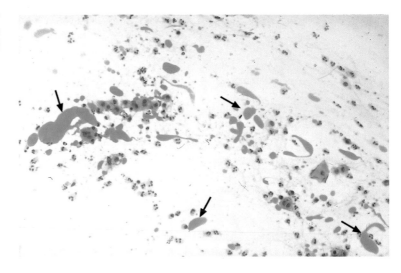

brinolysis by an enzyme locally produced in the mucosa; it does not coagulate (211). For this reason, **fibrin thrombi** are not detected under normal circumstances, but they may develop after injury or malignant erosion of blood vessels (Fig. 3.**47**). **Hemosiderin**, the intracellular storage form of iron, has the appearance of eosinophilic or blue-green granules (Fig. 3.**48a, b**) (368). Large histiocytes that incorporate blood components are called **siderophages** (Fig. 3.**49**).

▶ **Mucus:** Foamy cervical mucus is most likely caused by the ciliary action of endocervical cells (Fig. 3.**50**). The creation of **Curschmann spirals** is attributed to the same mechanism (Fig. 3.**51**) (151). These spirals resemble a twisted, knotted rope with a dark center and a light, fibrous periphery. Occasionally, the columnar secretory cells produce calcified, concentrically laminated, round structures, called **psammoma bodies** (418).

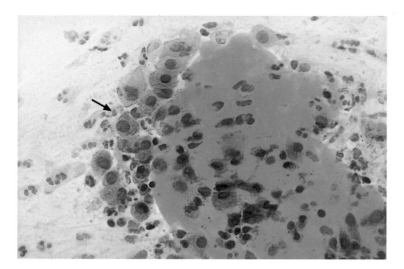

Fig. 3.**45** **Ointment components.** The ointment is being resorbed by surrounding macrophages (—►). ×250.

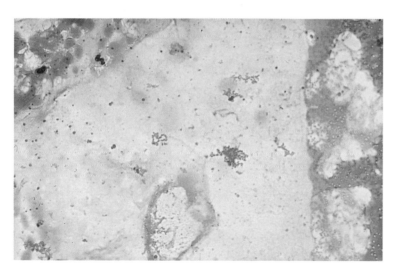

Fig. 3.**46** **Blood obscuring the cervical smear.** The blood is partly corpuscular (erythrocytes), partly lysed. ×250.

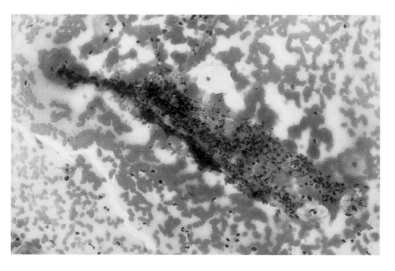

Fig. 3.**47** **Fibrin thrombus.** Leukocytic infiltration of a thrombus, surrounded by intact erythrocytes and caused by vascular erosion due to invasive cervical carcinoma. ×100.

Fig. 3.**48 a, b Hemosiderin.** Predominantly in granular form, with superficial cells visible at the bottom. ×250.
a Eosinophilic hemosiderin.

b Blue-green hemosiderin.

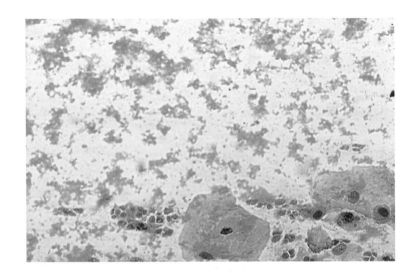

Fig. 3.**49 Siderophages.** Loose clusters of macrophages (⟶) that have incorporated blood components. Leukocytes and blue-green ointment residues (⟹) are visible nearby. ×400.

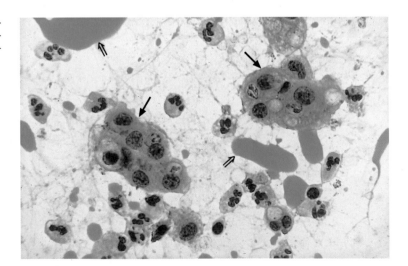

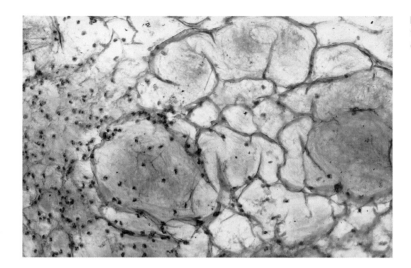

Fig. 3.**50** **Air bubbles in the cervical mucus.** The foam was probably created by the action of ciliary cells. ×100.

Fig. 3.**51** **Curschmann spiral. The coiled mucinous fiber** (⟶) was probably created by the action of ciliary cells. ×160.

Fixation Errors

Desiccation of the cells due to delayed fixation or dilution of the fixative frequently causes discoloration of the smear, **pseudoeosinophilia**, and **nuclear degeneration** (Figs. 3.**52**, 3.**53**) (368). The artifacts may be misinterpreted as keratinized cells or enlarged nuclei. The changes may affect only parts of the smear, particularly marginal areas, either because the entire smear was not homogeneously covered by spray fixative, or because the microscopic slide was not completely immersed in the fixation dish.

Staining Errors

Discoloration of the smear may be caused by inferior batches of stain, wrong staining periods, or excessive use of the same staining series (368).

On the other hand, prolonged staining with certain solutions, or the use of unsuitable batches, may result in **overstaining** of the cytoplasm, thus making it difficult to distinguish between cytoplasm and nucleus (Fig. 3.**54**). Central pseudoeosinophilia of cell clusters is also typical for overstained preparations (Fig. 3.**55**) (368).

Occasionally, hematoxylin may precipitate as a result of insufficient washing during the staining procedure; this is called **hematoxylin pigment** (Fig. 3.**56**) (368).

Fig. 3.**52 Fixation error.** Faint staining and pseudoeosinophilia of the smear. Cells and nuclei are severely degenerated as a result of desiccation. ×100.

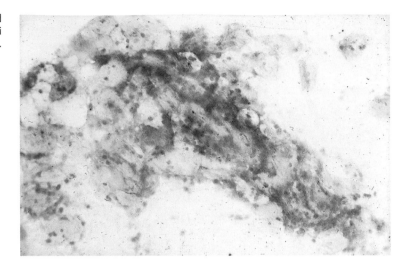

Fig. 3.**53 Fixation artifacts.** Caused by desiccation, these artifacts mimic cytoplasmic vacuolation or koilocytosis (——►). ×100.

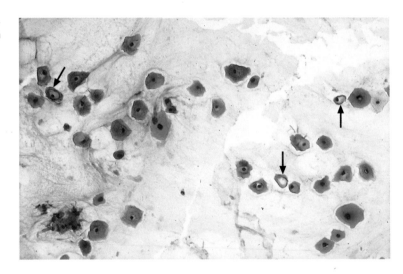

Fig. 3.**54 Overstaining.** Caused by staining error. The nuclei of parabasal cells are difficult to distinguish from the cytoplasm (——►). ×250.

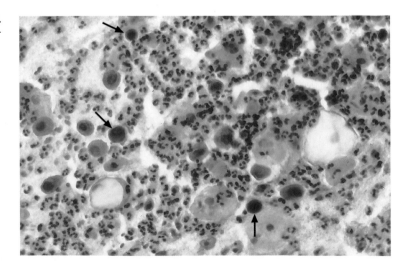

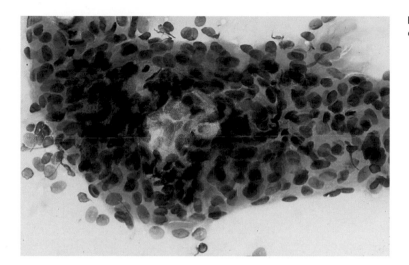

Fig. 3.**55** **Pseudoeosinophilia.** The center of the dense cell cluster is incompletely stained. ×400.

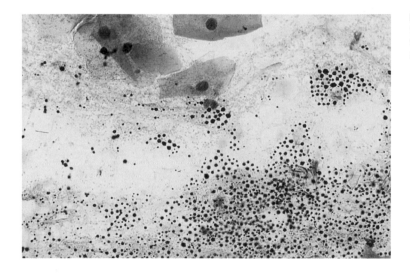

Fig. 3.**56** **Hematoxylin pigment.** The precipitation is probably due to incomplete washing during the staining procedure. Superficial cells are visible at the top. ×400.

Mounting Errors

The **cornflake artifact** is due to the formation of brownish, round granules on top of squamous epithelial cells, thus severely hampering assessment of the smear (Fig. 3.**57a, b**). This phenomenon probably results from the inclusion of air during the mounting process and is caused by water evaporation due to insufficient dehydration (368). The dark coloration is due to the interference of light reflected between air, mounting medium, and cell surface. This frequently observed phenomenon is therefore explained by exposure to water and air during improper or delayed mounting (101).

Assessment is not possible with large **inclusions of air**, or when the wrong side of the slide has been mounted, because the optical parameters have changed. Proper remounting of the smear is required (Figs. 3.**58**, 3.**59**).

Confusion of Material

Confusion of patients' names, resulting in the wrong assignment of cell material, may occur at the sender's end or in the cytological laboratory. The error is usually not recognized and is of no consequence as long as the cytological findings are negative. If, however, a mix-up is suspected and the findings are positive, the patients involved need to be recalled to the originating doctor's office or clinic for the smear preparation to be repeated.

Fig. 3.**57 a, b   Cornflake artifact.**

a Formation of fine brown granules on top of the cells, presumably as a result of insufficient dehydration during the mounting process. ×400.

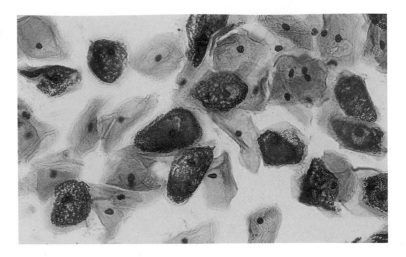

b The cornflake pattern overlying the cells may severely hamper the diagnosis. ×400.

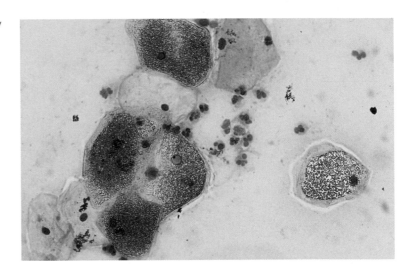

Fig. 3.**58   Inclusion of air** (upper right half of image). This often changes the cell image to such an extent that expert assessment of the smear is no longer possible. Remounting may help. ×100.

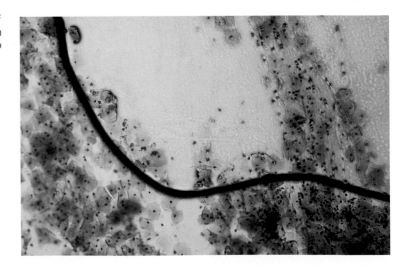

80

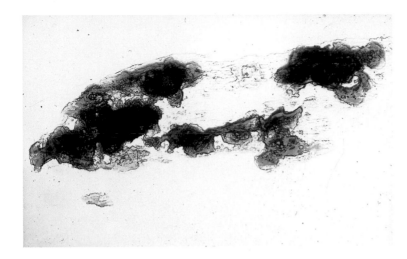

Fig. 3.**59 Mounting error.** Placing the coverslip on the wrong side of the microscope slide changes the optical parameters. Proper remounting is required. ×100.

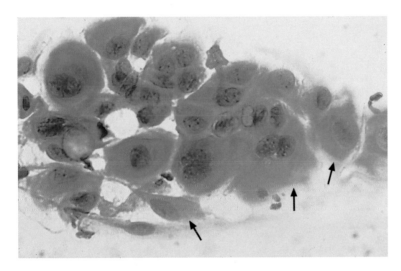

Fig. 3.**60 a–d Acute radiation effects.**
a Parabasal cells with enlarged and degenerated nuclei, amphophilic cytoplasm, and indistinct cell borders (⟶). ×400.

Radiation

The effects of radiation occur less often with the glandular epithelium than with the readily regenerating squamous epithelium. Immature cells, such as **parabasal cells** and **metaplastic cells**, are especially affected (Fig. 3.**60 a–d**) (38). Since ionizing radiation attacks the sensitive nucleic acids, it irreversibly damages the genetic material of the cells. For this reason, effects of radiation can still be recognized in cytological smears for years and even decades after radiation therapy (388). Radiation always leads to changes in the overall image of the smear, which shows initially acute, and later chronic, inflammatory reactions (121).

Initially, the changes affect mainly the immature cells, but after only a few weeks they are also pronounced in all other layers of the squamous epithelium (Figs. 3.**61 a, b**; 3.**62 a, b**; 3.**63 a, b**) (388). These changes consist mainly of irregular **vacuolation** as well as **phagocytosis** and **cannibalism**, a process leading to the formation of "bird's eye cells." The changes also include **macrocytosis, leukotaxis**, and changes in color and structure, such as **pseudoeosinophilia, amphophilia, bizarre cytoplasmic formations**, and **spindle-shaped cells**.

At the same time, the size of the nuclei increases without significantly shifting the nucleocytoplasmic ratio, thus resulting in **proportional nuclear enlargement**. The nuclear structure may become **hyperchromatic**, although without any atypia. Pronounced degenerative nuclear changes usually occur in the form of **karoylysis, karyorrhexis**, or **karyopyknosis**. Binuclear and polynuclear cells are

b Small intermediate cells with blurred cell borders and enlarged, degenerated nuclei. ×1000.

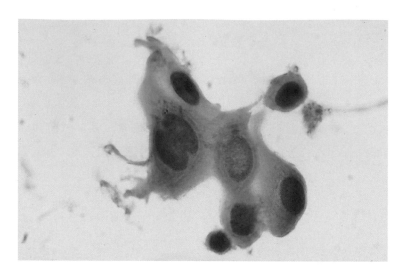

c A cluster of endocervical columnar epithelial cells (→) with enlarged, hyperchromatic nuclei. Some nuclei show distinct chromocenters. Superficial and intermediate cells are visible nearby. ×250.

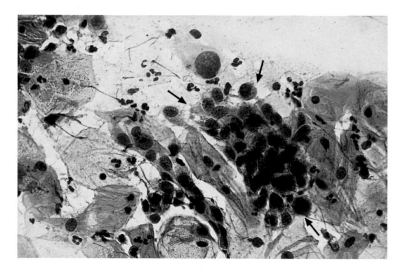

d A cluster of endocervical columnar epithelial cells (→) with enlarged, hyperchromatic nuclei. Nuclear and cytoplasmic structures are blurred. Superficial cells are visible nearby. The background of the preparations indicates inflammation. ×250.

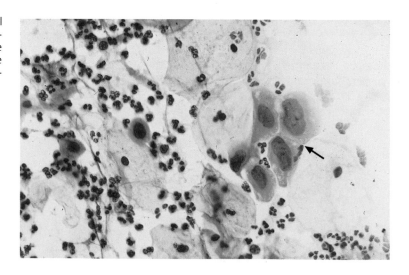

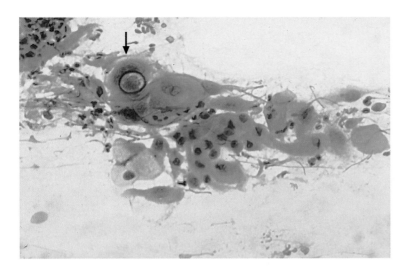

Fig. 3.**61 a, b   Chronic radiation effects.**
a Variable shape of the cytoplasm, indistinct cell borders, and vacuolation (⟶). The nuclei are degenerated. ×250.

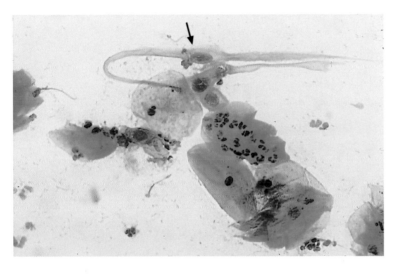

b Appearance of spindle-shaped cells. The nuclei are degenerated and hypochromatic (⟶). ×250.

common, and enlarged nucleoli as a sign of the ongoing epithelial regeneration are also observed.

Folic Acid Deficiency

Deficiency of vitamin B_{12} and/or folic acid may occur during pregnancy and in association with certain forms of anemia. It has a negative effect on all cells because these substances are required for DNA synthesis (258). A characteristic sign is **macrocytosis** with proportional enlargement of cells and nuclei of the squamous epithelium, and binuclear cells and cytoplasmic vacuolation are also found occasionally (Fig. 3.**64**).

Cytostatic Treatment

Similar changes to those of folic acid deficiency are observed after treatment with cytostatic agents, since many of these substances function as folic acid antagonists (368, 392). The resulting changes of the cell image are usually even more pronounced than in folic acid deficiency (Fig. 3.**65 a**). The nuclei may be **hyperchromatic** and **polymorphic**, and partial nuclear condensation is occasionally observed (Fig. 3.**65 b**).

Surgical Sequelae

Within the first 4 weeks after an intravaginal surgical intervention, such as conization or hysterectomy, **granulation tissue** usually de-

Fig. 3.**62 a, b   Radiation effect.**
a Macrocytosis, cannibalism, and phagocytosis, resulting in so-called bird's eye cells (⟶). ×630.

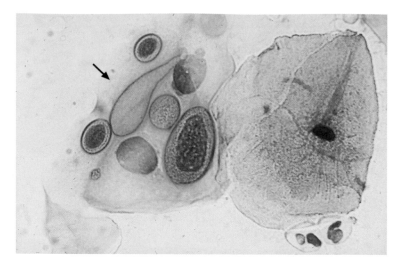

b Macrocytosis, cannibalism, and phagocytosis resulting in so-called bird's eye cells (⟶). 630×.

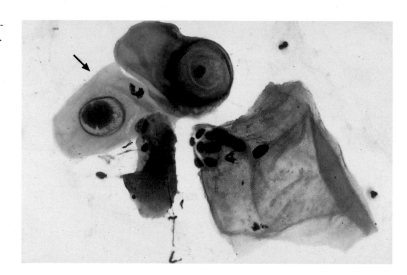

Fig. 3.**63 a, b   Polynuclear giant cells.** Caused by radiation effect.
a Giant cell with phagocytosed leukocytes. ×400.

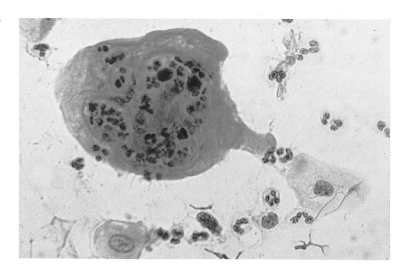

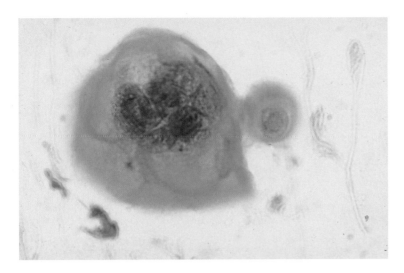

b Giant cell with vacuolation, pseudoeosinophilia, and indistinct cell borders. ×630.

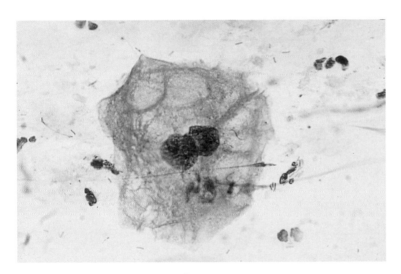

Fig. 3.**64** **Folic acid deficiency.** Macrocytosis and cytoplasmic vacuolation of a binuclear cell. ×630.

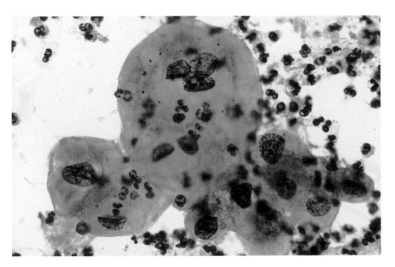

Fig. 3.**65 a, b** **Effect of chemotherapy.**
a Polynuclear cells with proportional nuclear enlargement and hyperchromatic, polymorphic nuclei. ×630.

3

b Cells with hyperchromatic, polymorphic nuclei and central nuclear condensation (———➤). ×400.

Fig. 3.**66 Granulomatous postoperative inflammation.** Cell image with polymorphic cells and nuclei. The background of the preparation indicates inflammation. ×400.

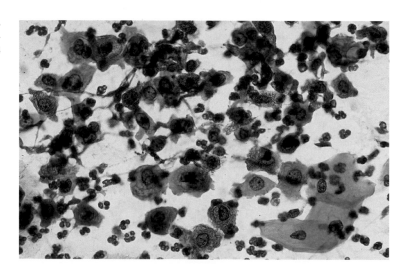

Fig. 3.**67 Intrauterine device.** A loose cluster of immature metaplastic cells, with an inflammatory cell image caused by the presence of an IUD. ×630.

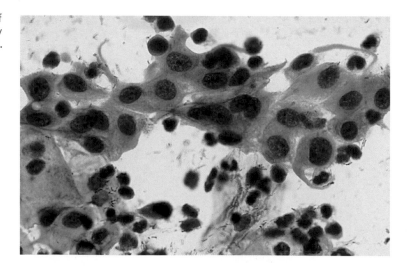

velops in the area of the surgical wound (81). Cytologically, this tissue consists of squamous and endocervical epithelia with inflammatory changes, regenerating epithelium, leukocytes, histiocytes, fibroblasts, and protein-rich cell debris (Fig. 3.**66**).

As a late sequel of hysterectomy, columnar epithelial cells may be present in the smear if the endocervix has not been completely removed, or if portions of the uterine tubes have been caught in the vaginal blind sac, a condition referred to as **tubal prolapse** (81).

Intrauterine Devices

Patients using an intrauterine device (IUD) often show reactive changes in the smear image. These include nuclear enlargement and vacuolation in the endocervical columnar epithelium, presence of metaplastic cells (Fig. 3.**67**) and endometrial cells, inflammatory cells in the background, and, in particular, signs of actinomycosis (78) (see also p. 92).

However, it is only this combination of changes that is characteristic for the presence of an IUD, not any of the individual changes. The chronic state of irritation caused by the foreign body seems to promote colonization of the cervix by pathogens.

Abnormal Microbiology

Facultative pathogenic microorganisms may be detected in the cervical smear, sometimes also together with isolated damaged cells, without inducing an inflammatory reaction visible in the overall cell image. Whether an inflammatory reaction occurs depends on the pathogen count and the aggressiveness of the organism, but, most importantly, on the body's state of immunity.

Most of the time, these microorganisms exist in a parasitoid, saprophytic state. They are nonaggressive but may lead to degenerative changes of cells (314). Many pathogens may be identified on the basis of the characteristic cytological changes they induce in the smear. The physiological lactobacillus flora is present in less than one third of all women.

Mixed Flora

A mixed vaginal flora consists mostly of cocci and rods, and the latter may temporarily include lactobacilli (Fig. 3.**68a, b**). As some bac-

teria measure only a few tenth of a micrometer, the 100× magnification commonly used for screening is not sufficient for microscopic differentiation of the flora. Identification requires a 400× magnification, although the exact assignment and differentiation of bacteria are not possible in a Papanicolaou smear (185). For proper identification, special staining procedures or cultures are required.

Cocci

Sometimes there is a pure coccal flora consisting of evenly spread, isolated, and punctiform bacteria. Such an infection frequently causes an increase in karyopyknosis, eosinophilia, or even pseudokeratinization of the squamous epithelial cells (Fig. 3.**69a, b**) (368). An inexperienced examiner may misinterpret these changes as the effect of increased follicular hormone activity, or a sign of HPV infection, or keratinizing dysplasia.

Gardnerella

A *Gardnerella* flora is relatively common and sometimes difficult to distinguish from a coccal flora (74). It occurs predominantly when the vaginal epithelium is highly differentiated (98).

The bacterium was formerly called *Haemophilus vaginalis*. Clinically, the infection is accompanied by a characteristic fishy odor caused by the release of amines. If there are signs of inflammation, this is referred to as **bacterial vaginosis** (80). Addition of potassium hydroxide solution to the vaginal discharge intensifies the odor, which is the diagnostic principle of the amine test (79).

The bacterial lawn in the cytological smear resembles dust-like particles with a tendency to accumulate like sand dunes. In addition, there are many squamous epithelial cells that are densely covered with bacteria due to a chemotactic effect; these are called "clue cells" (198). These changes are detected in native smears by phase contrast microscopy (Fig. 3.**70**) as well as in stained preparations (Figs. 3.**71**, 3.**72a, b**).

3

Fig. 3.**68 a, b   Mixed flora.**
a Predominantly rod-shaped bacteria and squamous epithelial cells with slightly swollen nuclei. No inflammatory reaction. ×630.

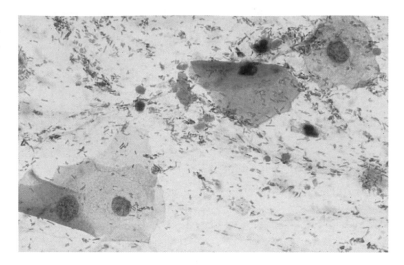

b Cocci and rods surrounding an eosinophilic superficial cell with pyknotic nucleus. ×1000.

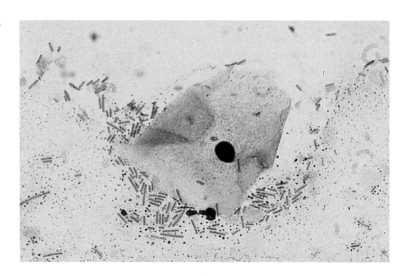

Fig. 3.**69 a, b   Coccal flora.**
a Punctiform bacteria. No inflammatory reaction of the squamous epithelium. ×630.

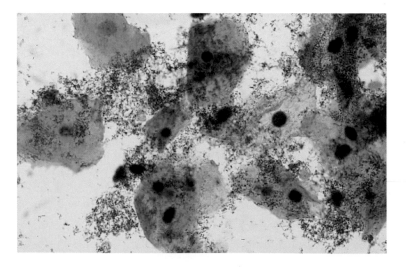

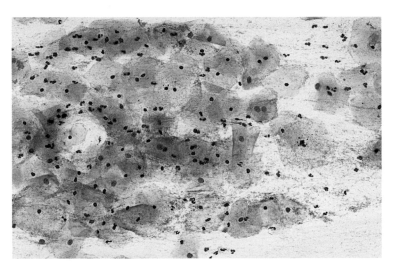

b Reactive pseudokeratinization of the squamous epithelium and an increase in pyknotic nuclei. No inflammatory reaction of the squamous epithelium. ×200.

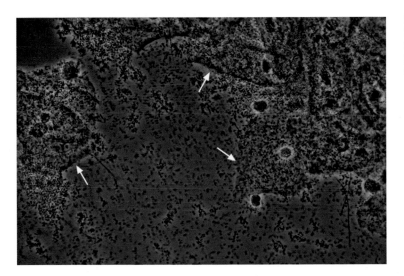

Fig. 3.**70 Clue cells associated with Gardnerella infection.** Phase contrast image of superficial cells overlaid by a dense bacterial lawn (→). The background shows the typical image of dustlike particles. ×630.

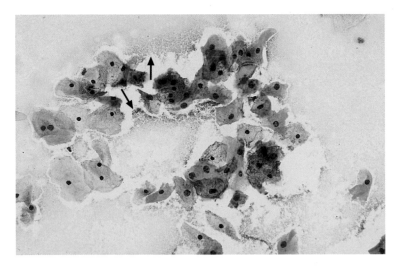

Fig. 3.**71 Gardnerella flora.** Typical sand-dune phenomenon (→). No inflammatory reaction of the squamous epithelium. ×250.

Fig. 3.**72 a, b** **Clue cells associated with Gardnerella infection.**

a Dense concentration of bacteria on some squamous epithelial cells (⟶). ×630.

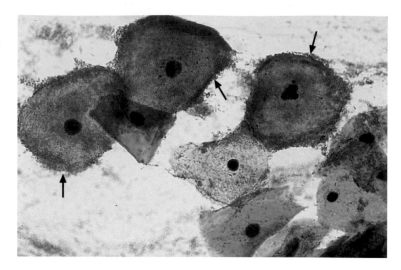

b Dense concentration of bacteria on an intermediate cell. ×1000.

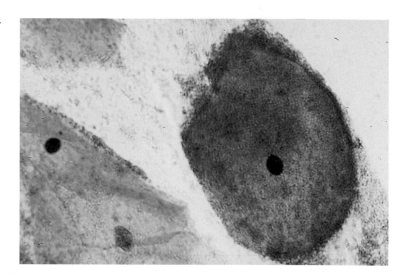

Chlamydia

Bacteria of the genus *Chlamydia* are very small and therefore difficult to recognize. Nevertheless, they may be detected in cervical smears by means of special methods (230). Because of their high pathogenicity, a special screening test is carried out during early pregnancy in the context of maternity care to ensure timely treatment of any infection (253). *Chlamydia* infection represents one of the most common causes of sterility, since it often leads to ascending inflammation with involvement of the uterine tubes, thus causing **salpingitis** (7). In the course of inflammation, scarring may block tubal patency.

Both the presence and the reproduction of *Chlamydia* depend on an energy-rich metabolism. These bacteria can therefore only survive as intracellular parasites. They preferentially infect parabasal cells, immature metaplastic cells, endocervical cells, and urothelial cells (99). In these cells, the pathogen forms cytoplasmic **vacuoles** with diffuse borders (inclusion bodies) of variable size without affecting the nuclei (Fig. 3.**73 a–c**) (118). The reproductive cycle takes place within these vacuoles, leading to the formation of **elementary bodies**, which show a bluish to reddish staining (352). Once the host cell is destroyed, the bacteria are released and may become infectious again. Detection of epithelial cells suspected to contain *Chlamydia* is too nonspecific to allow the assumption of an actual infection. At the most, such cells may be the reason for employing special detection methods using monoclonal antibodies to *Chlamydia*.

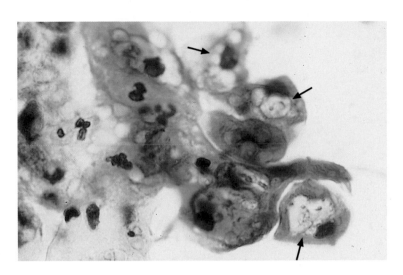

Fig. 3.**73 a–c Chlamydia infection.**

a Immature metaplastic cells with large vacuoles of variable size and indistinct margins (⟶). ×1000.

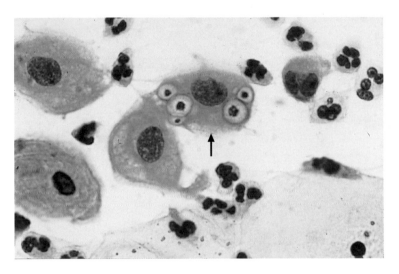

b Vacuoles of variable size and inclusion bodies within a parabasal cell (⟶). ×1000.

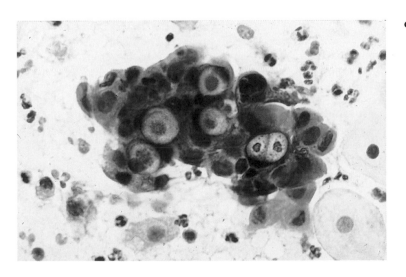

c A cluster of endocervical cells with vacuoles of variable size, indistinct margins, and eosinophilic inclusion bodies. The background of the preparation indicates inflammation. ×400.

Fig. 3.**74** **Leptothrix flora.** Long, piliform, loop-forming bacteria are surrounding superficial cells. ×630.

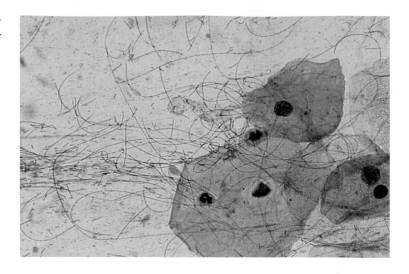

Fig. 3.**75** **Actinomycosis.** Bacterial lawn showing a dark, central concentration and peripheral filaments. ×360.

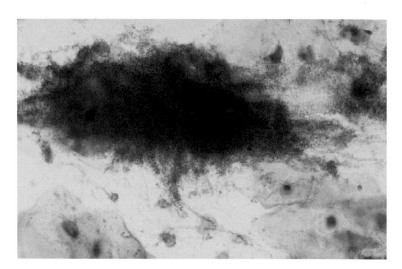

Leptothrix

These apathogenic **filamentous bacteria** are thin, piliform microbes measuring up to 100 µm in length and mostly forming curves or loops (Fig. 3.**74**) (55). In comparison with other bacteria, they are rather rare, although they tend to appear together with trichomonads.

Actinomycetes

Actinomycosis, an infection caused by actinomycetes, is observed in 10–20% of all **IUD users** (117). As a rule, the background of the smear indicates a severe inflammatory reaction, although the pathogenicity of this infection seems to be minor (81).

Actinomycosis is a bacterial infection and is recognized by dirty-looking, moldlike, filamentous **aggregations** of various sizes. Frequently, these mycelial structures have a central condensation from where fine, irregular filaments extend to the periphery. They are also called actinomycotic drusen, sulfur bodies, or Gupta bodies, and consist of bacteria, proteins, and polysaccharides (Fig. 3.**75**) (117).

Candida

The yeast *Candida albicans* is frequently detected in vaginal smears. This is especially true following antibiotic therapy, in diabetic patients with elevated glycosuria, and occasionally after taking high doses of estrogen (386). Macroscopically, candidiasis is often as-

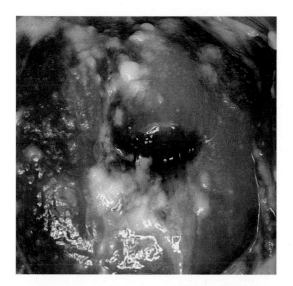

Fig. 3.**76** **Acute vaginal candidiasis.** Leukoplakia-like coating of the portio vaginalis, with intense redness of the squamous epithelium.

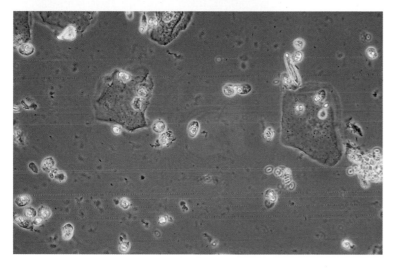

Fig. 3.**77** **Budding cells.** Phase contrast image of vaginal candidiasis. In addition to some superficial cells, numerous brightly reflecting, oval budding cells are present. ×630.

sociated with a whitish coat of the portio vaginalis (Fig. 3.**76**).

The detection of this fungus is much easier in wet mounts than in stained preparations because the pathogen shows higher contrast when viewed by phase contrast microscopy (Figs. 3.**77**, 3.**78**) (40). The exact identification of different fungal types in the cytological smear is not possible; it requires special culture media.

In stained preparations, fungal infections often yield a diffuse, milky background and are associated with pseudoeosinophilia and partial disintegration of individual squamous epithelial cells in the form of polka-dot cells or fenestrated cells (Fig. 3.**79**, and see Figs. 3.**12**, 3.**13** on pp. 60, 61).

The yeast either forms oval **budding cells** (**blastospores**), or develops tubular hyphae (**pseudomycelia**), both of which are either eosinophilic or do not stain at all (Figs. 3.**80 a–c**, 3.**81**, 3.**82**) (170). Budding yeast cells measure about 5 μm, thus being slightly smaller than erythrocytes; they often show budding protrusions in a "mother-and-child" configuration. By contrast, hyphae may reach a length of 100 μm and more. They consist of characteristic, double-contoured tubes, which are segmented and branched and show irregular, bamboo-like constrictions (Figs. 3.**78**, 3.**82**). These hyphae often form complex networks. Degenerating hyphae lose their double contours and are then difficult to distinguish from threads of mucus or elongated leukocytes (139).

Fig. 3.**78 Pseudomycelium.** Phase contrast image of vaginal candidiasis. In addition to some superficial cells, numerous double-contoured fungal hyphae (pseudomycelium) with typical segmentation are present. ×1000.

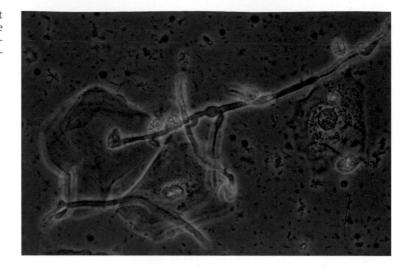

Fig. 3.**79 Fenestrated cell caused by candidiasis.** The cell is hardly recognizable due to partial dissolution of the eosinophilic cytoplasm (→). Irregular cytoplasmic ridges give the cell a segmented appearance. ×1000.

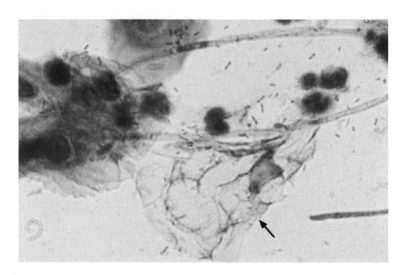

Fig. 3.**80 a–c Budding yeast cells indicating candidiasis.**
a Clusters of blastospores at the periphery of superficial cells (→). No inflammatory reaction in the background. ×400.

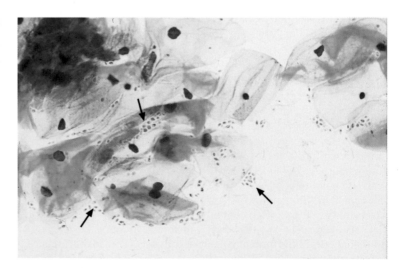

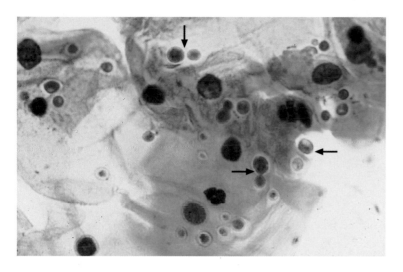

b Eosinophilic blastospores (⟶) of variable size and arrangement. No inflammatory reaction of the squamous epithelium. ×1000.

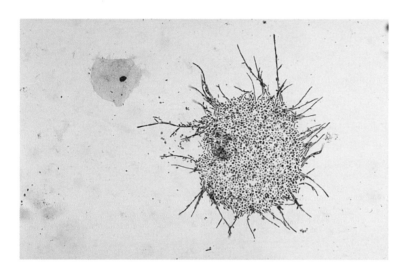

c Aggregation of blastospores which start to sprout and form a pseudomycelium. ×200.

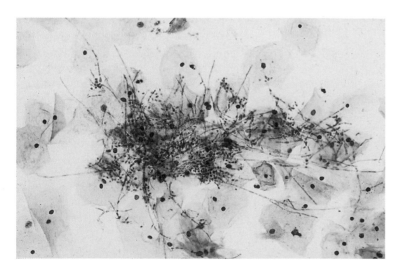

Fig. 3.**81** **A network of pseudomycelia indicating candidiasis.** Numerous budding cells and long hyphae are present. No inflammatory reaction of the squamous epithelium. ×250.

Fig. 3.**82 Typical Candida albicans pseudomycelium.** Well-preserved, branching hyphae with double contours and segmentation are seen between superficial cells. ×1000.

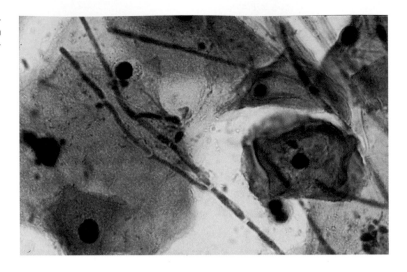

Fig. 3.**83 Spermatozoa in the vaginal discharge.** Phase contrast image. The tail and acrosome of each sperm appear dark, while the rest of the sperm head appears bright. ×1000.

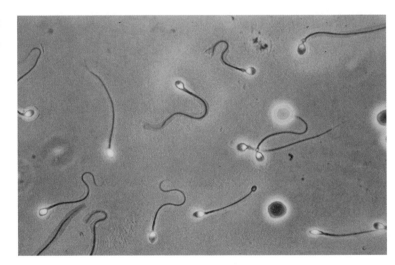

Spermatozoa

The detection of spermatozoa in native vaginal smears (Fig. 3.**83**), as well as their morphology and mobility, play an important role in fertility clinics and also in the forensic investigation of sexual offenses (338). In an already stained smear, chromosomal analysis is still possible by means of genetic fingerprinting.

Sperms are frequently detected in vaginal smears, and they are easy to recognize from their typical structure (Fig. 3.**84**) (338). They are faintly basophilic or do not stain at all. When degenerating, they lose their tails; the remaining sperm heads are difficult to distinguish from blastospores. However, whereas budding cells have a rather homogeneous structure, sperm heads are characterized by a faintly stained apical cap, the **acrosome**.

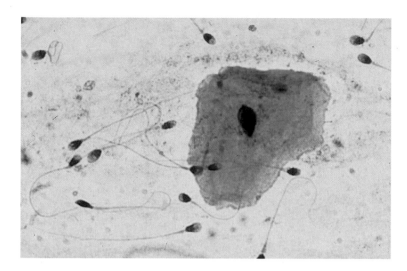

Fig. 3.**84 Well-preserved spermatozoa.** Note the undulating tail and apical translucency in the area of the acrosome which complements the dark area in the phase contrast image. ×1000.

Trichomonads

Trichomonal infection is one of the most common sexually transmitted diseases. The pathogens are not always easy to detect in the smear because of their faint staining (76). Because of their poor preservation in stained preparations, they are easily confused with degenerated parabasal cells, histiocytes, or mucus particles. In wet mounts, however, live trichomonads are usually reliably identified under the phase contrast microscope because of their mobility (Fig. 3.**85**).

Depending on the degree of maturation, trichomonads measure 10–30 µm in diameter, thus corresponding in size to parabasal cells or small intermediate cells. They are egg- or pear-shaped and faintly basophilic (Fig. 3.**86a, b**). The nucleus stains poorly as well. Occasionally, the four flagella are detected; these are used for locomotion. The cytoplasm often contains **reddish granules**, which are characteristic for these pathogenic protozoa (76).

Most of the time, the overall cell image indicates a severe inflammatory reaction, with the squamous epithelial cells showing enlarged, hyperchromatic nuclei and typical **perinuclear halos** (Fig. 3.**86c**) (160). Trichomonal infection is frequently associated with foamy cervical mucus, presumably generated by the action of the pathogen's flagella (Fig. 3.**87**) (161).

Fig. 3.**85 Trichomonal vaginitis.** Phase contrast image. In addition to many superficial cells, a cluster of pathogens with bright halos is visible in the center (→). The detection of erratic movements of trichomonads in the native preparation is evidence of the disease. ×630.

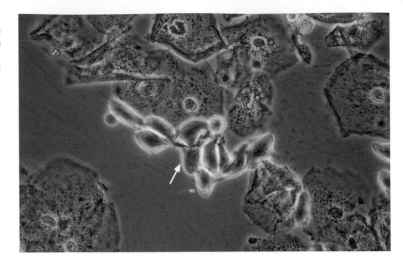

Fig. 3.**86 a, b Trichomonal infection.**

a Numerous basophilic pathogens (→) are present between squamous epithelial cells that show reactive swelling of their nuclei. ×400.

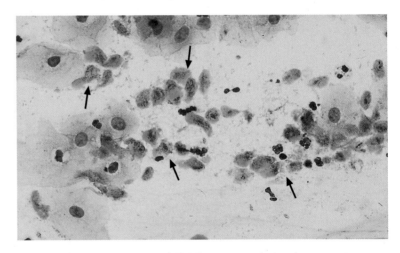

b In the lower right corner lies a single basophilic, round trichomonad (→) of the size of a para-basal cell. It shows typical eosinophilic cytoplasmic granules, a faintly stained nucleus, and an indistinct flagellum. Some intermediate cells with swollen nuclei are visible nearby. ×1600.

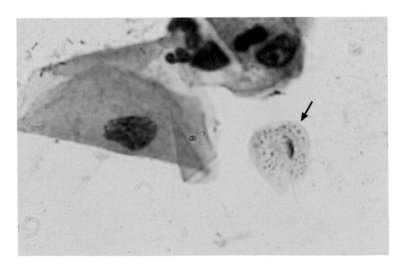

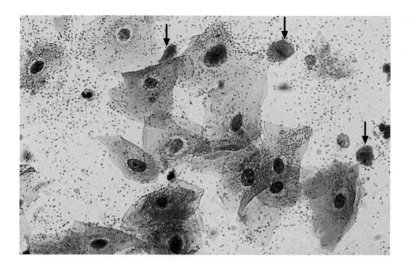

c Formation of typical perinuclear halos in squamous epithelial cells. Isolated pathogens are visible in the upper right corner (→). ×630.

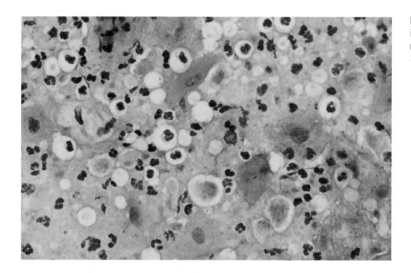

Fig. 3.**87 Trichomonal vaginitis.** The colorful cell image with foamy cervical mucus and inflammatory background of the preparation is typical. ×500.

Herpes Simplex Virus

Occasionally, the smear shows cytological changes that are caused by herpes simplex virus (HSV) type 2, less often by HSV type 1 (165). The infection mostly causes acute attacks, and the virus may remain dormant for many years without causing eruptions in the region of the nerves affected (414). Detection of the infection is possible by means of serology and cell culture. Since viral reproduction requires cells that are capable of dividing, these viruses preferentially infect parabasal cells, immature metaplastic cells, and endocervical cells (236).

HSV infection is usually associated with an inflammatory reaction, and the overall cell image reflects this. Since it affects the deeper cell layers where it causes cell destruction, further maturation of the squamous epithelial cells is no longer possible. The inflammatory reaction creates space-occupying lesions in the deep epithelial layers, which clinically manifest themselves as characteristic painful vesicles or ulcers. The content of these vesicles consists of inflammatory exudate as well as infected, disintegrating parabasal cells; it becomes infectious when the vesicle ruptures. After rupture of the vesicle, new infectious ulcers develop at the same site (Fig. 3.**88**).

The smear image exhibits enlarged nuclei of infected cells. The chromatin is arranged in clumps adhering to the nuclear envelope. This makes the nuclear envelope hyperchromatic,

Fig. 3.**88 Genital herpes simplex.** Numerous ulcers of variable size are present on the anterior lip of the external os of uterus.

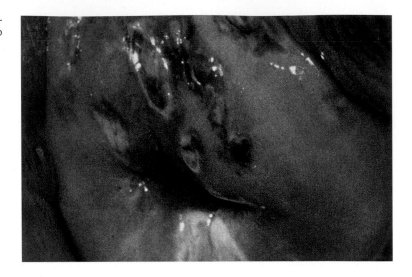

Fig. 3.**89 a, b Genital herpes simplex infection.**
a Development of numerous polynuclear cells with hyperchromatic nuclear envelope and central nuclear translucency ("ground-glass phenomenon"). ×400.

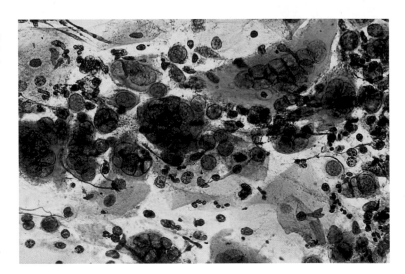

while the center of the nucleus appears without structure—an effect referred to as the "ground-glass phenomenon" (Fig. 3.**89 a, b**) (238). Simultaneous damage of the cytoplasm leads to syncytial confluence of several cells, thus forming **polynuclear cells** (Fig. 3.**90 a**). The individual nuclei compress one another (nuclear molding) and develop dark **inclusion bodies** in case of recurrent infection (Fig. 3.**90 b**) (254).

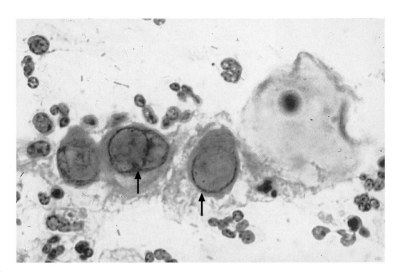

b Typical ground-glass phenomenon of the nuclei. The inside of the nuclei appears empty, and the nuclear envelope is evenly hyperchromatic (⟶). ×630.

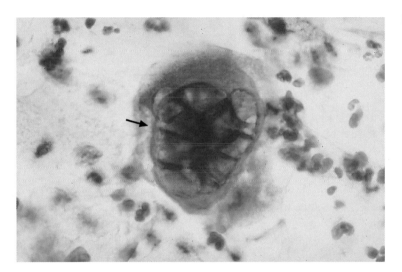

Fig. 3.**90 a, b   Genital herpes simplex infection.**
a A polynuclear cell (⟶), with the ground-glass nuclei compressing each other (nuclear molding). ×630.

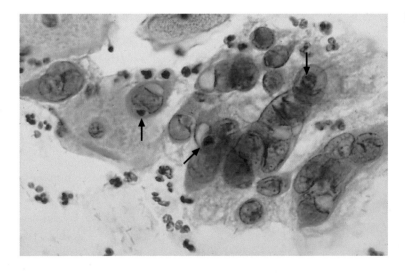

b Typical image of dark, intranuclear inclusion bodies (⟶). ×400.

Fig. 3.**91** **Acuminate condyloma.** A cluster of fine papillae on the anterior lip of the external os of uterus (→). The end of the thread of an intrauterine device is visible in the endocervical region (⇒).

Human Papillomavirus

Human papillomavirus (HPV) infection:
▶ May not be apparent
▶ Leads to condylomas
▶ Or leads to neoplasia

Inapparent HPV infection. This is very common, reflecting the fact that HPV is easily transmitted by sexual contact. The incidence is 30% or more (217). Unlike HSV infection, the course of this disease is rather chronic and may exist over months and years, depending on the body's state of immunity. The symptoms during the **clinically inapparent course** are nonspecific; sometimes only itching is reported. There are about 100 different types of HPV, most of which are neither pathogenic nor carcinogenic (294). HPV infection does not cause an inflammatory reaction in the overall cell image, although such a reaction may be present for other reasons.

Condylomas. Certain types of HPV, in particular types 6 and 11, may cause hyperkeratotic papillomas that manifest themselves clinically as **acuminate condylomas** or genital warts (Fig. 3.**91**) (102). Unlike HSV infection, the cellular damage due to HPV does not lead to cell death (103). On the contrary, the virus causes acute cell proliferation, which leads to thickening of the parabasal layer on the one hand, and to enhanced maturation with keratinization of the superficial layer on the other. Although the DNA-containing virus needs immature, dividing cells for its reproduction, the cytopathic effect is delayed and affects only the intermediate and superficial epithelial layers.

In the cytological smear, the infected cells usually exhibit a hyperchromatic nucleus with degenerative changes, and binuclear cells are common. Certain viral metabolites are released into the cytoplasm and damage it, thus creating a clear area. Cells altered in this way are called **koilocytes** (Greek: koilos, hollow) and are pathognomonic for HPV infection (Fig. 3.**92 a, b**) (15, 181). Some authors assume that the detection of distinct koilocytes indicates an increased risk for the patient to develop dysplasia (220). On the other hand, only about half of all carcinomas in which koilocytes had been detected were also found to be positive for HPV (81).

Less characteristic are some examples of cytoplasmic damage referred to as **indirectly caused HPV changes**. These lead either to aberrant staining, especially **amphophilia**, or to structural changes, such as **fissures** (cracked cytoplasm) or **polka-dot cells** and **fenestrated cells** (Fig. 3.**93 a–e**) (341). Rapid maturation of the cells may lead to imperfect keratinization and, therefore, detection of **parakeratotic or dyskeratotic cells** with degenerated nuclei (Fig. 3.**94 a, b**) (399). Although such cells are often found in cases of HPV infection, they are not specific for this infection; horn cells may be present due to other causes, such as bacterial inflammation, pressure-induced damage in patients using a vaginal diaphragm, or superficial reactions after inflammation.

Exfoliated infected cells finally undergo natural degeneration with disintegration of both cytoplasm and nucleus, thus releasing infectious virus particles from the nucleus (428).

Neoplasia. The various types of HPV can be distinguished by means of special detection methods, particularly DNA hybridization, thus facilitating a prognosis regarding the potential **carcinogenicity** of the virus (159). Only a few types, particularly 16 and 18, have carcinogenic potential. It should be borne in mind, however, that HPV infection has high rates of both remission and recurrence, and that presence of the virus does not necessarily lead to disease. General HPV screening is therefore not justified because of the high costs and the controversy regarding the clinical consequences (81) (see also pp. 169 ff.).

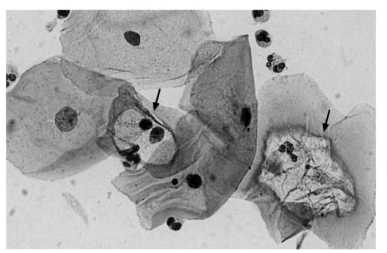

Fig. 3.**92 a, b Koilocytosis due to HPV infection.**
a Superficial squamous epithelial cells with typical sharp-edged punch holes in the cytoplasm (moth-eaten appearance). The perinuclear halos have dense cytoplasmic demarcations (⟶), and the nuclei undergo karyopyknosis. Binuclear cells are characteristic. ×630.

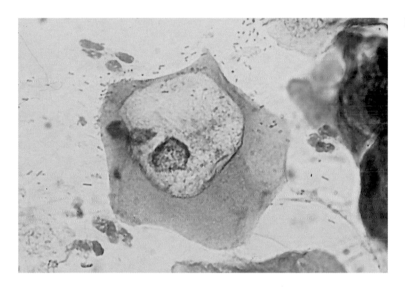

b A superficial cell showing a large perinuclear halo and degenerated nucleus with eccentric position. ×630.

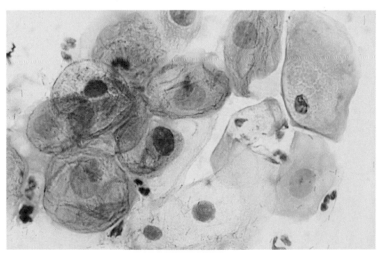

Fig. 3.**93 a–e Indirect signs of HPV infection.**
a Amphophilic cytoplasm with signs of koilocytosis. ×630.

b Fissures in the degenerating cytoplasm (**cracked cytoplasm**) (→). In addition, the cell exhibits koilocytosis. ×1000.

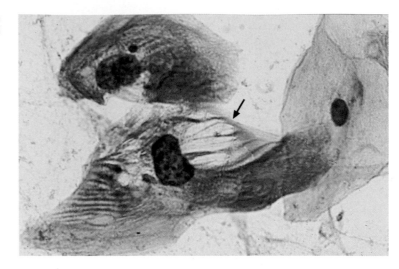

c Eosinophilic cytoplasmic degeneration with dot formation (**polka-dot cells**) (→). ×250.

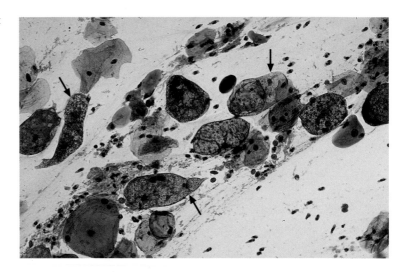

d Eosinophilic disintegration of the cytoplasm and dot formation (**polka-dot cell**). ×1000.

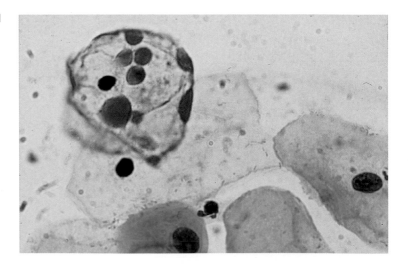

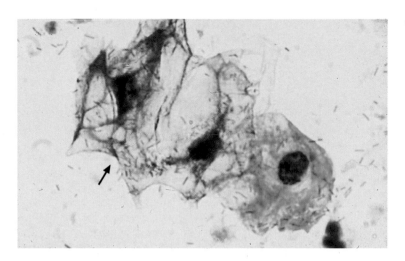

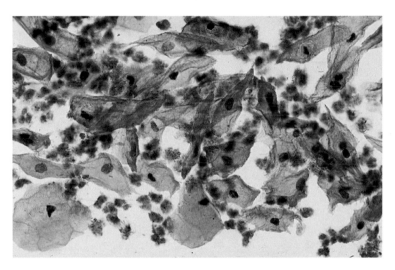

Fig. 3.**94 a, b Dyskeratocytes.** Horn cells with some nuclear abnormalities are an indirect sign of HPV infection.

a The background of the preparation indicates inflammation. ×400.

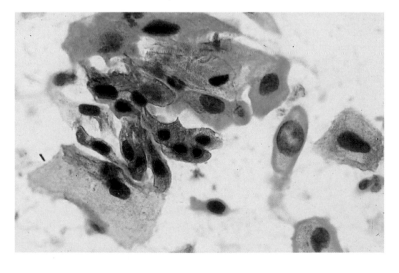

b The background of the preparation is clean. Normal intermediate cells are visible nearby. ×790.

Table 3.**1** Effects of acute inflammation on the portio and vagina

Macroscopic effects

▶ Redness
▶ Discharge

Histological effects

▶ Hyperemia
▶ Leukocytic infiltration

Cytological effects

On the cytoplasm
▶ Amphophilia
▶ Pseudoeosinophilia
▶ Keratinization
▶ Perinuclear halo
▶ Vacuolization
▶ Structural changes
▶ Heterolysis

On the nucleus
▶ Nuclear swelling and disturbance of the nucleocytoplasmic ratio
▶ Hyperchromatic nucleus with coarse chromatin structure
▶ Hyperchromatic nuclear membrane
▶ Formation of chromocenter and nucleolus
▶ Karyolysis, karyorrhexis, karyopyknosis

Vaginitis

Inflammatory reactions observed in the vaginal smear may be induced by bacterial or viral infections, protozoosis, or mycosis, as well as by mechanical, chemical, radiogenic, or thermal factors. As a rule, these reactions start with an acute phase, followed by a chronic phase and a regenerative phase. In recurrent infection, the different phases may even coexist.

Occasionally, specific changes are detected that are characteristic for the agent causing the inflammation, such as inclusion bodies, cytopathic effects, radiation damage, etc.

Acute Phase

Acute inflammation of the portio and vagina induces the changes shown in Table 3.**1**.

Macroscopy and Colposcopy

The intense redness of the vaginal epithelium caused by acute inflammation is easily recognized as the stromal papillae become very pronounced, thus causing **red punctuation of the epithelium** (223). The acetic acid test further enhances this effect because the very thin parts of the epithelium overlying the papillae react less intensely than the thicker areas between the papillae (Fig. 3.**95**). This creates a benign punctuation pattern (Fig. 3.**96 a**) (54). A more severe inflammation may have the opposite effect, thus creating a mosaic pattern (Fig. 3.**96 b**). In contrast to areas of abnormal punctuation, which are always circumscribed and localized, inflammatory punctuation is diffuse and usually manifests itself on the entire vaginal epithelium. Simultaneously, there is an increased secretion of fluid (**discharge**) from the swollen squamous epithelium as well as from the endocervix.

Histology

The inflammatory reaction is triggered by epithelial damage, to which the body responds with humoral defenses. There is a vascular and a cellular reaction. The blood vessels of the stroma dilate, and the blood flow increases (**hyperemia**). At the same time, granulocytes migrate out from the blood vessels. They penetrate first the connective tissue and eventually the epithelium (**leukocytic infiltration**) (Fig. 3.**97 a**). Furthermore, a proliferative stimulus acts on the basal cell layer, thus ensuring rapid exfoliation of the diseased epithelial layers. This quickly moves cells from the deep layers of the epithelium up to the surface, thus preventing their normal maturation (Fig. 3.**97b**) (160).

Cytology

Leukocytic exudate. In acute inflammation, phase contrast cytology may be used for immediate diagnosis (Fig. 3.**98**). A leukocytic exudate is often found in both native and stained preparations. However, the number of granulocytes does not necessarily correlate with the severity of the inflammation, because leukocytes may be present in noninflam-

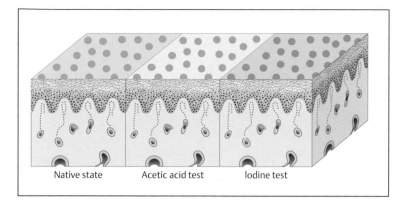

Fig. 3.**95 Colposcopic findings in case of acute vaginitis.** The acetic acid test results in enhancement of the red punctuation pattern on the epithelial surface. The iodine test causes a change in color due to the presence of glycogen ("leopard pattern").

Native state   Acetic acid test   Iodine test

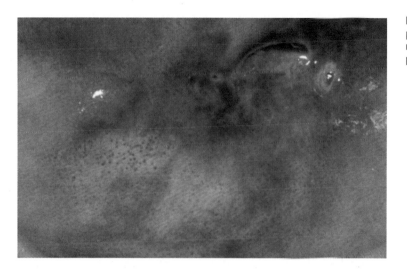

Fig. 3.**96 a Vaginitis.** Colposcopic image after applying the acetic acid test, showing diffuse inflammatory punctuation with peripheral redness at the posterior lip of the external os of uterus.

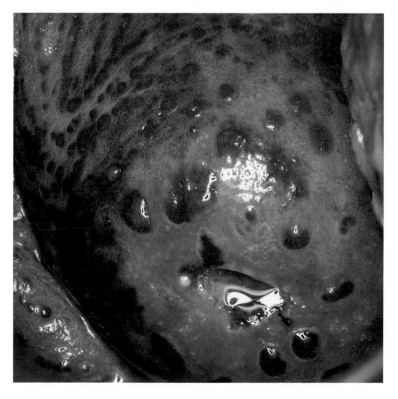

Fig. 3.**96 b Trichomonal vaginitis.** Colposcopic image after applying the acetic acid test, showing severe inflammation with diffuse, mosaic-like structure of the portio vaginalis and the adjacent vaginal epithelium.

3

Fig. 3.**97 a, b   Acute inflammation of the vaginal epithelium.**
a The connective tissue (stroma) exhibits increased vascularization and leukocytic infiltration. The squamous epithelium shows a reduced number of cell layers (→) due to the increased proliferation and exfoliation of cells. HE stain, ×100.

b Increased proliferation and enhanced exfoliation prevent normal maturation, thus leading to fewer cell layers. The nuclei are enlarged and show nucleoli, though no atypia. HE stain, ×400.

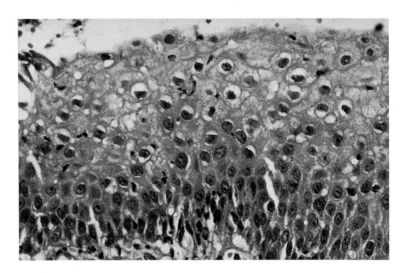

Fig. 3.**98   Acute vaginitis.** Phase contrast image with numerous granulocytes, parabasal cells, and immature metaplastic cells (→). A mixed flora is visible in the background. ×630.

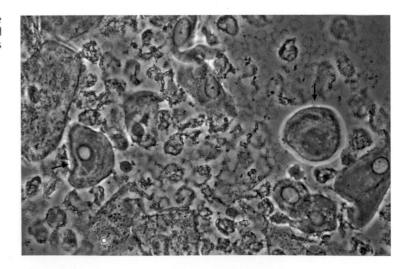

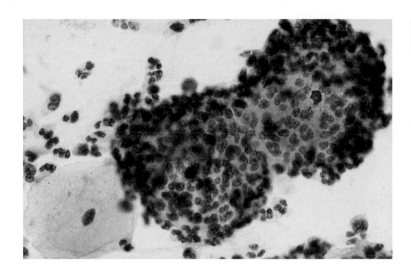

Fig. 3.**99 a Acute inflammation with leukotaxis.** Superficial cells show a dense accumulation of leukocytes on their surface. ×630.

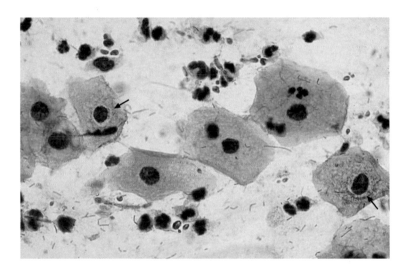

Fig. 3.**99 b Mild inflammation.** Swollen and hyperchromatic nuclei of superficial cells with perinuclear halos (⟶). ×630.

matory processes as well—for example, during menstruation (368) (see also p. 66). The accumulation of leukocytes on top of epithelial cells is due to a chemotactic effect and is referred to as **leukotaxis** (Fig. 3.**99 a**).

Cellular exfoliation. Inflammatory reactions of the squamous epithelium may be detected even in the absence of leukocytes. In mild inflammation, especially when associated with fungal infections, only the superficial and intermediate cells react (Fig. 3.**99 b**). By contrast, in severe inflammation the strong proliferative stimulus causes exfoliation of basal and parabasal cells, which usually show the same reactive changes (Fig. 3.**100**) (160). The background of the preparation is dirty, showing an increase

in mucus, cell debris, erythrocytes, hemosiderin, or precipitated proteins. The endocervical epithelium is usually also involved in the inflammatory reaction (see pp. 127 ff.).

Inflammatory reactions of the nucleus. The nuclei of squamous epithelial cells may either be active or degenerated.

▶ **Activation:** The nuclei are swollen and hyperchromatic, exhibiting hyperchromatic nuclear membranes as well as chromocenters or nucleoli (Fig. 3.**101 a, b**).
▶ **Degeneration:** This involves karyolysis, karyorrhexis, or karyopyknosis (Fig. 3.**102 a, b**).
▶ **Differential diagnosis:** Both activation and degeneration often cause a shift of the nucleocytoplasmic ratio in favor of the nu-

Fig. 3.100 Severe inflammation. Exfoliation of deep cell layers, with enlarged and hyperchromatic nuclei, hyperchromatic nuclear envelopes, distinct nucleoli, and evenly distributed chromatin. A normal superficial cell is visible on the left. ×1000.

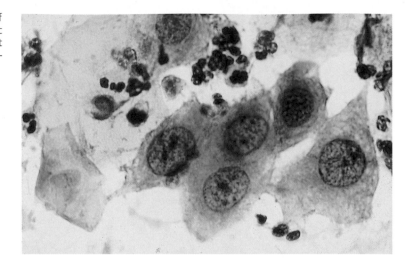

Fig. 3.101 Inflammation with activation of cell nuclei.

a Uniformly enlarged and hyperchromatic nuclei with formation of chromocenters, associated with amphophilic cytoplasm. ×630.

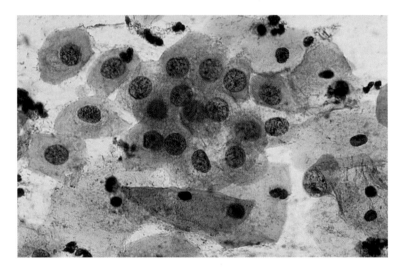

b Extremely enlarged, hyperchromatic nuclei of intermediate cells, with hyperchromatic nuclear envelope, formation of chromocenters, and irregular chromatin structure. ×1000.

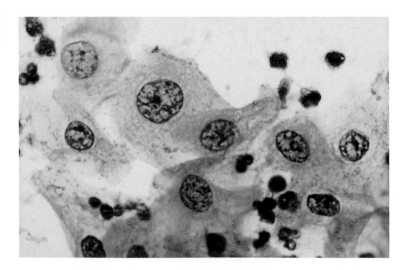

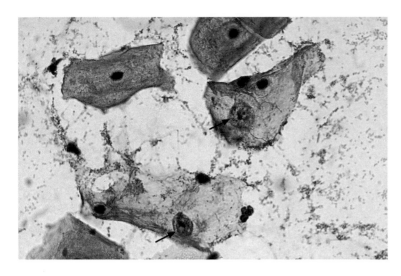

Fig. 3.**102** **Inflammation with nuclear degeneration.**
a First signs of karyorrhexis (→) and karyopyknosis, associated with reactive keratinization of the cytoplasm. ×630.

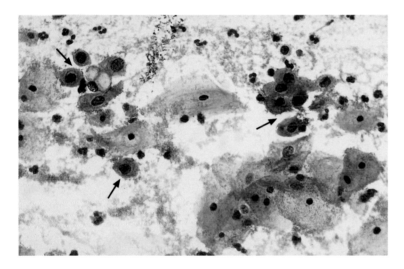

b Karyopyknosis of superficial cells and nuclear condensation of parabasal cells (→). ×400.

cleus. The nuclear changes—which may be associated with an **irregular, coarse chromatin pattern**, irregularities of the nuclear membrane, polymorphism in nuclear size and shape, as well as anisocytosis—make it difficult to distinguish between inflammatory reactions and malignant processes. However, the diffuse expression of these criteria and the presence of nucleoli are more typical for inflammatory reactions than for dysplastic neoplasia (368).

Inflammatory reactions of the cytoplasm. The cytoplasm also responds to inflammatory stimuli and may exhibit different staining properties, such as staining red as well as blue (**amphophilia**). Frequently, there are also **pseudoeosinophilia, keratinization, perinuclear halos, vacuolation,** or **structural changes** together with cytoplasmic disintegration (polka-dot cells or fenestrated cells) (Figs. 3.**103**–3.**105**; see also Figs. 3.**12**, 3.**13**, pp. 60, 61). Furthermore, bacterial toxins may completely or partially lyse the cells (**heterolysis**), thus generating indistinct cell borders and naked nuclei (Fig. 3.**106**) (368).

Fig. 3.**103 Inflammation with cytoplasmic reaction.**
a Nuclear swelling and cytoplasmic vacuolation of parabasal cells. Numerous leukocytes are visible in the background. ×400.

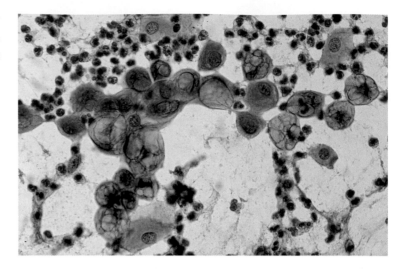

b Nuclear enlargement and karyopyknosis of superficial cells, associated with pseudo-eosinophilia and keratinization of the cytoplasm (—▶). ×1000.

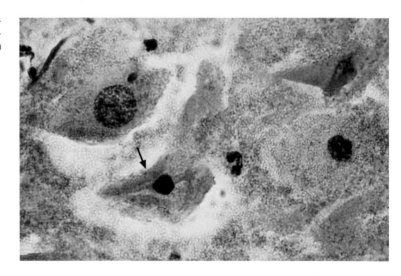

Fig. 3.**104 Inflammation with cytoplasmic reaction.** Nuclear swelling and perinuclear halo (—▶). ×1000.

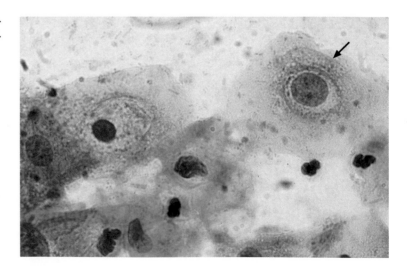

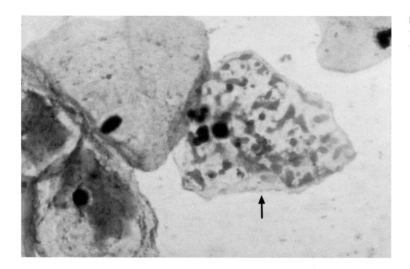

Fig. 3.**105** **Inflammation with cytoplasmic reaction.** Disintegration of the cytoplasm (polka-dot cell) (→). ×630.

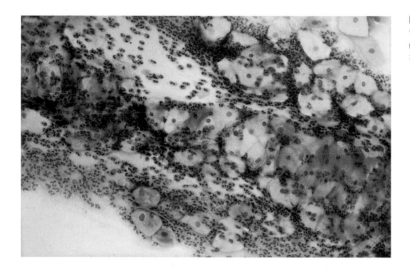

Fig. 3.**106** **Inflammation with cytoplasmic reaction.** Pseudoeosinophilia, indistinct cell borders (due to heterolysis), and leukocytic background. ×100.

Table 3.**2** Effects of chronic inflammation on the portio and vagina

Macroscopic effects	Histological effects	Cytological effects
▶ Few characteristic changes	▶ Granulation tissue	▶ Histiocytes
		▶ Plasma cells
		▶ Lymphocytes
		▶ Fibroblasts
		▶ Smooth muscle cells

Chronic Phase

Chronic inflammation of the portio and vagina is characterized by the changes listed in Table 3.**2**.

Macroscopy

The native cell image is less characteristic during the chronic phase of inflammation, and the redness of the vaginal epithelium and the amount of discharge have usually decreased.

Fig. 3.**107** **Chronic inflammation of the vaginal epithelium.** The connective tissue (stroma) shows granulation tissue, increased vascularization, and reticular cells that partly reach the epithelial surface (——➤). The epithelial stratification begins to normalize. HE stain, ×100.

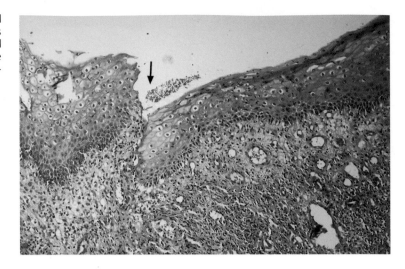

Fig. 3.**108** **Histiocytes.** Foamy cytoplasm, indistinct cell borders, and kidney-shaped, partly degenerated nuclei. ×400.

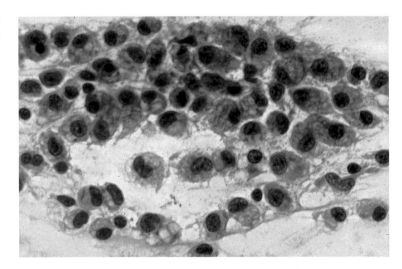

Histology

The stroma shows an increase in both vascularization and connective tissue formation, as reflected by increased numbers of fibroblasts, collagen fibers, and round cells (histiocytes, plasma cells, lymphocytes). Connective tissue with this kind of organization is referred to as **granulation tissue** (Fig. 3.**107**).

Cytology

Chronic inflammation is characterized by an increase in reticulocytes (111). Simultaneously, there is a reduction in the number of granulocytes and a decline in acute inflammatory reactions in both the nuclei and the cytoplasm of epithelial cells. The background of the preparation appears increasingly clean.

Histiocytes. Histiocytes have a chromophobic cytoplasm with poorly defined borders and a kidney-shaped, usually eccentric nucleus, which becomes condensed and pyknotic when it degenerates (Figs. 3.**108**, 3.**109 a**) (240). They sometimes show hyaline cytoplasmic degeneration and then stain intensely eosinophilic, thus making it difficult to distinguish them from small parakeratotic cells (Fig. 3.**109 b**). Histiocytes may merge to form polynuclear **giant cells** (Fig. 3.**110**). These vary in size, and they phagocytose other cells and cellular debris, in which case they are referred to as **macrophages** (Fig. 3.**111 a, b**) (275).

Plasma cells. Plasma cells are round and are about the same size as small histiocytes. Their cytoplasm is chromophobic to basophilic, their cell borders are more defined than those of histiocytes, and their round nucleus is usually

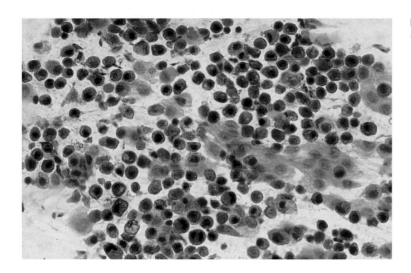

Fig. 3.**109 a, b   Degenerated histiocytes.**
a Nuclear condensation and lysis of the cytoplasm. ×400.

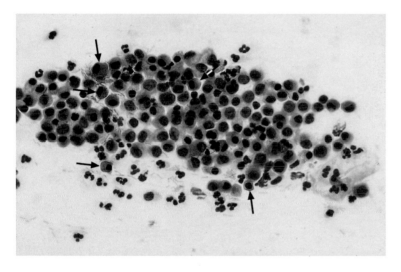

b Nuclear condensation, karyopyknosis, and hyaline degeneration of the cytoplasm (⟶). ×400.

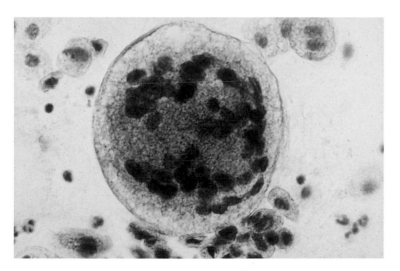

Fig. 3.**110   Histiocytic giant cells.** Numerous irregularly arranged nuclei. ×400.

3

Fig. 3.**111** **Macrophages** (phagocytic histiocytes).
a Incorporation of leukocytes (——►). Intermediate cells are visible in the background. ×400.

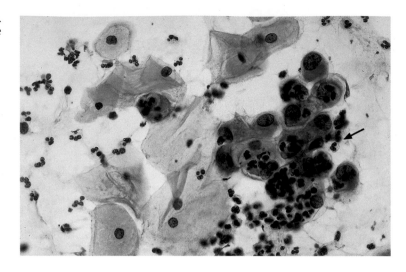

b Vacuolation (——►). Leukocytic background. ×250.

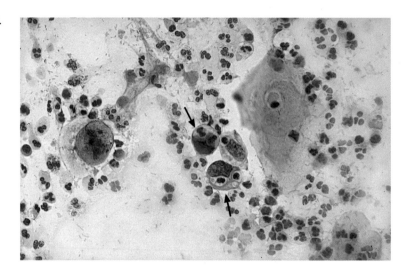

Fig. 3.**112** **Plasma cells.** Linear arrangement of plasma cells (——►), with marginal nuclei showing a typical cartwheel structure. Superficial cells in the upper half of the micrograph. ×250.

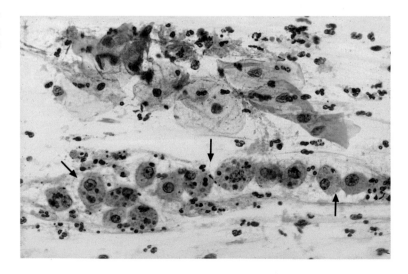

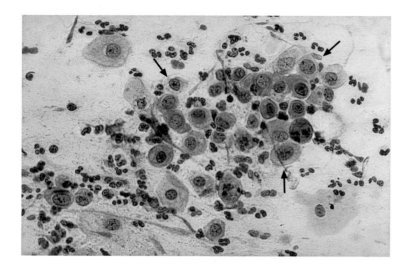

Fig. 3.**113** **Lymphocytes.** A loose cluster of lymphocytes (→) surrounded by parabasal cells and granulocytes. ×250.

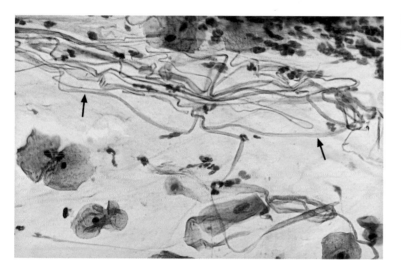

Fig. 3.**114** **Smooth muscle cells.** The cytoplasm is extremely elongated (→). Superficial cells are visible nearby. ×250.

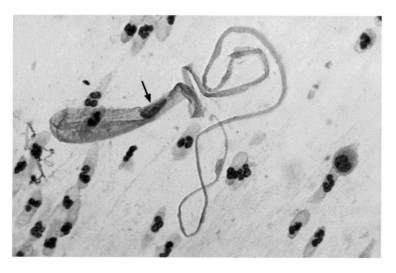

Fig. 3.**115** **A smooth muscle cell.** The cell has the shape of a tadpole and contains a pyknotic nucleus (→). ×400.

Fig. 3.**116** **Fibroctes.** A loose aggregation of narrow, spindle-shaped cells with pyknotic nuclei (→). Intermediate cells are visible in the center. ×400.

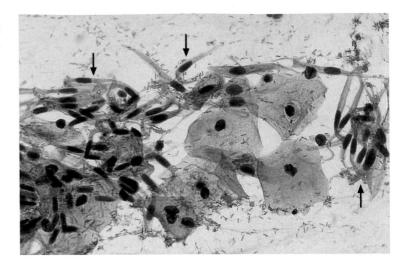

Fig. 3.**117** **Fibrocyte.** An isolated narrow and extremely elongated cell with pyknotic nucleus. Superficial and intermediate cells are visible nearby. ×500.

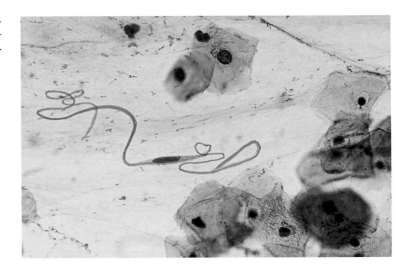

Fig. 3.**118** **Fibroblasts.** A loose cluster of spindle-shaped, voluminous cells (→) with active, vesicular nuclei. ×400.

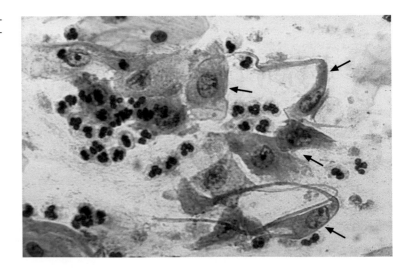

118

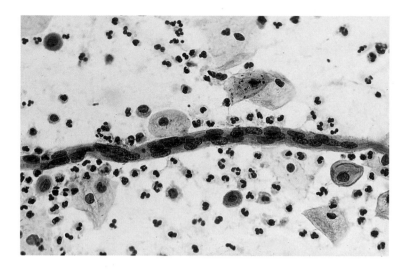

Fig. 3.**119** **Polynuclear fibroblast.** An extremely elongated, isolated cell with numerous nuclei lined up in a row. ×400.

Table 3.**3**   Effects of epithelial regeneration on the portio and vagina

Macroscopic effects	Histological effects	Cytological effects
▶ Few characteristic changes, occasionally leukoplakia	▶ Basal cell hyperplasia ▶ Superficial keratinization ▶ Regenerative epithelium	▶ Basal cells ▶ Surface reaction ▶ Regenerative epithelium

found at the cell margin, exhibiting a typical cartwheel structure (Fig. 3.**112**) (349).

Lymphocytes. Lymphocytes are easily recognized by their small size and their round nucleus (Fig. 3.**113**).

Fibrocytes, fibroblasts, and smooth muscle cells. These are rarely found in the smear; they usually derive from inflammatory traumatism of the connective tissue (stroma) or from a submucosal myoma (Figs. 3.**114**–3.**119**) (237).

Regenerative Phase

Epithelial regeneration after inflammation of the portio and vagina exhibits the changes listed in Table 3.**3**.

Macroscopy

Regeneration rarely causes typical macroscopic changes of the squamous epithelium. Occasionally, **leukoplakia** is observed; as it presents itself as regular and flat, it does not cause suspicion (Fig. 3.**120**).

Histology

Histological examination usually reveals **basal cell hyperplasia**, which causes a thickening of the basal cell layer (Fig. 3.**121**) (81). During the healing phase, normal epithelial maturation resumes, and superficial cells reappear (Fig. 3.**122**). Occasionally, **superficial keratinization** may occur as the result of overreaction (Fig. 3.**123**) (355).

After severe inflammation or trauma, both squamous epithelium and columnar epithelium show tissue repair in the form of **regenerative epithelium** that grows horizontally across the defective epithelium (Fig. 3.**124**). Whereas the squamous epithelium is regenerated by basal cells, the glandular epithelium develops from reserve cells. In both cases, repair by proliferation begins in the defective area adjoining the still intact epithelium; here, voluminous cells with enlarged nuclei, containing nucleoli, develop within a short period (178). These epithelial cells proliferate rapidly until they completely cover the defective surface. This is followed by the development of basal cells, or reserve cells, from the regenerative epithelium and, subsequently, by gradual development of all stages of normal squamous, or columnar, epithelial maturation (Fig. 3.**125**) (81).

3

Fig. 3.**120** **Leukoplakia.** A circumscribed, flat, and benign area found on the portio vaginalis as a result of chronic irritation.

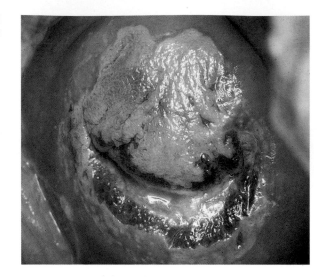

Fig. 3.**121** **Basal cell hyperplasia.** Regenerative inflammatory phase of the vaginal epithelium. The normal epithelial stratification has been completely lost and has been replaced by a layer of evenly distributed hyperchromatic, isomorphic basal cells. HE stain, ×400.

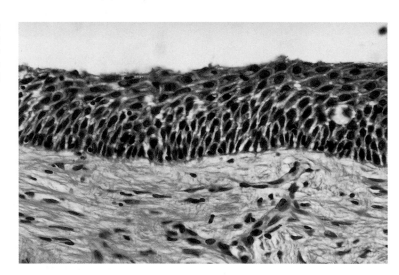

Fig. 3.**122** **Basal cell hyperplasia.** Acute inflammatory phase of the vaginal epithelium leads to increased proliferation, thus steadily widening the basal cell layer. After the inflammation has healed, the hyperplastic basal cell layer regresses, and cell maturation normalizes.

	Super-ficial-cell	Inter-mediate cell	Para-basal cell	Basal cell	Para-basal cell	Inter-mediate cell	Super-ficial-cell
Super-ficial cell layer							
Inter-mediate cell layer							
Para-basal cell layer							
Basal cell layer							

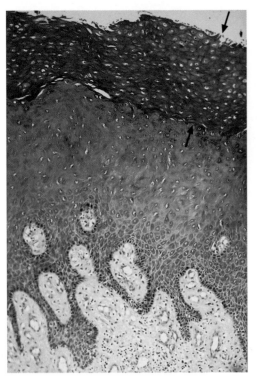

Fig. 3.**123 Surface effect.** Regenerative inflammatory phase of the vaginal epithelium. The epithelial stratification has normalized, but the superficial zone of keratinization is relatively wide (➞). HE stain, ×100.

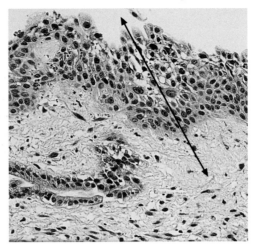

Fig. 3.**124 Regenerative epithelium.** Regenerative inflammatory phase of the vaginal epithelium. The epithelium shows no signs of maturation; it consists of immature cells, the nuclei of which are enlarged and polymorphic and contain macronucleoli. The nuclear axes show a certain polarization in one direction, as indicated by the arrows. HE stain, ×200.

	Super-ficial cell	Tissue defect	Regene-rative epi-thelium	Basal cell	Para-basal cell	Inter-mediate cell	Super-ficial-cell
Super-ficial cell layer							
Inter-mediate cell layer							
Para-basal cell layer							
Basal cell layer							

Fig. 3.**125 Regeneration of a tissue defect.** In case of complete loss of the squamous epithelium, the adjacent basal cells undergo proliferative metamorphosis and transform into a regenerative epithelium. After the latter has filled the tissue defect, they change back into basal cells from where normal epithelial maturation starts.

Fig. 3.**126 a, b   Basal cell hyperplasia.**
a Isolated basal cells (⟶) lie between superficial cells with degenerated nuclei. ×400.

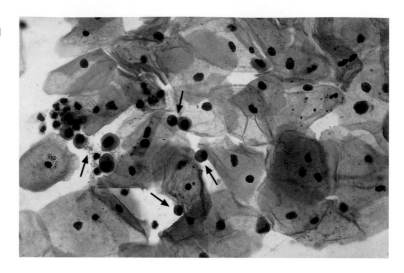

b Several basal cells (⟶) with degenerated nuclei. Superficial cells are visible in the lower half of the micrograph. ×630.

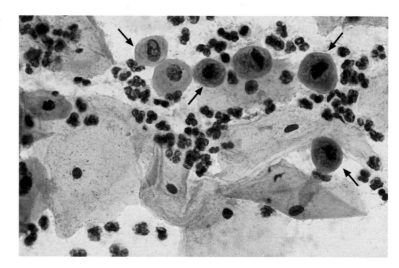

Cytology

Basal cell hyperplasia. In the cytological smear, basal cells are frequently found during the healing phase of inflammation, and this is referred to as **basal cell hyperplasia** (81) (see also pp. 19 ff.). The basal cells are distinctly smaller than parabasal cells and have a round, active or degenerated, and often hyperchromatic nucleus. Their cytoplasm has distinct borders and stains darkly cyanophilic (Fig. 3.**126 a, b**, and Figs. 2.**19**–2.**21**, pp. 19, 20). Basal cells are almost always isolated and diffusely distributed in the smear; they rarely form small clusters. If they show nuclear degeneration, they may be confused with cells derived from a small-cell carcinoma in situ. In the context of physiological cell renewal, basal cell hyperplasia leads to complete healing and subsequent maturation of the squamous epithelium.

Surface effect. Hasty maturation may cause superficial keratinization, resulting in parakeratotic cells. These are often spindle-shaped and elongated, and they may even form keratin pearls (Figs. 3.**127**, 3.**128**) (282). These phenomena are called **surface reaction**. They are difficult to distinguish from mild dysplasia or from changes caused by HPV.

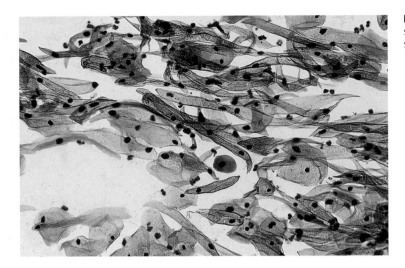

Fig. 3.**127** **Surface effect.** Keratinized, spindle-shaped superficial cells are often arranged like a shoal of fish. ×200.

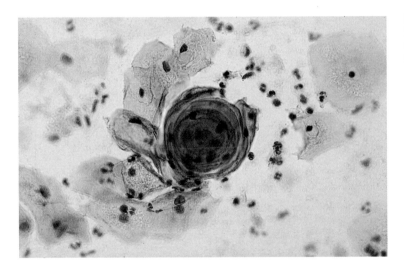

Fig. 3.**128** **A benign keratin pearl due to the surface effect.** The cells are arranged like the layers of an onion and have pyknotic nuclei. Some superficial cells are visible nearby. ×400.

Regenerative epithelium. Severe epithelial damage—such as that caused by necrotic inflammation, cryotherapy, electrical or surgical intervention, injury, or radiation—may completely destroy parts of the squamous or columnar epithelium. This triggers rapid epithelial repair by means of a syncytial tissue, underneath which the original epithelium is then rebuilt (178). This syncytial tissue represents the **regenerative epithelium**, which usually appears in the form of two-dimensional cell clusters in the vaginal smear. Because the new epithelium covering the defect grows in a tangential direction, the cellular and nuclear axes within these cell clusters tend to have a uniform direction; this is known as polarization (Fig. 3.**129a, b**). The enlarged, hyperchromatic nuclei almost always contain one or more prominent macronucleoli. Regeneration of the squamous epithelium is characterized by sharply outlined, compact clusters with homogeneous cytoplasm, whereas regeneration of the endocervical epithelium exhibits loose clusters with indistinct margins and vacuolated cytoplasm (see pp. 142 ff.) (100).

Fig. 3.**129 a, b Squamous epithelial regeneration.**

a A cell cluster showing the polarization of cells and nuclei (as indicated by the arrows), nuclear enlargement, and formation of macronucleoli. ×400.

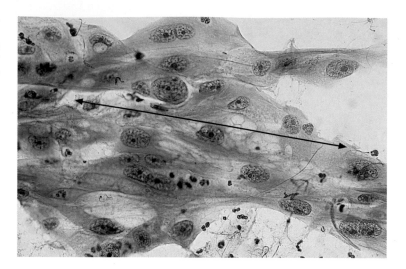

b A two-dimensional epithelial sheet viewed from above, showing uniformly arranged nuclei with nuclear enlargement and distinct nucleoli. Isolated superficial cells are visible nearby. ×400.

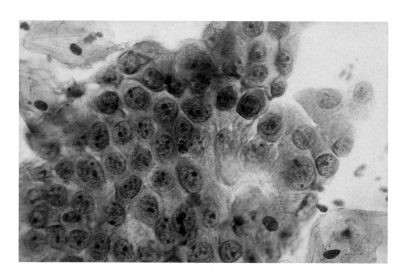

Therapy

Physiological colonization with lactobacilli is found in only about one third of all women. A facultative pathogenic flora is therefore not considered abnormal, as long as there are no complaints. Even atrophy in the vaginal region is not usually associated with any clinical symptoms.

Atrophic, infectious, and inflammatory changes of the vagina are therefore treated only when the patient complains of certain symptoms, such as pain, itching, burning, discharge, or offensive odor, or when it is feared that the lack of treatment might cause permanent damage, such as sterility resulting from salpingitis. Treatment may also become necessary if interfering factors prevent the unambiguous assessment of the cytological smear.

Epithelial atrophy. Atrophy of the vaginal epithelium, which is due to estrogen deficiency after the menopause, may lead to dryness and vulnerability of the vaginal epithelium and also to susceptibility to infections. These conditions are easily treated with hormone replacement therapy, as is commonly done today. Systemic (oral, parenteral, or transdermal) application of hormone is preferred to topical treatment, although the latter is often used as an additional measure (see also p. 24).

Bacterial and fungal inflammations. These are usually effectively treated with any of the many topical antibiotics and antimycotics that are commercially available. Tetracyclines are primarily used for cocci and *Chlamydia*, metronidazoles for *Gardnerella* and trichomonads, and nystatins for fungal infections. If the pathogen is resistant, broad-spectrum antibiotics or antimycotics may be used as well. It may be necessary to treat the sexual partner as well. Furthermore, the colon should also be treated in case of fungal infection, because it often represents a persistent reservoir of pathogens. Although topical therapeutic application is usually not possible during menstruation, treatment should be resumed during the early days of the cycle (e.g., by applying the ointment to the tampon), since the vaginal flora may be disturbed by the accumulation of blood and the body's resistance is diminished during this time.

Topical treatment may be supplemented by systemic (usually oral) therapy. As this will destroy the natural vaginal flora, treatment should be followed by acidification using lactate in combination with lactobacilli (79). Sometimes the symptoms recur after symptom-free intervals, in which case the therapy needs to be repeated. In such cases, lasting therapeutic success may be achieved only by prophylactic replacement of lactic acid and lactobacilli over a longer period of time. Recently, prophylactic vaccination with altered lactobacillus strains has been recommended.

Herpes infection. Nowadays this infection may be successfully treated with aciclovir (topically, orally, or parenterally), although relapses cannot be excluded because of the neurotropic persistence of the virus.

Papillomavirus infection. Treatment of HPV infection poses certain problems. Endemic infection with this virus is high, and it is assumed that up to 80% of the population become infected at least temporarily (see also p. 169 ff.). Nevertheless, symptoms are rarely reported.

Certain types of HPV (high-risk types) are thought to be responsible for the development of cancer. Hence, it is desirable to combat these viruses on a wide scale by means of vaccination (see also p. 212 ff.) (387). Vaccines against certain papillomaviruses (types 6, 11, 16, and 18) have passed the stage of clinical trials, and are now commercially available (392a). Future developments will show whether these relatively expensive products will be accepted by the public, and whether possible viral mutations will ruin all expectations.

Condylomas, which are usually caused by low-risk types of HPV, may be removed by chemical or mechanical means. Swabbing iso-lated skin eruptions with podophyllum resin, or rather the application of purified podophyllotoxin, has proved best (115). The mechanism of action of this compound is based on the inhibition of cell division. In the case of diffuse condylomas, the body's own immune defenses may be stimulated by imiquimod, and topical chemotherapy with 5-fluorouracil may be used as well (186, 226). Laser treatment is very effective but requires a well-equipped center and a therapist with extensive experience (115).

The consistent use of condoms might provide a prophylactic protection against HPV infection, although large parts of the population are not easily convinced of the need to take such measures. Since the infection seems rapidly to become carcinogenic when a person has sexual contact very early in life, it makes sense to use condoms at least during the teenage years.

Vaginal Adenosis

Circumscribed glandular alterations within the vaginal epithelium, which may give rise to exfoliation of columnar epithelial cells, are very rare. They usually result from embryonic malformations, whereby the **remains of the fetal mesonephros** (wolffian duct and Gartner duct) manifest themselves as cysts (81).

Typically, however, vaginal adenosis also occurs in women whose mothers have been treated with **diethylstilbestrol** during pregnancy (132). These women have an increased risk of developing cancer.

Foci of endometriosis are occasionally detected in the vaginal wall, particularly in the area of the posterior vaginal fornix toward the portio vaginalis (Douglas space) (296). Here, these foci form bluish nodules as a result of regular hemorrhaging during menstruation.

Fig. 3.**130 a Endocervical naked nuclei.** A loose cluster of nuclei, partly with edematous swelling (→), with a regular chromatin pattern and smooth nuclear envelopes. ×400.

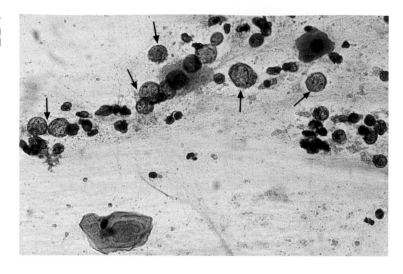

Fig. 3.**130 b Endocervical cells.** Some cells are still intact, others are only present as naked nuclei (→) within a stretch of mucus. ×400.

Fig. 3.**131 Ciliocytophthoria.** Cytoplasmic degeneration of endocervical glandular cells, showing detaching ciliary borders. Cilia and their terminal plates are visible in the upper half of the micrograph (→). ×1000.

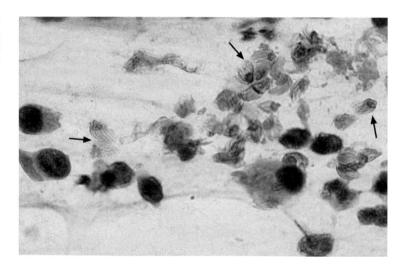

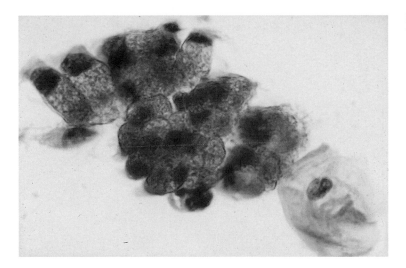

Fig. 3.**132** **Eosinophilic cytoplasmic degeneration.** Endocervical goblet cells. ×1000.

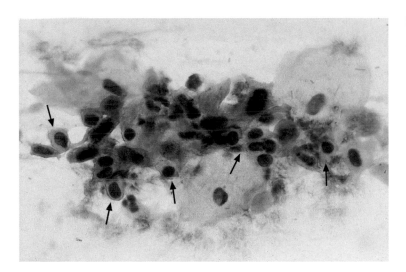

Fig. 3.**133** **Eosinophilic cytoplasmic degeneration.** Endocervical glandular cells with a tendency to shrink (⟶). ×400.

Endocervix

Degeneration

The endocervical epithelium is often affected by degeneration. The changes are essentially the same as those of the squamous epithelium. Because of the strong intercellular adhesion, endocervical cells in the smear are often found as clusters of **naked nuclei** which predominantly lie within strings of mucus (Fig. 3.**130 a, b**) (85).

Detachment of the ciliary border of ciliated cells is called **ciliocytophthoria** (Fig. 3.**131**) (232). It is not known whether this affects only endocervical cells or perhaps also cells of the tubular epithelium (50).

Degenerated endocervical cells may show pseudoeosinophilia, severe shrinkage in case of nuclear condensation, or hyaline cytoplasmic degeneration (Figs. 3.**132**, 3.**133**). Nonepithelial cells, such as histiocytes, exhibit similar degenerative changes. Degenerated leukocytes may look like whiplashes (Fig. 3.**134**).

Fig. 3.**134** **Degenerated leukocytes** showing a whiplash-like elongation (→) and nuclear debris. Parabasal and intermediate cells are also visible. ×400.

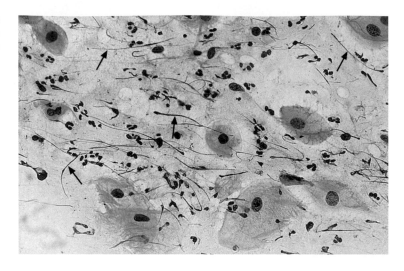

Fig. 3.**135** **Cervicitis.** Inflammation of the endocervical columnar epithelium. The connective tissue (stroma) contains granulation tissue. The columnar epithelium (→) is characterized by an increase in secretory vacuoles and by nuclear enlargement, thus causing pseudostratification in some areas. HE stain, ×200.

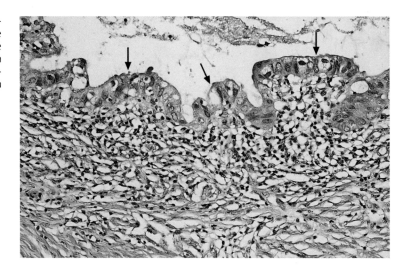

Cervicitis

Acute and Chronic Phases

The effects of acute and chronic phases of cervicitis on the columnar epithelium of the endocervix are listed in Table 3.**4**.

Macroscopy

Cervicitis may exist in isolation or in connection with vaginitis. In the case of an ectopia, the macroscopic signs include intense **redness** and an increased **tendency to bleed** upon mechanical contact, in particular. Furthermore, there is increased secretion in the form of **vaginal discharge**.

Histology

Histological examination reveals **hyperemia** of the stromal tissue associated with **leukocytic infiltration**. The columnar epithelium shows intense **vacuolation** of the cytoplasm (Fig. 3.**135**).

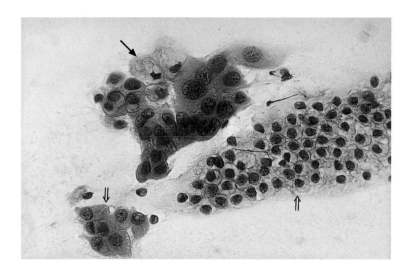

Fig. 3.**136 Cervicitis.** The enlargement of both nuclei and cytoplasm of the endocervical cells is associated with the formation of chromocenters and with cytoplasmic pseudoeosinophilia and vacuolation (⟶). A sheet of normal endocervical columnar epithelium (⟹) lies nearby. ×250.

Table 3.**4** Effects of acute and chronic inflammation on the columnar epithelium of the endocervix

Macroscopic effects

▶ Redness
▶ Discharge
▶ Hemorrhagic diathesis

Histological effects

▶ Hyperemia
▶ Leukocytic infiltration
▶ Increase in secretory vacuoles

Cytological effects

On the nucleus
▶ Proportional enlargement of nucleus
▶ Hyperchromatic nucleus with coarse chromatin structure
▶ Formation of chromocenter and nucleolus
▶ Polynuclear cells
▶ Degeneration

On the cytoplasm
▶ Amphophilia
▶ Pseudoeosinophilia
▶ Vacuolization
▶ Macrocytosis

Cytology

Inflammation leads to increased exfoliation of endocervical cells, with isolated cells being detected more often than cell clusters.

There may be significant **nuclear enlargement**, with the cytoplasm usually being increased as well; this is referred to as proportional nuclear enlargement (Fig. 3.**136**) (160). In case of severe cervicitis, there are also changes in the morphology of both cytoplasm and nuclei (Fig. 3.**137**). The nuclei are then often **hyperchromatic**, showing a **coarse chromatin structure** as well as **chromocenters** or **prominent nucleoli**. **Polynuclear cells** may also be detected (Fig. 3.**138**). **Degenerative factors** may lead to the appearance of large, naked nuclei which are difficult to distinguish from nuclei of cancer cells (160).

The cytoplasm may exhibit discoloration, such as **amphophilia** or **pseudoeosinophilia**, and the cells may show **vacuolation** or an increase in size; this is referred to as macrocytosis.

3

Fig. 3.**137** **Severe cervicitis.** Disorganized arrangement of endocervical cells and pronounced nuclear and cytoplasmic polymorphy (──►). ×250.

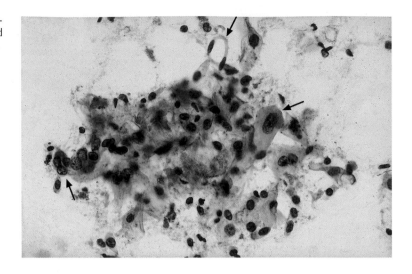

Fig. 3.**138** **Severe cervicitis** (in the same patient as in Fig. 3.**137**). Proportional nuclear enlargement, binuclear cells, hyperchromatic nuclei with irregular chromatin structure and distinct nucleolus. ×250.

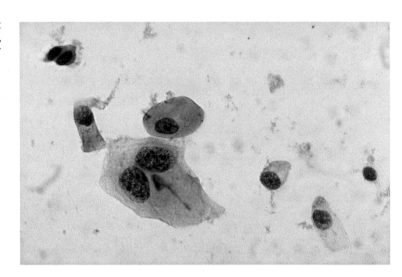

Tab. 3.**5**   Effects of regeneration and humoral factors on the endocervical epithelium

Colposcopic effects	Histological effects	Cytological effects
▶ Loss of transparency	▶ Hyperplasia of reserve cells	▶ Reserve cells
▶ Open glands	▶ Metaplasia	▶ Metaplastic cells
▶ Secretory cysts	▶ Regenerative epithelium	▶ Regenerative epithelium

Regenerative Phase

Two processes completely independent of each other have a significant impact on the endocervical epithelium: the regenerative phase of inflammation on the one hand, and hormonal factors on the other (Table 3.**5**).

Colposcopy

Loss of transparency. The transition of the endocervical epithelium into the more resistant squamous epithelium is called **transformation** in colposcopy; in morphological terms, it is known as **metaplasia** (52). The process may be induced by inflammatory as well as hormonal factors, and it occurs mainly during sexual maturation. Over the years, it gives rise

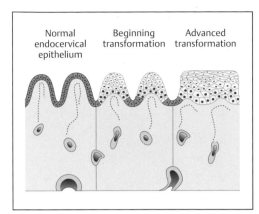

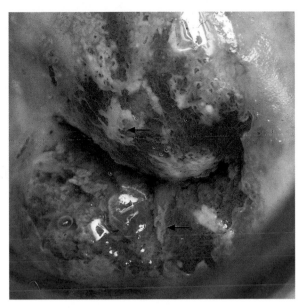

Fig. 3.**139 Epithelial transformation.** The diagram shows the effect of the transformation process on the surface of the endocervical epithelium. At the beginning of the metaplastic transformation, only the superficial epithelium of the reddish villi is transformed, leading to partial loss of transparency. Once the intervillous spaces are also transformed, the epithelium flattens and shows complete loss of transparency.

Fig. 3.**140 Early transformation zone in the area of an ectopia.** The metaplastic (whitish) squamous epithelium (——▶) slides like a tongue over the columnar epithelium.

to a shift of the squamous/columnar epithelial boundary in direction of the endocervix, thus obscuring parts of the endocervical epithelium (264). When viewed at higher magnification, the metaplastic epithelium exhibits a **loss of transparency** at the tips of the villi of the glandular epithelium (Fig. 3.**139**).

Epidermization. Once the metaplastic transformation also includes the intervillous spaces, the villi merge and the epithelium flattens. In colposcopy, the process causing white discoloration of the surface epithelium is called **epidermization**. The epithelium seems to slide over the columnar epithelium like a tongue (Fig. 3.**140**) (130).

Open glands. The above process often spares the openings of glands, and these appear as reddish craters within the whitish metaplastic epithelium; they are referred to as **open glands** (Fig. 3.**141**).

Secretory cysts. Occasionally the new epithelium overgrows the ducts of the glands, thus preventing the mucus from flowing out. This results in the formation of secretory cysts, called **Nabothian cysts**, which may protrude from the surface like bumps (223). Since the cysts are formed within the stroma, they bring the surrounding blood vessels to the surface, where they become clearly visible because of

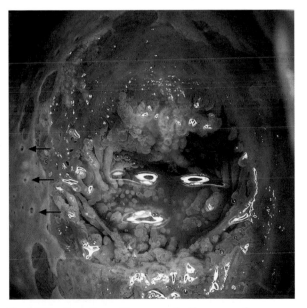

Fig. 3.**141 Transformation zone in the area of the cervical glands.** The metaplastic squamous epithelium covers the columnar epithelium. This process of epidermization initially spares the glandular orifices, thus creating open glands (——▶).

3

Fig. 3.142 Transformation zone in the area of the cervical glands. Once the metaplastic squamous epithelium has covered the glandular orifices, secretory cysts (Nabothian cysts) are formed (→).

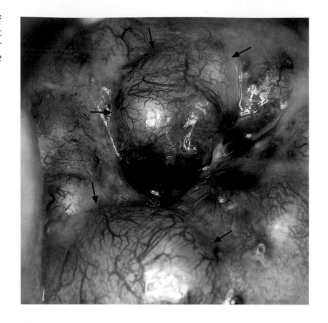

Fig. 3.143 Epithelial transformation in the area of the cervical glands. The diagram shows the effect of the transformation process on the endocervical glands. The metaplastic transformation occurs almost exclusively in the area of the superficial epithelium and spares the glandular canals, thus creating open glands. When the epithelium covers also the glandular orifices, secretion accumulates and temporary cysts (nabothian cysts) are formed.

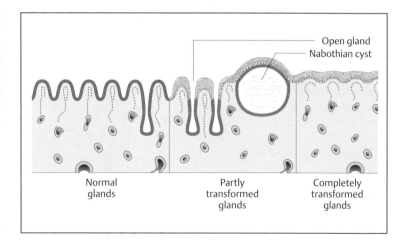

Open gland
Nabothian cyst

Normal glands

Partly transformed glands

Completely transformed glands

Fig. 3.144 Reserve cell hyperplasia. During the regenerative phase of inflammation, the endocervical columnar epithelium shows no mature glandular cells. It rather consists of several irregular layers of undifferentiated reticular cells that exhibit indistinct cell borders and enlarged, vesicular nuclei (→). HE stain, ×200.

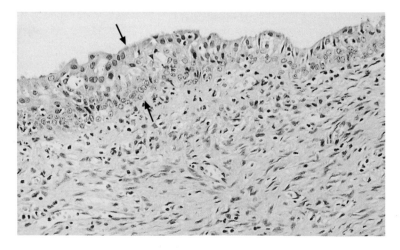

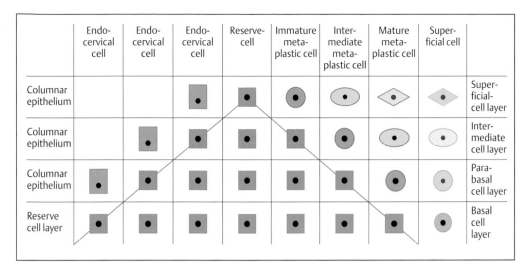

	Endo-cervical cell	Endo-cervical cell	Endo-cervical cell	Reserve-cell	Immature meta-plastic cell	Inter-mediate meta-plastic cell	Mature meta-plastic cell	Super-ficial cell	
Columnar epithelium			▪	▪	●	⬭	◇	◇	Super-ficial-cell layer
Columnar epithelium		▪	▪	▪	▪	●	⬭	⬭	Inter-mediate cell layer
Columnar epithelium	▪	▪	▪	▪	▪	▪	●	○	Para-basal cell layer
Reserve cell layer	▪	▪	▪	▪	▪	▪	▪	●	Basal cell layer

Fig. 3.**145** **Metaplasia.** The metaplastic transformation of the endocervical columnar epithelium into ectocervical squamous epithelium passes through a stage of **reserve cell hyperplasia,** during which the columnar epithelium exfoliates. The reserve cells give rise to metaplastic cells which gradually mature until a normal squamous epithelial stratification is achieved.

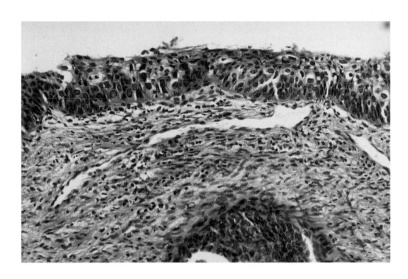

Fig. 3.**146** **Immature metaplasia.** The endocervical columnar epithelium shows no signs of maturation. It consists of several irregular layers of hyperchromatic cells with distinct cell borders and enlarged, mostly isomorphic nuclei. HE stain, ×200.

the thin epithelial layer (Figs. 3.**142**, 3.**143**). At the end of the transformation process, the metaplastic squamous epithelium can no longer be visually and morphologically distinguished from the original squamous epithelium (complete transformation).

Histology

Reserve cell hyperplasia. Histologically, metaplasia begins as reserve cell hyperplasia (Fig. 3.**144**) (122). The pluripotent reserve cells multiply and increasingly assume the character of squamous epithelial cells as they mature.

Metaplasia. After shedding of the columnar epithelium, metaplastic cells exfoliate until the maturation of the squamous epithelium is complete (Fig. 3.**145**). Initially this involves the development of immature metaplasia; the cytoplasm shows few signs of maturation, and the nuclei lie relatively close to one another (Fig. 3.**146**) (52). This is followed by moderate metaplasia and mature metaplasia, with the voluminous cytoplasm pushing the nuclei further apart. The cells gradually flatten longitudinally, and the epithelium begins to resemble the normal differentiated squamous epithelium (Fig. 3.**147**).

3

Fig. 3.**147 Moderate metaplasia.** The endocervical columnar epithelium increasingly shows signs of maturation, including the beginning of squamous epithelial stratification. The cells have distinct cell borders, with partly thickened cell margins (ectoplasm) (⟶). The nuclei are enlarged and hyperchromatic, but still isomorphic. HE stain, ×400.

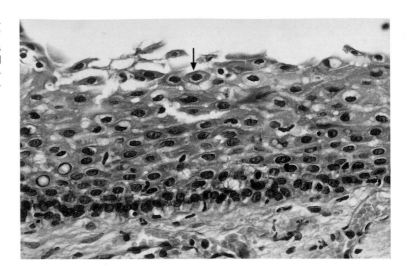

Fig. 3.**148 Glandular regenerative epithelium.** During the regenerative phase of inflammation, the endocervical columnar epithelium shows no mature glandular cells. Rather, it consists of several irregular layers of immature cells that exhibit plenty of foamy cytoplasm and enlarged, polymorphic nuclei with nucleoli (⟶). There is no recognizable polarization of the nuclear axes. HE stain, ×200.

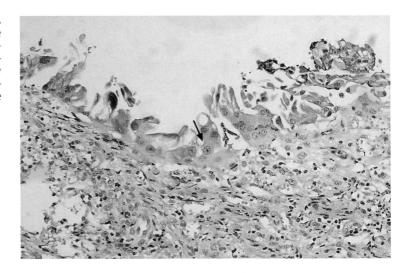

Regenerative epithelium. As in squamous epithelial erosion, a tissue defect causes the formation of **regenerative epithelium**, though it has a glandular character in the endocervical area (Fig. 3.**148**) (81).

Cytology

Reserve cells. Reserve cells are difficult to distinguish cytologically from small histiocytes. They usually exfoliate in loose clusters, have indistinct cell borders, and may contain an eccentric nucleus (Fig. 3.**149 a, b**) (122).

Metaplastic cells. These cells exhibit cytological signs of both columnar and squamous epithelia (282). Their exfoliation occurs usu-

ally in groups or clusters. Because of their endocervical origin, normal or inflammatory endocervical cells are frequently found nearby.

The cytoplasm of **immature metaplastic cells** is dark blue, purple, or blue-green in color, and the cells are round with sharply outlined borders (Fig. 3.**150 a, b**) (301). The cells may vary in shape (spider cells or tadpole cells) and exhibit vacuolation, phagocytotic activity, or cannibalism (Fig. 3.**151 a–e**). Exfoliation of metaplastic cells usually occurs in the form of loose cell clusters. The nuclei are often abnormal in shape, usually as the result of shrinkage. This is accompanied by bright cytoplasmic areas where the previously round nucleus has retracted (Fig. 3.**152**) (344). The resulting nuclear polymorphism often gives rise to confusion with dysplastic processes. For

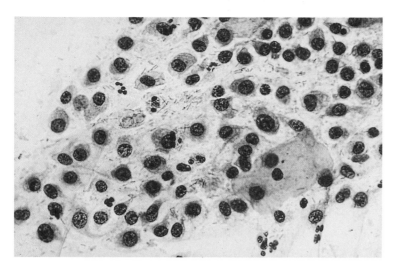

Fig. 3.**149 a, b   Reserve cells.**

a Reserve cell hyperplasia is characterized by loose clusters of reticular cells that have a foamy cytoplasm, indistinct cell borders, and active, hyperchromatic, evenly structured nuclei. An intermediate cell is visible on the lower right. ×400.

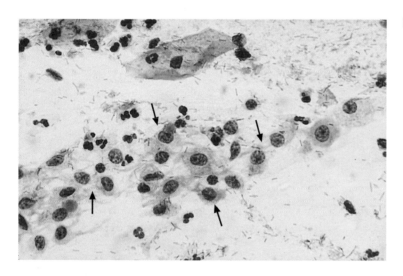

b A loose cluster of cells (→) with indistinct cell borders, foamy cytoplasm, and active nuclei. Two isolated intermediate cells are visible at the top. A lactobacillus flora is also present. ×400.

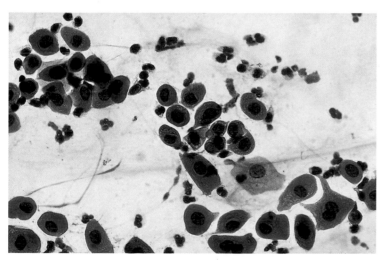

Fig. 3.**150 a, b   Immature metaplastic cells.**
a The loose clusters of oval cells show well-defined cell borders and deep blue cytoplasm. Some nuclei are active, while most nuclei are condensed. ×400.

b A cluster of deep blue cells exhibiting partly vacuolated cytoplasm and active, hyperchromatic nuclei with irregular chromatin structure and hyperchromatic nuclear envelope. ×1000.

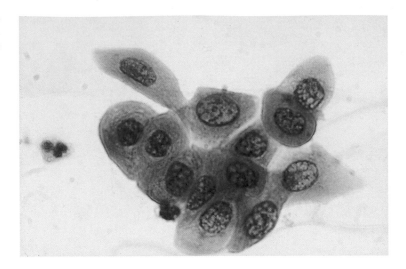

Fig. 3.151 a–e   Immature metaplastic cells.

a A loose group of star-shaped, immature metaplastic cells (spider cells) with active nuclei. Normal superficial cells are visible on the upper left. ×400.

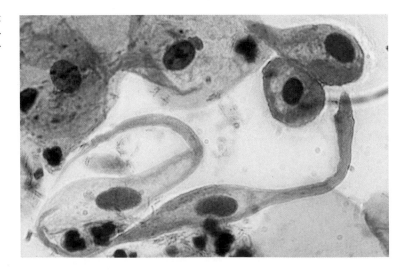

b Oval or spindle-shaped immature metaplastic cells with degenerated nuclei (tadpole cells). Normal intermediate cells are visible nearby. ×1000.

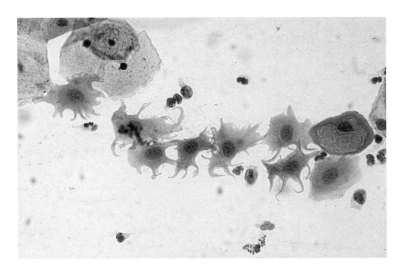

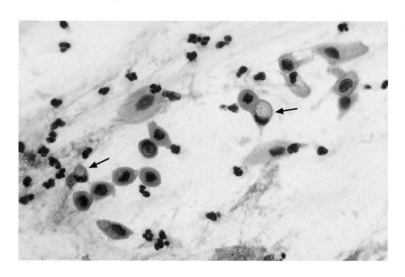

c Irregular, small, blue-green immature metaplastic cells with degenerated nuclei and, in part, cytoplasmic vacuolation (→). ×500.

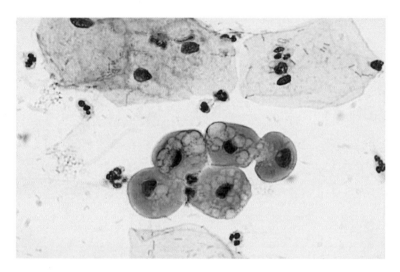

d Severe cytoplasmic vacuolation of blue-green immature metaplastic cells with well-defined borders and degenerating nuclei. Intermediate cells are visible at the top. ×630.

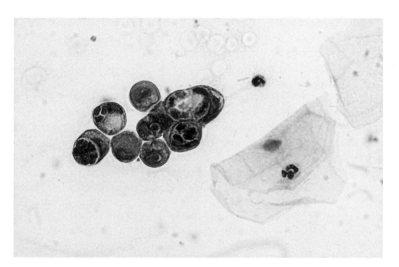

e Severe signs of degeneration of immature metaplastic cells include cytoplasmic vacuolation and pseudoeosinophilia as well as karyorrhexis. Superficial cells are visible on the lower right. ×630.

Fig. 3.152 Immature and intermediate metaplastic cells. A loose group of cells with hyperchromatic nuclei exhibiting the first signs of degeneration, such as nuclear condensation and hyperchromatic nuclear envelope. Some nuclei show a concave curvature opposite a corresponding light area of the cytoplasm (⟶). ×630.

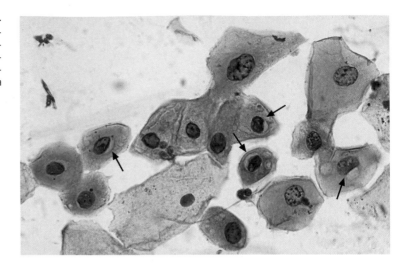

Fig. 3.153 Immature and intermediate metaplastic cells. A loose cluster of cells with thickened cell margins (ectoplasm) and a light center (endoplasm). The degenerated nuclei are usually condensed. ×500.

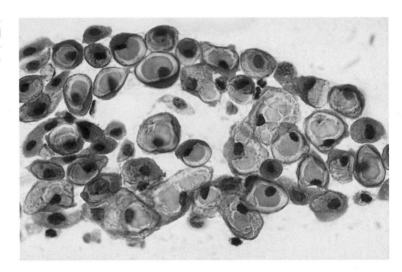

differential diagnosis, it is helpful to know that metaplasia preferentially leads to concave curving of the nucleus, whereas dysplasia tends to cause convex curving.

Intermediate metaplastic cells are larger than immature ones, usually stain violet or green, and have a transparent endoplasm and a dense ectoplasm, thus causing a thickening of the cell margin (Figs. 3.**153**, 3.**154a**) (368). The round nuclei may be active or degenerated, and the cytoplasm shows amphophilia, pseudoeosinophilia, and vacuolation (Figs. 3.**154b, c**, 3.**155**, 3.**156**).

Mature metaplastic cells are clearly larger and stain brighter than intermediate ones. They already resemble normal superficial cells and, in their terminal stage, can no longer be distin-

guished from them (Figs. 3.**157**, 3.**158a–c**) (368). Metaplastic cells are still detected as relicts long after the inflammation has subsided.

Regenerative epithelium. The rapid regeneration of the columnar epithelium of the endocervix after a surface defect is characterized by the appearance of loose cell clusters with indistinct cell borders and foamy cytoplasm, as well as enlarged nuclei and the formation of macronucleoli (100). Polarization of the nuclear axes in a certain direction, as commonly found in squamous epithelial regeneration, is usually absent (Fig. 3.**159a–d**). If the cells show secretory activity at the same time, the cytoplasm contains vacuoles (Fig. 3.**160a, b**).

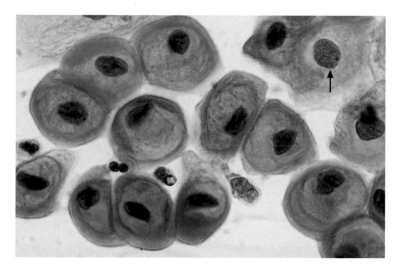

Fig. 3.**154 a–c Intermediate metaplastic cells.** Typical color and structure of ectoplasm and endoplasm.

a Degenerated nuclei predominate, while some nuclei are still active (⟶). ×1000.

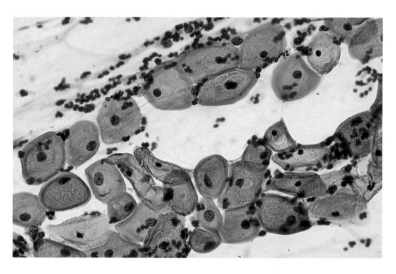

b Hyaline (yellow) cytoplasmic degeneration. ×400.

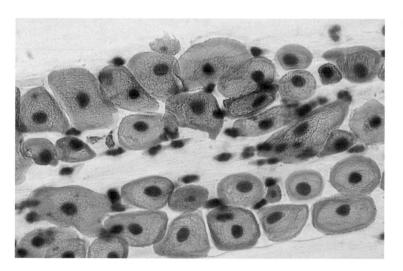

c Amphophilic cytoplasm. ×400.

Fig. 3.**155 Intermediate metaplastic cells.** A compact cluster of almost completely hyalinized cells with still recognizable ectoplasm and endoplasm. Normal superficial cells are visible on the upper left. ×250.

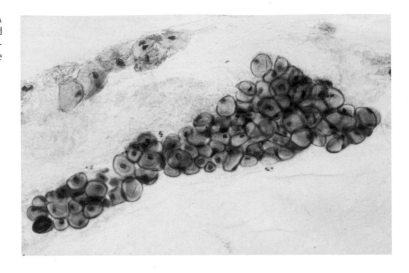

Fig. 3.**156 Vacuolar degeneration of metaplastic cells.** Formation of signet-ring cells, in which the nucleus is pushed to the cell border and therefore hardly recognizable (⟶). The background of the preparation indicates inflammation. ×400.

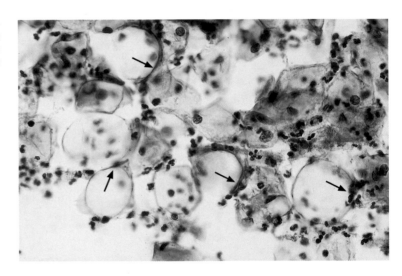

Fig. 3.**157 Intermediate and mature metaplastic cells.** A cluster of mostly green cells, in which the thickening of cell margins is still recognizable (⟶). The nuclei are partly vesicular, partly pyknotic. ×250.

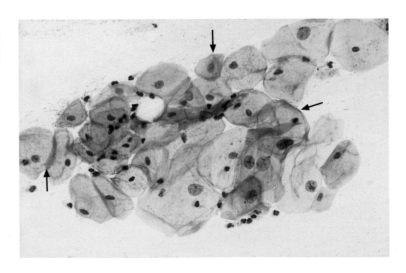

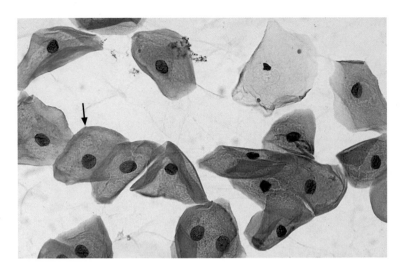

Fig. 3.**158 a–c Mature metaplastic cells.**

a Blue-violet staining of the cytoplasm, with the thickening of cell margins just recognizable (——➤). ×400.

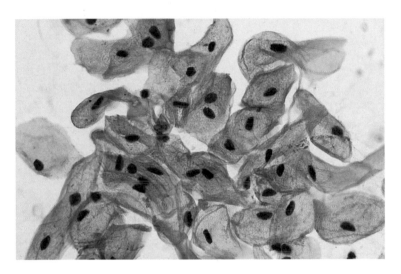

b The cells begin to assume a polygonal shape, thus clearly resembling normal intermediate cells. The still recognizable thickening of cell margins reflects the metaplastic origin of the cells. ×400.

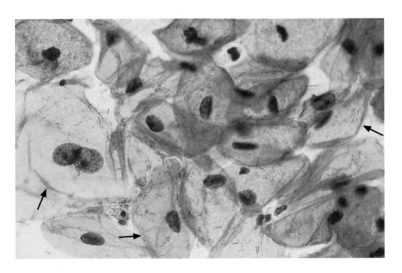

c The cells begin to assume a polygonal shape, while the cell margins are still slightly thickened (——➤). ×630.

Fig. 3.**159 a–d Glandular regenerative epithelium.**

a A cluster of cells with foamy cytoplasm and indistinct cell borders. The polarization of nuclei is hardly noticeable, macronucleoli are prominent, and the cell cluster has a flat, two-dimensional structure. ×250.

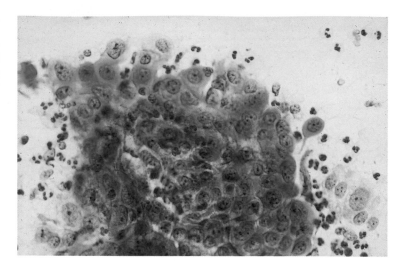

b A two-dimensional cell cluster with enlarged nuclei and prominent nucleoli. A normal intermediate cell is visible at the top. ×630.

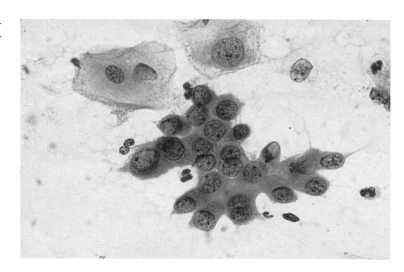

c Palisade arrangement of cells with enlarged basal nuclei and prominent nucleoli. Intermediate cells are visible nearby. ×630.

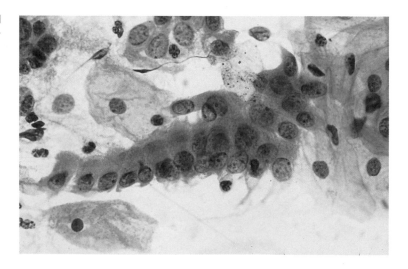

d An irregular cell cluster of spindle-shaped cells with enlarged, nucleolus-containing nuclei. ×630.

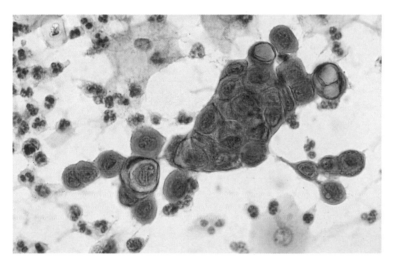

Fig. 3.**160 a, b Glandular regenerative epithelium.**

a A cluster of vacuolated degenerated cells. The background of the preparation indicates inflammation. ×630.

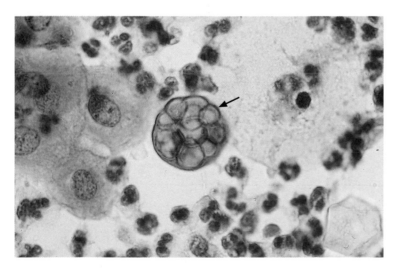

b Complete vacuolar degeneration of a cell (⟶). Some regenerative epithelial cells are visible on the left. The background of the preparation indicates inflammation. ×1000.

Fig. 3.**161 Follicular cervicitis.** The connective tissue (stroma) exhibits increased vascularization and lymphocytic infiltration. Enlarged lymph nodules are visible on the lower left (—→), and one such nodule has reached the epithelial surface on the upper right (⟹). The glandular epithelium is well developed, and the surface epithelium shows signs of metaplastic transformation. HE stain, ×160.

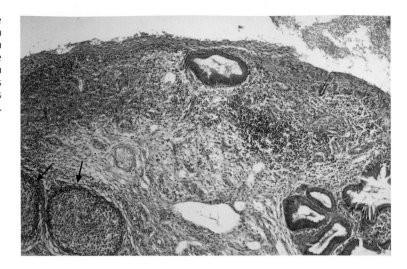

Fig. 3.**162 a, b Follicular cervicitis.**

a The germinal center of a lymph nodule contains huge numbers of lymphoblastic and reticular cells. The nuclei are partly intact, partly degenerated. ×320.

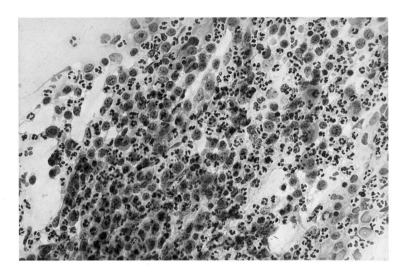

b Reticular cells with loose cytoplasm and indistinct cell borders. The nuclei exhibit a regular and hyperchromatic structure. Isolated intermediate cells are also present. ×500.

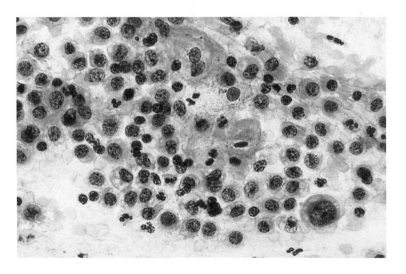

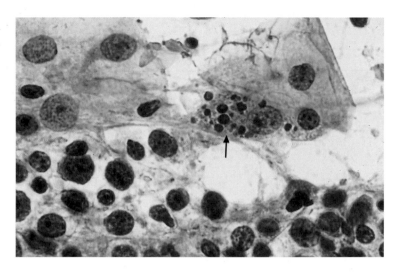

Fig. 3.**163 A starry sky cell associated with follicular cervicitis.** Numerous reticular cells are seen in the lower half of the micrograph, and one of these (⟶) shows characteristic nuclear debris ("starry sky cell" with tingible bodies). Intermediate cells are visible in the upper half of the micrograph. ×1000.

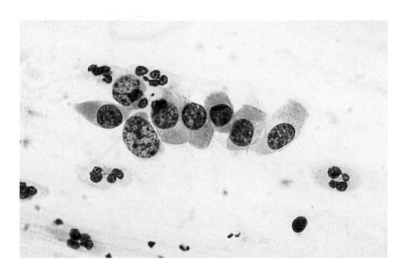

Fig. 3.**164 Tubal metaplasia.** Palisade arrangement of endocervical ciliary cells. The ciliary border is no longer visible. The nuclei vary in size and morphology and are hyperchromatic with irregular chromatin structure and formation of chromocenters. ×1000.

Follicular Cervicitis

A special form of endocervical inflammation is follicular cervicitis. Here, lymph follicles changed by the inflammation reach the surface of the glandular epithelium (Fig. 3.**161**). Exfoliation usually results in large, loose, and uniformly structured cell clusters, known as **germinal centers**. These are often found only in a few areas of the smear (315).

The smear image is typical for the lesion; it shows a large number of **reticular cells**, mainly lymphoblasts and histiocytes. The cells are round, and the foamy cytoplasm has blurred margins (Fig. 3.**162a, b**). The nuclei are relatively large and hyperchromatic, with coarse chromatin and chromocenters or micronucleoli. The more mature cell types are smaller and resemble lymphocytes; the imma-

ture cells may reach the size of macrophages and show phagocytotic activity. Typically, nuclear debris (**tingible bodies**) may be present within the cytoplasm; cells with these inclusions are referred to as "starry sky cells" (Fig. 3.**163**) (77).

Tubal Metaplasia

The presence of foci of **ciliated endocervical cells** in the smear indicates tubal metaplasia. These cells are not metaplastic cells in the usual sense but merely resemble the ciliated cells of the tubal mucosa; characteristic signs are enlarged nuclei and polynuclear cells (Fig. 3.**164**). If degenerative nuclear changes are also detected, the cells are frequently confused with those typical for glandular pre-

Fig. 3.**165** **Cervical polyp.** A smooth benign polyp has developed from the endocervical glandular epithelium and is now visible in the external os of the uterus.

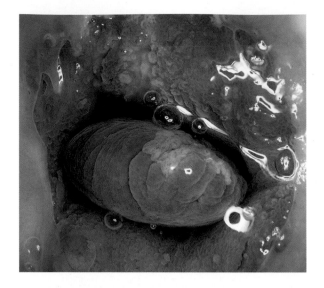

Fig. **3.166** **Cervical polyp.** The connective tissue base of the polyp has a papillary structure. The cervical columnar epithelium (at the bottom of the micrograph) exhibits many folds and villi, and also numerous secretory vacuoles (→). HE stain, ×160.

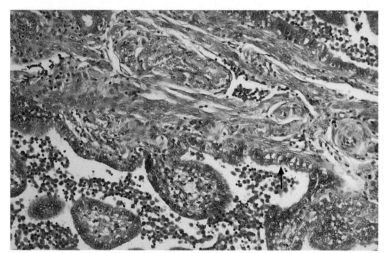

cancerous changes or adenocarcinoma of the endocervix (262).

Microglandular Hyperplasia

This involves the benign proliferation of endocervical glands and is attributed to hormonal overstimulation. The smear contains cell clusters from papillary glands, with the cells exhibiting mild nuclear enlargement and, occasionally, cytoplasmic vacuolation and degeneration.

Cervical Polyps

Cervical polyps are easy to recognize macroscopically by their intense red color and round shape (Fig. 3.**165**). Histological examination reveals a papillary structure of the connective tissue scaffold and a peduncular elongation of the stroma (Fig. 3.**166**). The cytological image of the smear shows regular clusters of endocervical columnar epithelium with cytoplasmic vacuolation and slight nuclear enlargement.

Therapy

Cervicitis. The treatment of cervicitis basically corresponds to that of vaginitis (see pp. 124 f.) as both organs are normally affected simultaneously. Occasionally, however, there may be isolated inflammation of the endocervical epithelium, particularly in patients using an IUD or in connection with chlamydial infec-

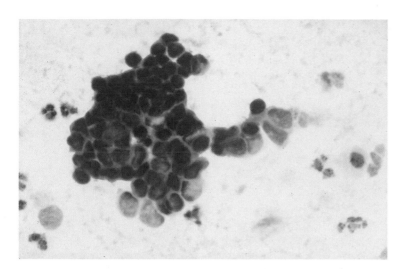

Fig. 3.**167 a, b Sheets of degenerated endometrial cells.**
a Nuclear enlargement by karyolysis. ×630.

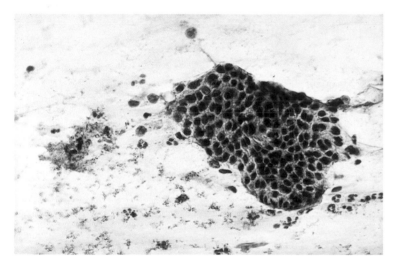

b Pseudoeosinophilic cytoplasm and shrinkage of nuclei due to karyopyknosis. ×250.

tion. In either case, antibiotic treatment is required.

In general, treatment focuses only on the acute and chronic phases of inflammation, less so on the regenerative phase. In particular, when there is metaplastic transformation of the epithelium due to hormonal factors, rather than inflammatory ones, treatment is not required because this type of transformation is a natural process.

Cervical polyps. These are usually removed immediately after their discovery in gynecological practice (by an outpatient procedure using dressing forceps). Polyps may cause discomfort because of their tendency to bleed, and there is also a certain risk of malignant degeneration. After removal, the wound is cauterized to prevent bleeding and regeneration

of the polyp. The intervention may be done without general anesthesia, and usually also without local anesthesia, since the endocervical mucosa is not sensitive to pain.

Endometrium

Degeneration

Endometrial cells in the smear usually show degenerative changes and are then difficult to distinguish from degenerated endocervical cells (see also p. 41). They often occur in sheets or clusters, and their nuclei are either enlarged due to karyolysis or condensed due to karyopyknosis (Fig. 3.**167 a, b**) (359).

Table 3.**6**   The effects of endometritis

Macroscopic effects	Histological effects
▶ Hemorrhagic diathesis	▶ Hyperemia ▶ Leukocytic infiltration ▶ Vacuolization
Cytological effects on the cytoplasm	**Cytological effects on the nucleus**
▶ Leukocytic infiltration ▶ Vacuolization	▶ Absolute and relative nuclear enlargement ▶ Hyperchromatic nucleus with coarse chromatin structure ▶ Formation of chromocenter and nucleolus ▶ Degeneration

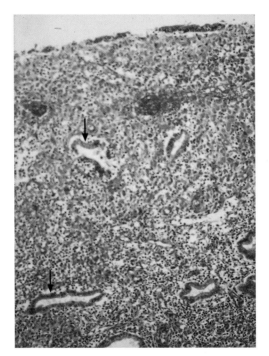

Fig. 3.**168** **Endometritis.** The connective tissue (stroma) exhibits increased vascularization and leukocytic infiltration. The glandular epithelium is well developed (⟶). HE stain, ×100.

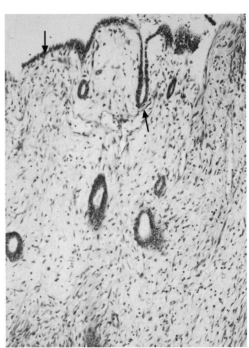

Fig. 3.**170** **Endometrial atrophy.** The connective tissue (stroma) shows a reduction in cell numbers and an increase in sclerosis. The glandular epithelium is poorly developed (⟶). HE stain, ×100.

Endometritis

The various stages of inflammation are more difficult to determine in the endometrium than in the squamous epithelium or endocervical epithelium. Changes caused by endometritis are listed in Table 3.**6**.

Macroscopy

Endometritis usually manifests itself clinically either as prolonged menstruation (hypermenorrhea) or as midcycle bleeding (metrorrhagia).

Histology

Histologic examination reveals **hyperemia** of the blood vessels and **leukocytic infiltration** of stroma and epithelium. Increased vascularization may also be present (Fig. 3.**168**).

Cytology

On the cellular level, endometrial cells with inflammatory changes show **enlarged** and **hyperchromatic** nuclei with a **coarse chromatin structure** and sometimes also the **formation**

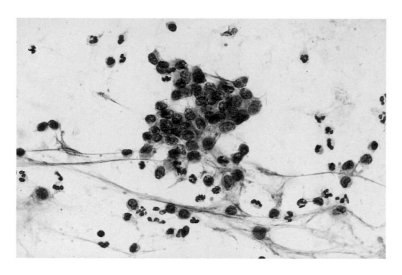

Fig. 3.**169 a, b   Endometritis.**
a A loose cluster of hyperchromatic endometrial cells with sparse cytoplasm and regular arrangement. ×400.

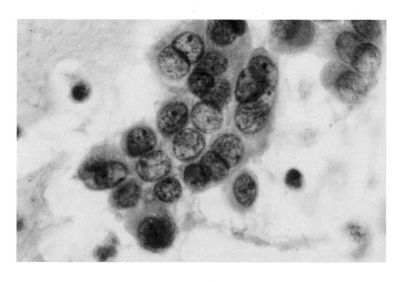

b A sheet of endometrial cells showing enlarged nuclei with chromocenters and sparse cytoplasm with indistinct cell borders. ×1000.

of chromocenters or **nucleoli** (Fig. 3.**169 a, b**) (81). This may cause significant problems in differentiating endometritis from uterine carcinoma. The cell clusters often exhibit cytoplasmic **vacuoles** and **leukocytic infiltration**.

The occurrence of endometrial cells in the smear is mostly associated with a rejection reaction within the uterine cavity, and this is why nonepithelial elements—such as granulocytes, histiocytes, blood, and precipitated proteins—are often found nearby (67). Smears taken during menstruation or midcycle bleeding may be completely obscured by the exudate. Smears taken directly from an extracted

IUD often show reactive inflammatory changes in the sense of **endometritis**, with the endometrial cell clusters usually being densely infiltrated by leukocytes (83).

Endometrial Atrophy

After the menopause, the atrophic endometrium (Fig. 3.**170**) often shows sclerosing processes of blood vessels, which may traumatize the vessels and cause bleeding (68). As well as the blood, atrophic endometrial cells usually appear in the smear too, either isolated

Fig. 3.**171 Endometrial atrophy.** A compact cell sheet of cells showing small condensed nuclei. ×250.

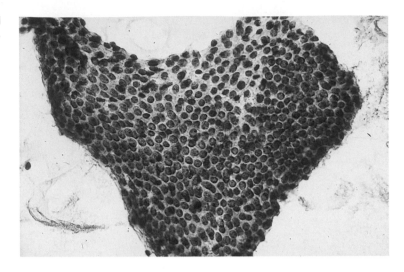

Fig. 3.**172 Endometrial polyp.** Hysteroscopic image of a smooth-walled endometrial polyp protruding on the left side of the image (→). Mild hemorrhage is visible in the fundus of the uterus (⇒).

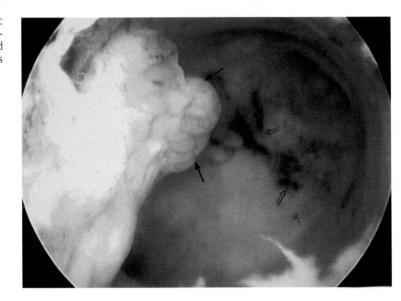

Fig. 3.**173 Endometrial polyp.** Sonographic image of a smooth-walled endometrial polyp in the fundus of the uterus, showing a highly-reflective internal structure (→).

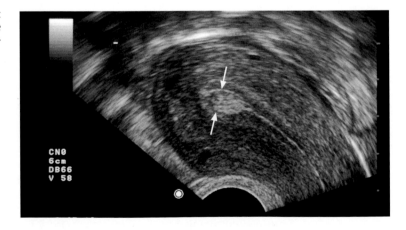

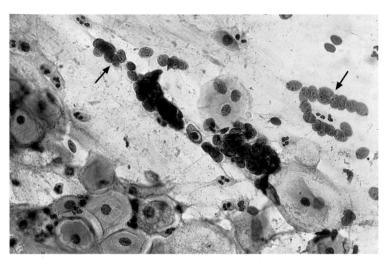

Fig. 3.**174 a, b Endometrial polyp.**
a Degenerated endometrial cells in the smear are widely distributed and isolated. The naked nuclei have a uniform, bipolar shape (→). Parabasal cells are visible in the lower half of the micrograph. ×400.

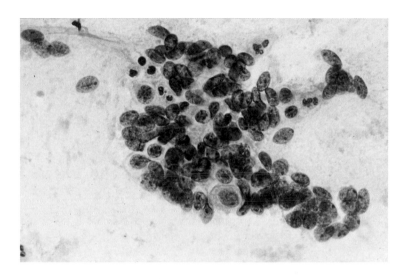

b The polyp contains a loose cluster of hyperchromatic endometrial cells. The nuclei are bipolar in shape and have a smooth nuclear membrane and partly contain nucleoli. ×630.

or in sheets (Fig. 3.**171**). They are often difficult to distinguish from highly differentiated abnormal endometrial cells. Their detection should be viewed with the same suspicion as postmenopausal bleeding and therefore requires further evaluation.

Endometrial Polyps

Endometrial polyps tend to bleed and are sometimes discovered during hysteroscopy Fig. 3.**172**) or sonography (Fig. 3.**173**).

On the cellular level, polyps are characterized by the occurrence of regular cell clusters in the smear, and there may be slightly enlarged, hyperchromatic nuclei with nucleoli (81). In case of cytoplasmic degeneration, uniformly **bipolar, naked nuclei** are found, and

these may occur isolated (Fig. 3.**174 a**) or in loose cell clusters (Fig. 3.**174 b**).

Glandular (Simple) Endometrial Hyperplasia

On the histological level, glandular endometrial hyperplasia is associated with increased numbers of coiling glands, and often also with cystic dilation of the glandular lumina, thus creating a Swiss cheese pattern (Fig. 3.**175**).

The cytological image usually resembles that of endometrial polyps (Fig. 3.**176**) (219). However, nuclear changes may be so pronounced that differentiation from malignant processes becomes impossible.

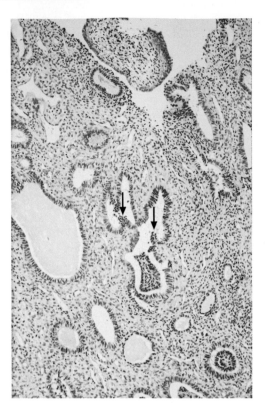

Fig. 3.**175 Glandular (simple) endometrial hyperplasia.** Glandular lumina are increased in numbers and they vary in size (Swiss cheese pattern). In parts, the glandular epithelium exhibits intraluminar budding (→). The single-layered epithelium does not show any atypia. HE stain, ×100.

Therapy

Treatment of endometritis or endometrial hyperplasia, which during menopause manifests itself mostly in the form of acyclic vaginal bleeds, usually involves administration of high doses of estrogen to promote rapid and full development of the inflamed mucosa (110, 391). Gestagens are administered either simultaneously or subsequently to facilitate rapid bleeding of the artificially developed mucosa (hormonal abrasion). However, continuous bleeding after treatment raises suspicions of other causes of bleeding—such as endometrial polyps, atypical endometrial hyperplasia, uterine myoma, or even uterine carcinoma—and requires curettage of the uterine cavity (instrumental abrasion). After the menopause, instrumental treatment of vaginal bleeding is preferred to facilitate early recognition of a potential carcinoma.

Ovary

Degeneration

Normal glandular cells derived from the ovaries are rarely seen in the cytological smear (84). Because of the great distance they travel from the ovaries to the vagina, they are subjected to severe degenerative influences and are therefore no longer distinguishable from other degenerating cells.

Cells derived from ovarian carcinoma, however, are occasionally better preserved and are then identified as **abnormal glandular cells** (Fig. 3.**177**; see also pp. 237 ff.).

Salpingitis and Ovarian Cysts

Inflammatory changes of the ovary are usually caused by pelvic inflammatory disease (PID) or, more specifically, salpingitis. These changes manifest themselves as purulent abscesses. Benign cysts are also common (see also p. 45).

Endoscopy and Sonography

Inflammatory, cystic, or tumorous changes of the ovaries manifest themselves clinically as lower abdominal pain and may be palpable during the gynecological examination. The changes sometimes show up in sonograms and may be macroscopically visible during laparoscopy (Figs. 3.**178**, 3.**179**) (267).

▶ **Salpingitis** is characterized by intense redness, purulent secretion, or scar formation.
▶ **Follicular cysts** usually have one cavity (unilocular cysts) and contain a clear, serous fluid.
▶ **Lutein cysts** often bleed and then contain thin blood.
▶ **Endometrial cysts**, by contrast, are filled with thick, dark blood and are therefore also known as chocolate cysts.
▶ **Cystic tumors** usually have several cavities (multilocular cysts). They are associated with a certain risk of malignant degeneration. In **serous cystadenoma** the cavities are filled with a clear fluid, whereas **mucinous cystadenomas** have mucin-filled cavities.
▶ **Dermoid cysts** are benign teratomas and are usually filled with sebum, hairs, other skin appendages, or constituents of other organs.

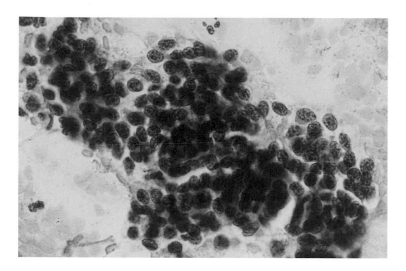

Fig. 3.**176 Glandular endometrial hyperplasia.** An irregular, loose cluster of hyperchromatic endometrial cells. The nuclei are isomorphic, yet their chromatin structure is more coarse and irregular than usual. ×630.

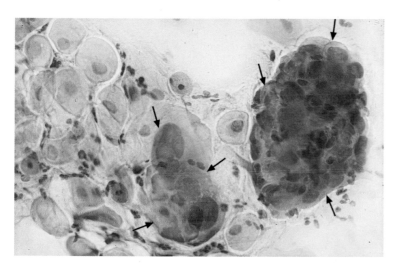

Fig. 3.**177 Degenerated clusters of abnormal glandular cells.** The clusters are derived from an ovarian carcinoma; they show peripheral retraction lines (⟶) and an irregular, three-dimensional arrangement of nuclei. Parabasal cells are visible nearby. ×250.

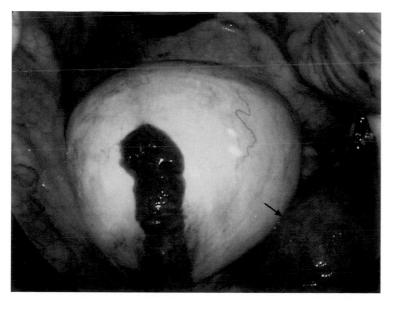

Fig. 3.**178 Ovarian cyst.** Laparoscopic image of a benign, smooth-walled left ovarian cyst, with the infundibulum of the uterine tube lying on top. The fundus of the uterus is visible on the lower right (⟶).

Fig. 3.179 Ovarian cyst. Sonogram of a benign, smooth-walled, unilocular left ovarian cyst, 4.9 × 6.5 cm in size.

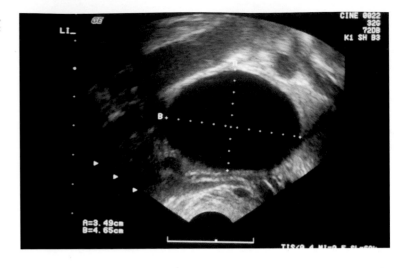

Fig. 3.180 Follicular cyst of the ovary. The follicular lumen is visible on the upper right. The wall of the cyst consists of a wide layer of undifferentiated cells (—▶). These **granulosa cells** have little cytoplasm, indistinct cell borders, and large isomorphic nuclei with nucleoli. Masson–Goldner stain, ×200.

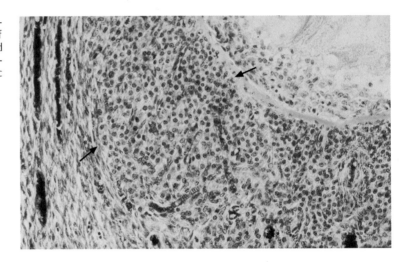

Fig. 3.181 Lutein cyst of the ovary. The lumen of the cyst is filled with numerous differentiated cells (—▶). These **granulosa-lutein cells** have plenty of granular cytoplasm, well-defined cell borders, and isomorphic nuclei with nucleoli. Masson–Goldner stain, ×200.

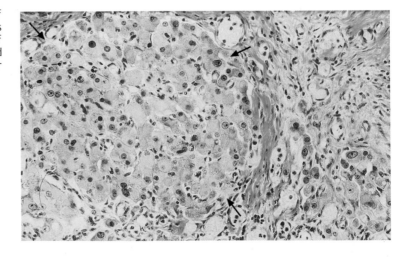

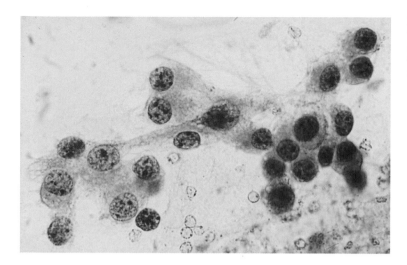

Fig. 3.**182** **Granulosa cells.** Puncture fluid from a follicular cyst, showing a loose cluster of granulosa cells, which have a sparse, foamy cytoplasm with indistinct cell borders and relatively large, hyperchromatic nuclei. Nucleoli are sometimes visible. ×790.

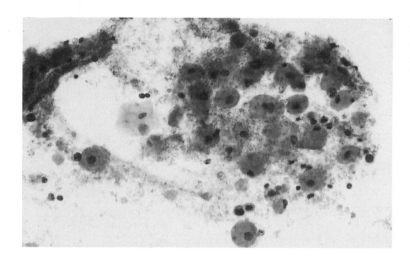

Fig. 3.**183** **Granulosa-lutein cells.** Puncture fluid from a lutein cyst, showing granulosa-lutein cells, which have a voluminous cytoplasm with relatively well-defined cell borders. Their nuclei are mostly small, degenerate, and pyknotic. ×320.

Histology

Histological examination usually reveals abscess cavities filled with masses of leukocytes due to the inflammation.

The following types of cells or epithelia are found in the walls of these cysts:

▶ **Follicular and lutein cysts: granulosa cells** and **granulosa-lutein cells** (Figs. 3.**180**, 3.**181**)
▶ **Endometrial cysts:** endometrium
▶ **Cystadenomas:** columnar epithelium
▶ **Dermoid cysts:** squamous cell epithelium

Cytology

Puncture fluid. Persistent ovarian cysts of benign appearance are usually punctured, and the puncture fluid is sent off for cytological assessment and hormone analysis (263).

The puncture fluid predominantly contains precipitated proteins, blood, and histiocytes with vacuolated cytoplasm, which are called macrophages or **foam cells**. Other cells detected in the various types of cysts include:

▶ **Follicular cysts:** granulosa cells (Fig. 3.**182**)
▶ **Lutein cysts:** granulosa-lutein cells (Figs. 3.**183**, 2.**68**)
▶ **Endometrial cysts:** siderophages (foam cells with phagocytosed hemosiderin) (Fig. 3.**184**) (81)
▶ **Cystadenomas:** columnar epithelial cells
▶ **Dermoid cysts:** squamous epithelial cells and horn cells

Hormone analysis. In ovarian cysts of questionable appearance, estrogen levels of more than 3700 pmol/L suggest a benign, functional cyst. Values below 1200 pmol/L tend to suggest a neoplastic change, and extirpation of the cyst should be considered (6).

3

Fig. 3.**184 Siderophages.** Puncture fluid from an endometrial cyst, showing histiocytes with phagocytosed hemosiderin (→). ×790.

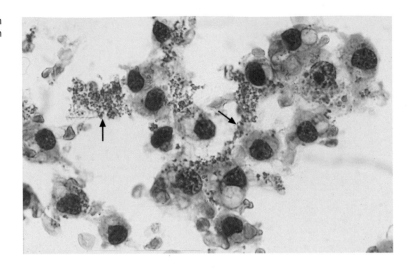

Fig. 3.**185 A cluster of dyskeratotic cells.** In an atrophic cervical smear, the presence of dyskeratotic cells (→) derived from a vulvar carcinoma is due to retrograde transmission. Intermediary cells and basal cells are visible in the background. ×400.

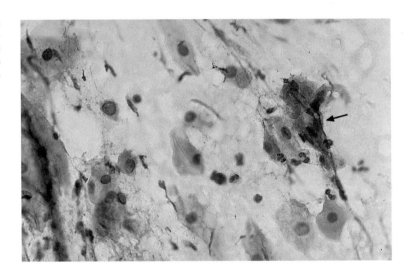

Therapy

Salpingitis. Salpingitis is usually treated with antibiotics, anti-inflammatory agents, and physical rest. Early and adequately prolonged therapy is important to prevent long-term complications. Late sequelae include scarring, which interferes with tubal patency and is one of the most common causes of female sterility.

Follicular cysts. Small follicular cysts are usually only monitored by ultrasound; larger cysts are punctured. The puncture fluid is then subjected to cytological examination and hormone analysis, the results of which may prompt extirpation of the cyst (6).

Endometrial cysts. Once diagnosed, they are usually removed. This is followed by temporary gestagen treatment so that the ectopic endometrium will not be exposed to the proliferative effect of estrogen (173). This therapy aims at the desiccation of focal endometriosis.

Cystadenomas, dermoid cysts, and potentially malignant cysts. To ensure thorough histological processing of the tissue, such cysts are almost always extirpated (see also p. 240), since puncture may involve the risk of spreading tumor cells into the abdominal cavity or abdominal wall (374).

Vulva

Degeneration

The horn cells exfoliated from the stratum corneum of the vulva are very resistant to environmental influences because of their stable keratin scaffold. It is therefore very rare that they exhibit signs of degeneration (108). Sometimes they are subject to retrograde transmission into the vaginal cavity and are then still well preserved even when the overall cell image shows severe degeneration (Fig. 3.**185**) (245).

Vulvitis

Inflammation is the most common cause of benign lesions of the vulva. It is mostly due to fungal infection, but may also result from bacterial infection. Another inflammatory manifestation is infection with herpes simplex virus (HSV).

Macroscopy

Inflammatory diseases of the vulva are normally associated with intense redness, which is often accompanied by a whitish coating in case of candidal infection (Fig. 3.**186**). Diffuse localization of the changes suggests a bacterial or fungal infection; the formation of circumscribed vesicles or lesions suggests that they are caused by HSV.

Inspection of the vulva—unlike that of the portio vaginalis—is hampered by problems

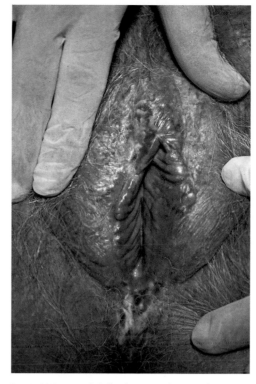

Fig. 3.**186** **Candidal vulvitis.** Redness, edematous swelling, and whitish coating on the labia majora and in the perianal region.

posed by two natural light filters present in the skin: the superficial horn cells of the stratum corneum and the melanin pigment of the basal layer (Fig. 3.**187**) (244). These give rise to different hues and structures of the skin surface, without being typical for a specific dis-

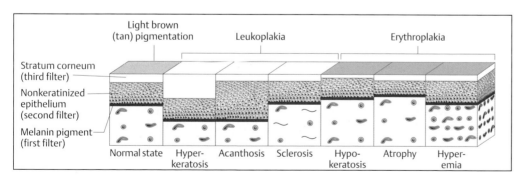

Fig. 3.**187** **Vulvar diseases.** Diagram explaining the colposcopic appearance of different epithelial reactions in response to various disease processes.

Fig. 3.**188** **Acute candidal vulvitis.** The connective tissue (stroma) exhibits increased vascularization and leukocytic infiltration. The enhanced proliferation of the epithelium causes a loss of stratification and a horny layer containing nuclei (parakeratosis) (→). ×100.

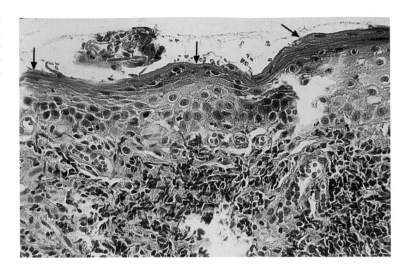

Table 3.**7** Descriptive terms for the documentation of macroscopic findings related to vulvar diseases

Description of color
▶ Leukoplakia
▶ Erythroplakia
▶ Others (e. g., pigmentation)

Description of type
▶ Thickening or tumor
▶ Atrophy
▶ Defect or ulcer

Description of consistency
▶ Soft
▶ Hard

Description of expansion
▶ Size
▶ Diffuse or not diffuse

Description of localization

ease. This complicates the differential diagnosis and often leads to incorrect interpretations (157). Nevertheless, prominent leukoplakia should always be considered potentially malignant (as is the case with lesions of the portio vaginalis), whereas lesions within or beneath the skin are mostly benign. A strictly descriptive record of changes is best suited for the documentation of macroscopic findings (Table 3.**7**).

Histology

Acute inflammatory disease stimulates epithelial proliferation, thus leading to hasty, and therefore parakeratotic, cornification

(Fig. 3.**188**). Chronic inflammation causes additional thickening of the stratum corneum, which is called hyperkeratosis.

Cytology

In the presence of inflammatory lesions of the vulva, the smear contains variable amounts of orthokeratotic and parakeratotic cells, although the proportion of parakeratotic cells is usually above 40% (Fig. 3.**189a**) (248). Furthermore, the nuclei are often swollen, making it sometimes difficult to distinguish the changes from dysplastic processes (Fig. 3.**189b**). If vaginitis with increased amounts of discharge is also present, vaginal epithelial cells that have been altered by inflammation may also be detected in the smear. Horn cells of the vulva do not exhibit the type of inflammatory reactions known to occur in cervical cells.

Vulvar Dystrophy

Dystrophic lesions of the vulva are divided into an atrophic form, **lichen sclerosus**, and a hyperplastic form, **squamous epithelial hyperplasia** (124). It is thought that local deficiencies in testosterone or testosterone receptors are responsible for their development.

Macroscopy

Vulvar dystrophy manifests itself almost always as whitish changes of the skin (**leukoplakia**). These occur either within the skin or as prominent skin lesions (**hyperkeratosis**). However, the color of the skin as well as the

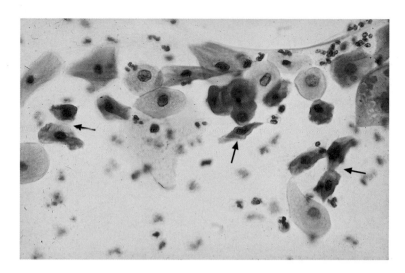

Fig. 3.**189 a Parakeratotic cells associated with candidal vulvitis.** Dissociated isomorphic cells with slightly swollen, condensed nuclei (→). Also present are several vaginal superficial and intermediate cells. The background of the preparation indicates inflammation. ×250.

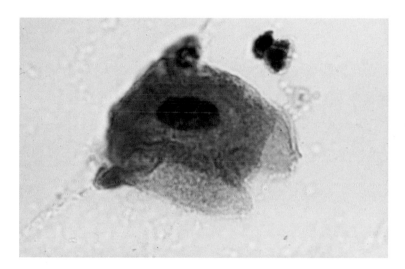

Fig. 3.**189 b Parakeratotic cell associated with vulvitis.** The nucleus is degenerated and slightly enlarged. ×630.

mode of localization and spreading may be very different in each particular case.

Physiological senile atrophy of the vulva frequently occurs after the onset of menopause and is characterized by pronounced atrophy of the labia minora (Fig. 2.**70**, p. 47). Depigmentation of the skin, such as **vitiligo**, causes benign white spots.

Lichen sclerosus. In its classic manifestation, this atrophic dystrophy is arranged in a butterfly pattern. It is characterized by leukoplakia within the skin and a mother-of-pearl gloss, and it is associated with atrophy and waxy degeneration of the labia minora (Fig. 3.**190**).

Squamous epithelial hyperplasia. In hyperplastic dystrophy, the skin appears thickened and hyperkeratotic. The cells exhibit weak positive staining in the toluidine blue test.

At an advanced state, the vaginal orifice may be constricted and vulnerable, and rhagades, fissures, and microhematomas may occur. The damaged skin is prone to fungal and bacterial superinfection. Subjective symptoms include itching, burning, and painful intercourse (Fig. 3.**191**).

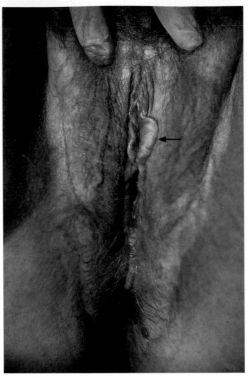

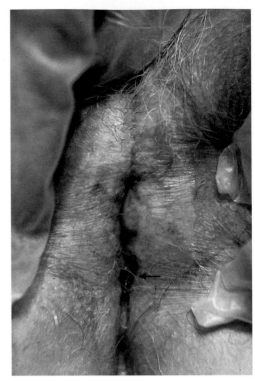

Fig. 3.190 Lichen sclerosus. A butterfly-shaped leukoplakia within or beneath the skin is visible in the perianal region and at the interlabial folds. The labia minora undergo waxy degeneration (⟶).

Fig. 3.191 Advanced vulvar dystrophy. Complete atrophy (the labia minora and the clitoris are no longer visible), leukoplakia, lichenification of the skin, and stenosis of the vaginal orifice. The skin's vulnerability leads to fissures of the posterior labial commissure during palpatory examination (⟶).

Fig. 3.192 Lichen sclerosus of the vulva. The connective tissue (stroma) shows a reduced content of cells and severe sclerosing. The epithelium is poorly developed and exhibits parakeratotic cornification. HE stain, ×50.

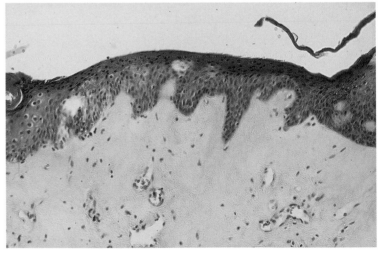

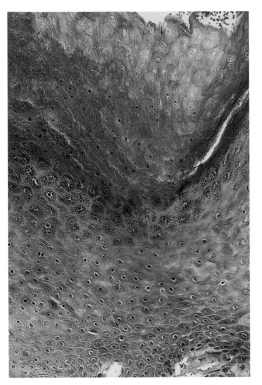

Fig. 3.**193 Squamous epithelial hyperplasia of the vulva.** The epithelium has become wider, and the papillary body is prominent. The surface shows severe hyperkeratosis, in parts also parakeratosis. HE stain, ×50.

Histology

Lichen sclerosus. The epithelium is atrophic and covered by an orthokeratotic or parakeratotic stratum corneum of varying thickness. The rete ridges are flattened, and the parvocellular stroma is sclerosed and infiltrated by round cells (Fig. 3.**192**) (168).

Squamous epithelial hyperplasia. Epithelial hyperplasia is characterized by a wide epithelium with reinforced rete ridges (Fig. 3.**193**). Dysplasia may occur—and this is why epithelial hyperplasia, unlike lichen sclerosus, is considered a potentially precancerous lesion.

Cytology

Cytological smears taken from dystrophic vulvar lesions normally show exclusively orthokeratotic and parakeratotic cells in varying amounts. Parakeratotic cells dominate during inflammatory episodes (Fig. 3.**194**). If dysplasia is present, dyskeratocytes may be detected as well.

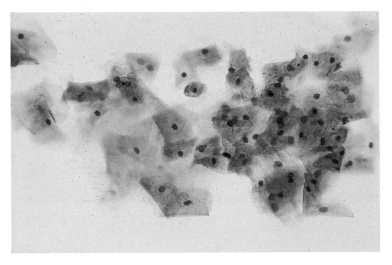

Fig. 3.**194 Parakeratotic cells in case of vulvar dystrophy.** The isomorphic nuclei are degenerated. ×250.

3

Vulvar Condylomas _____

Condylomas of the vulva represent a benign manifestation of HPV infection.

Macroscopy

Vulvar lesions caused by HPV may be diffusely distributed and invisible in the native state. After application of acetic acid, evenly spread skin changes that resemble drops of lime can be recognized macroscopically or under the colposcope (Fig. 3.**195**). More frequently observed are circumscribed warts of varying sizes (genital warts or condylomas) (Fig. 3.**196**).

Histology

Koilocytes and isolated horn cells are detected in the upper epithelial layers of condylomas, and there is a widening of the spinous layer (acanthosis) (Fig. 3.**197**). Frequently, papillomatous growths with branching connective tissue foundations are present (Fig. 3.**198**).

Cytology

Even in the presence of prominent condylomas, koilocytotic cells (Fig. 3.**199 a**) are rarely present in the smear because the superficial horny layer prevents their exfoliation (245).

By contrast, **parakeratotic and dyskeratotic cells** are found in large numbers (Fig. 3.**199 b**). This makes it difficult to differentiate condylomas from dysplastic lesions.

Fig. 3.**195** **Signs of HPV infection of the vulva.** Diffuse whitish stippling in response to the acetic acid test.

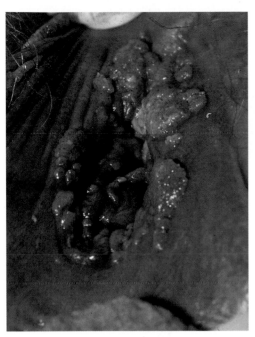

Fig. 3.**196 Acuminate condylomas.** Numerous lesions of various sizes near the vaginal orifice.

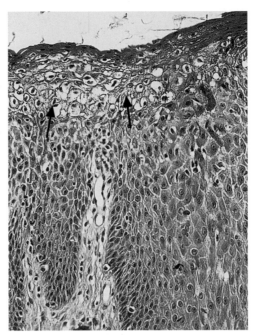

Fig. 3.**197 Flat condyloma of the vulva.** Beneath the horny layer (stratum corneum), in the stratum granulosum, are numerous koilocytes. They usually contain degenerated, pyknotic nuclei (—►). HE stain, ×100.

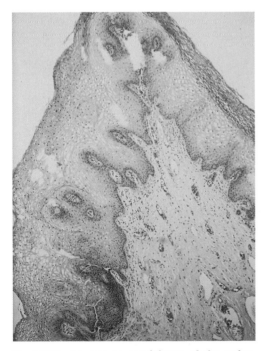

Fig. 3.**198 Acuminate condyloma of the vulva.** The connective tissue (stroma) forms a papillary base. The epithelium has become much wider and contains koilocytes. The superficial stratum corneum exhibits intense parakeratosis and, in parts, mild dyskeratosis. HE stain, ×50.

Fig. 3.**199 a A koilocytotic parakeratotic cell.** The cell is derived from a vulvar condyloma and contains a degenerated nucleus. ×630.

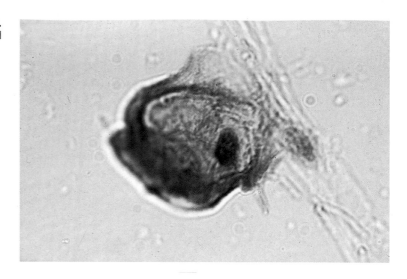

Fig. 3.**199 b   Slightly dyskeratotic horn cells.** The cells are derived from a vulvar condyloma. They show an increase in cytoplasm, nuclear enlargement, and anisonucleosis. ×250.

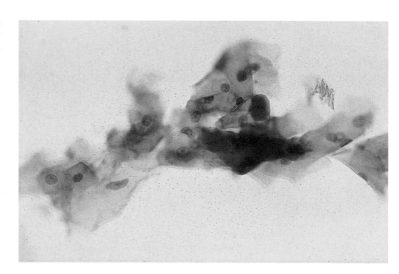

Therapy

Inflammatory diseases of the vulva. In principle, the same therapeutic strategy as for vaginitis applies here as well (see pp. 124 f.).

Dystrophic lesions of the vulva. Unlike the treatment of atrophic vaginal epithelium, androgens (in particular, testosterone propionate) rather than estrogens are used for topical treatment (248). Combination with corticosteroids (in particular, triamcinolone and clobetasol) and application of progesterone are also advantageous.

Condylomas. Topical treatment with podophyllotoxin, imiquimod, or 5-fluorouracil is used here (32, 115) (see p. 125). Circumscribed lesions may be removed by surgical or electrical ablation or destroyed by laser treatment (357).

Malignant
Changes
of the Female
Genital Tract

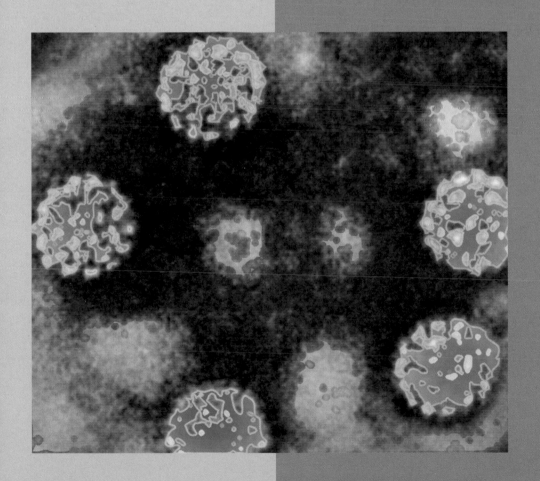

4

4 Malignant Changes of the Female Genital Tract

Portio and Vagina

Epidemiology

Thanks to the effectiveness of cancer screening programs, invasive cervical carcinoma is rare today (0.1–0.2‰). Severe precancerous stages, however, occur roughly 10 times as frequently, and less severe stages as much as 100 times as frequently.

Incidence. Currently, at the start of the 21st century, the **relative** frequency of cervical carcinoma worldwide is 0.1–0.2‰. This means that one or two out of 10 000 women examined each year are affected by this disease (355). In the late 1960s, this figure was still 0.3–0.4‰, more than twice as high as it is today (Fig. 4.1) (7a, 118a).

Cytological examinations aimed at the early detection of cervical cancer have been carried out since the early 1950s (51), and prescriptive screening has been available in Germany since 7 July 1971. This means that women aged 20 and above are entitled to have such an examination once a year at the expense of the national health insurance scheme (166).

Prevalence. The **absolute** frequency of cervical carcinoma is higher than the incidence because it includes all women, not just those who have undergone cytological testing. Currently, the prevalence is 0.3–0.4‰ worldwide (81).

Early detection. The population participation rate for prescriptive screening for early detection of cancer varies from country to country; in Germany it is 50–60% (316, 366). It should be taken into account, however, that women who undergo cytological testing include not only those who participate in prescriptive screening but almost all those who have a gynecological examination for other reasons such as contraception, pregnancy, or menopausal symptoms. It is therefore difficult to estimate the actual percentage of women who undergo cytological testing more or less regularly, and this percentage is probably much higher than that for early detection screening. Nevertheless, failure to participate in regular screening for cancer prevention is the highest risk factor possible for the development of cervical carcinoma (62). The risk is six times higher in this group than for women who are regularly monitored.

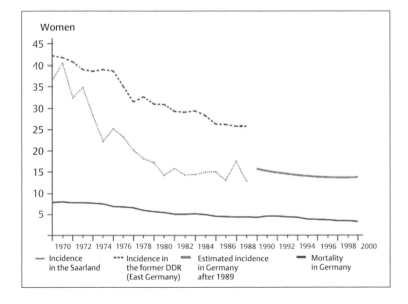

Fig. 4.1 Incidence and mortality of cervical carcinoma in Germany between 1970 and 2000. Age-adjusted incidence and mortality per 100 000 women screened each year. (Robert Koch Institute, Berlin, Germany, see ref. 7a.)

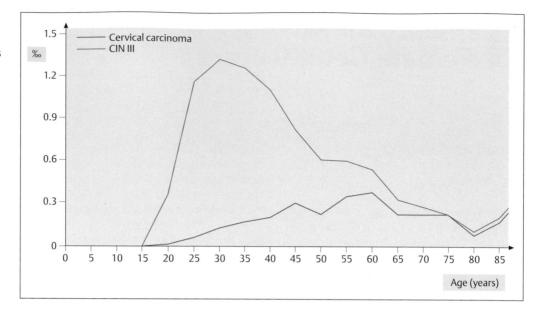

Fig. 4.**2** **Age distribution of cervical carcinoma** as compared to CIN III lesions (Cytological Laboratory of the Swiss Cancer Society).

Age distribution. The age distribution of women affected by cervical carcinoma shows a peak at age 60, and the frequency is about the same at ages 40 and 70 (355). However, grade III cervical intraepithelial neoplasia (CIN III) occurs four times more often before the age of 40 than after age 70 (Fig. 4.**2**).

Precancerous stages. The incidence of precursors of cancer differs significantly from that of invasive cervical carcinoma, being about 100 times higher (312). The incidence of mild to moderate dysplasia is about 1%, and that of severe dysplasia and carcinoma in situ is about 2‰ (Fig. 4.**3**). The peak of precancerous stages occurs at age 30, which is 30 years earlier than for invasive carcinoma (327). There is a steep increase starting at age 15 and a slow, steady decline after age 30 and into old age (Fig. 4.**2**).

Factors influencing epidemiology. Over the last 50 years, the epidemiology of cervical carcinoma—and that of its precursors, in particular—has been clearly influenced by two factors: the successful screening for early detection, on the one hand, and the increasing endemic infection of the population with human papillomavirus (HPV), on the other. Timely detection of severe precancerous stages leads to their therapeutic elimination before a carcinoma develops (355). The increased frequency of HPV infection, however, has caused a dramatic increase in mild and moderate dysplasia, and these stages are usually

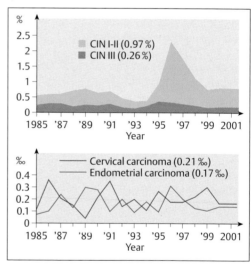

Fig. 4.**3** **Frequency distribution of positive Pap smears,** based on 1.14 million cytological smears evaluated between 1985 and 2001 (Cytological Laboratory of Prof. Nauth, Stuttgart, Germany). The peak of the CIN I–II findings during 1996/1997 is attributed to a change in the method of smear collection at this time (using a Szalay spatula instead of a cotton swab), which caused a temporary increase in positive findings (particularly of CIN I–II lesions) due to overrating.

not removed right away but only monitored at regular intervals. These circumstances have resulted in a distinct shift of incidence rates within the last decades.

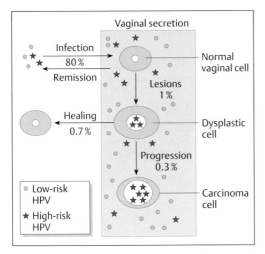

Fig. 4.**4** **Viral carcinogenesis**. High-risk types of HPV cause lesions in about 1 % of infected women and cancer in 0.3 %.

Morphogenesis

The current view is that cervical carcinoma is caused by HPV infection. Carcinogenesis takes place via several intermediate stages known as **dysplasia** or **cervical intraepithelial neoplasia** (CIN). The last of these stages (CIN III) is difficult to distinguish from carcinoma in situ (CIS).

In terms of **chronological development**, we distinguish:

▶ Mild dysplasia (CIN I)
▶ Moderate dysplasia (CIN II)
▶ Severe dysplasia (CIN III)
▶ Carcinoma in situ (CIN III/CIS)
▶ Invasive carcinoma

In terms of **morphological development**, we distinguish:

▶ Nonkeratinizing dysplasia
▶ Metaplastic dysplasia
▶ Keratinizing dysplasia
▶ Small-cell dysplasia
▶ Glandular dysplasia (see pp. 214 ff.)

Several forms of dysplasia often exist simultaneously, side by side. The morphogenesis of squamous cell carcinoma and that of endocervical adenocarcinoma cannot be evaluated in isolation from each other.

Invasive squamous cell carcinoma of the cervix shows the following growth patterns, which may exist simultaneously:

▶ Large-cell squamous carcinoma
▶ Keratinizing squamous carcinoma
▶ Small-cell squamous carcinoma

HPV Infection

There are numerous hypotheses for the development of cervical carcinoma, and they involve factors ranging from promiscuity, parity, genital hygiene, nutritional habits, and addiction—particularly cigarette smoking—to viral infections (208).

Transmission. Current discussions focus on HPV infection, which is the most likely candidate for being involved in cervical carcinogenesis (147, 381). Certain HPV subtypes, particularly types 16 and 18, are thought to be the real cause of cervical cancer (82). Since this virus is transmitted by sexual contact, specific risk factors include an early start of sexual activity, a high number of sexual partners, and low socio-economic status (142).

It is now assumed that 80 % of the population come into contact with HPV for shorter or longer periods of time during their lives (382), and that about 10 % of all women become infected with high-risk types of HPV (see also p. 102). However, only 1 % of this group develop lesions, and about 70 % of these undergo spontaneous healing (Fig. 4.**4**).

Route of infection. The virus turns virulent when penetrating the epithelial barrier, thus being able to infect cells that are still capable of dividing, namely, basal cells, immature metaplastic cells, and reserve cells (81). The virus is most effective when the epithelium has already been damaged by bacterial or fungal infections, as this facilitates the penetration of virus particles into the epidermis (221). This pathway of infection is particularly facilitated on the boundary between squamous epithelium and columnar epithelium of the portio vaginalis where the metaplastic process of transformation predominates (342).

Viral replication. The shell of protein that protects the viral genome, the capsid, disintegrates upon entrance of the virus into the host cell, and the naked viral DNA begins to replicate (productive infection). Because of the cell's greater metabolic needs, the nucleus increases in size and becomes polyploid. Cytologically, this change correlates with mild to moderate dysplasia. During the terminal phase of squamous epithelial differentiation, the capsid proteins, or **late** proteins (L1 and L2) are synthesized, thus ensuring survival and infectiousness of the virus after exfoliation and disintegration of infected superficial cells (8, 221a). Occasionally, however, the viral DNA becomes integrated into

4

the genome of the host cell; this amounts to a genetic "accident" as the virus is now unable to replicate (nonproductive infection). Now, only some early proteins are synthesized, but late proteins are suppressed, thus evading the host's immune response. The **early** proteins E6 and E7, specifically, act as **oncoproteins** by inactivating **tumor suppressor proteins** and stimulating cell proliferation (429). Cytologically, these changes correlate with severe dysplasia or carcinoma in situ. At this stage of carcinogenesis, certain components of tobacco smoke—when excreted at high concentration into the cervical mucus—possibly act as cocarcinogens by further raising the rate of cell division (252).

Controlled carcinogenesis. For variable periods of time, the host cell's regulatory mechanism is able to recognize genomic damage caused by mutation. The damaged cells are then eliminated by means of DNA repair and subsequent controlled cell death, or **apoptosis**, mediated by the tumor suppressor genes *p16* and *p53*. This process is called **controlled carcinogenesis** (Fig. 4.5) (73a, 173a and b, 350).

Uncontrolled carcinogenesis. Over longer periods of time, the mutagenic effects of the virus may disturb certain short nucleotide sequences, so-called microsatellites, in the host cell's DNA repair genes. The resulting **microsatellite instability** interferes more and more with the normal function of these repair genes. A high degree of instability finally leads to uncontrolled cell proliferation, thus promoting the induction of secondary mutations that are no longer eliminated through DNA repair. Thus, the development of cancer can no longer be prevented, even if the virus were to be eliminated at this stage. This process is called **uncontrolled carcinogenesis** (Fig. 4.6) (350).

Koilocytes. The virus-infected basal cells mature into superficial cells during their migration through the epithelium. At the intermediate and superficial cell stages, viral metabolites are released from the nuclei and cause damage in the cytoplasm, thus creating irregular perinuclear halos. Cells affected in this way are called **koilocytes** (15, 181) (see also p. 102).

Dyskeratocytes. The basal cells respond to HPV infection with enhanced proliferation. This results in accelerated epithelial maturation and may be accompanied by temporary keratinization on the surface. Koilocytic intermediate and superficial cells mature into atypical horn cells, called **dyskeratocytes**. These finally exfoliate

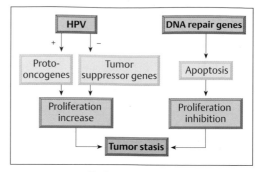

Fig. 4.**5 Controlled carcinogenesis.** Long-term dysplasia is associated with **tumor stasis**, a state in which the increase in HPV-stimulated cell proliferation is inhibited by the host cell's defense mechanism, through the activation of DNA repair genes.

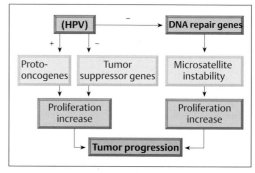

Fig. 4.**6 Uncontrolled carcinogenesis. Tumor progression** occurs when HPV damages the body's repair genes, thus preventing the inhibition of increased cell proliferation.

and release HPV particles from their nuclei when they undergo cytolysis (see also p. 102). Reinfection, or the infection of a sexual partner, is therefore possible (220).

Contributing factors. The aggressiveness of HPV depends on the subtype, but also on certain factors that are not yet well understood. The body's immune status probably plays a major role (184). This explains why cervical neoplasms occur more frequently in connection with diseases that weaken the immune system and, in particular, with HPV infection (85a). As maturation of the squamous epithelium usually ceases after the menopause, as a result of hormonal deficiency, the virus is no longer able to attack during this period of life. Hence, the frequency of cervical carcinoma declines with age (355).

Types of HPV. Infection with certain subtypes of the virus, particularly types 6 and 11, causes pronounced thickening of the superficial zone of keratinization, thus creating warts, or **con-**

Cytology	Super-ficial cell	Mature dysplas-tic cell	Interme-diate dys-plastic cell	Immature dysplas-tic cell	Squamous carcinoma in situ cell	Squamous carcinoma cell
Superficial-cell layer						
Intermediate cell layer						
Intermediate cell layer						
Parabasal cell layer						
Basal cell layer						
Histology	Normal	Mild dysplasia	Moderate dysplasia	Severe dysplasia	Carcinoma in situ	Invasive carcinoma

Fig. 4.**7 Stages of carcinogenesis in nonkerat-inized squamous epithelium.** The dysplastic basal cell layer thickens with each increase in dysplasia. The epithelium has less and less time to mature, thus causing exfoliation of more and more imma-ture epithelial cells.

dylomas. These are epithelial proliferations without precancerous potential (218, 381) (see also pp. 162 ff.). Other types of HPV, mainly types 16 and 18, stimulate the basal cells so intensely that they respond not only with increased proliferation but also with nuclear abnormalities.

Dysplasia

Development of dysplasia. The layer of abnormal basal cells thickens over the years, and the dysplastic basal cells have less and less time for maturation (Fig. 4.**7**) (187). They are therefore still at a relatively low level of differentiation when they reach the epithelial surface. Over the years, exfoliated cells are more and more immature, and koilocytic features become rare as they only occur in mature cells. The condition associated with the increased thickness of the abnormal basal cell layer and the resulting disturbed maturation in the upper cell layers of the squamous epithelium is called **dysplasia** (313).

Grades of dysplasia. Depending on the severity of the precancerous stage, the condition is classified as **mild, moderate, or severe dysplasia (CIN I, II, or III)** (311). The German nomenclature distinguishes an additional stage, **carcinoma in situ (preinvasive carcinoma)** (48).

Types of dysplasia. The morphology of dysplasia differs depending on which type of cells—basal cells, immature metaplastic cells, or reserve cells—has been infected (246a, 281). The further differentiation of these cell terms in different types of carcinomas may also play a role (Fig. 4.**8**).

In about 70% of patients with dysplasia, basal cells or immature metaplastic cells are infected, and this results in **nonkeratinizing** or **metaplastic dysplasia** (Fig. 4.**9**).

In about 20% of cases, the dysplastic stage is skipped due to infection. This possibly happens when preexisting inflammation causes basal cell hyperplasia and the concurrent HPV infection prevents the epithelium from maturing. The stage of abnormal basal cell hyperplasia thus leads directly to the development of **small-cell carcinoma in situ** (Fig. 4.**10**).

In another 7%, the infection causes superficial keratinization, or dyskeratosis, of the abnormal epithelium. This is called **keratinizing dysplasia** (328), and its structure resembles vulvar dysplasia (see Fig. 4.**105**, p. 241).

In the remaining 3%, the reserve cells of the columnar epithelium are infected. The resulting abnormal reserve cell hyperplasia then leads to **glandular dysplasia** and, finally, to endocervical **adenocarcinoma in situ** (Fig. 4.**11**) (403) (see pp. 214 ff.). However, the actual frequency of adenocarcinoma and its precursors seems to be higher (about 10%) than can be detected by cytological means (248a). Because glandular cells tend to exfoliate very little, and their criteria for malignancy are difficult to detect, glandular dysplasia is probably not always recognized cytologically.

The various morphological types of dysplasia may develop independently of one another, and their morphogenesis is often mixed (104).

Regression of dysplasia. In principle, dysplastic lesions are reversible, and regression probably depends on the body's immune status. About 60–70% of all mild dysplasia will regress within a year and turn into normal epithelium (239). For higher grades of dysplasia, however, regression is less likely (371). Nevertheless, the frequency of regression for CIN III lesions is assumed to be up to 30% (267a).

4

Fig. 4.8 Morphogenesis of cervical carcinoma. Diagram. Carcinogenesis takes place either in the **basal cells** of the squamous epithelium or in the **reserve cells** of the columnar epithelium. The final stage is usually a nonkeratinizing—less frequently a keratinizing or small-cell—squamous carcinoma, or an adenocarcinoma of the endocervix. (According to S.F. Patten, modified.)

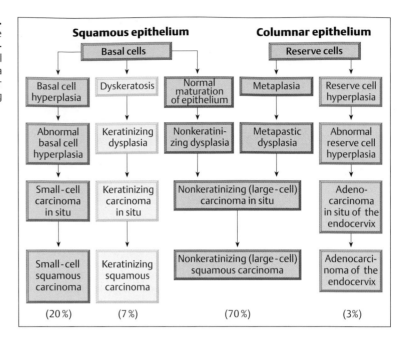

Fig. 4.9 Development of metaplastic dysplasia. When HPV lesions occur in the endocervical epithelium during metaplastic transformation, abnormal reserve cells develop first, followed by **abnormal metaplastic cells** of various degrees of maturation. This finally leads to fully developed dyskaryosis of the squamous epithelium.

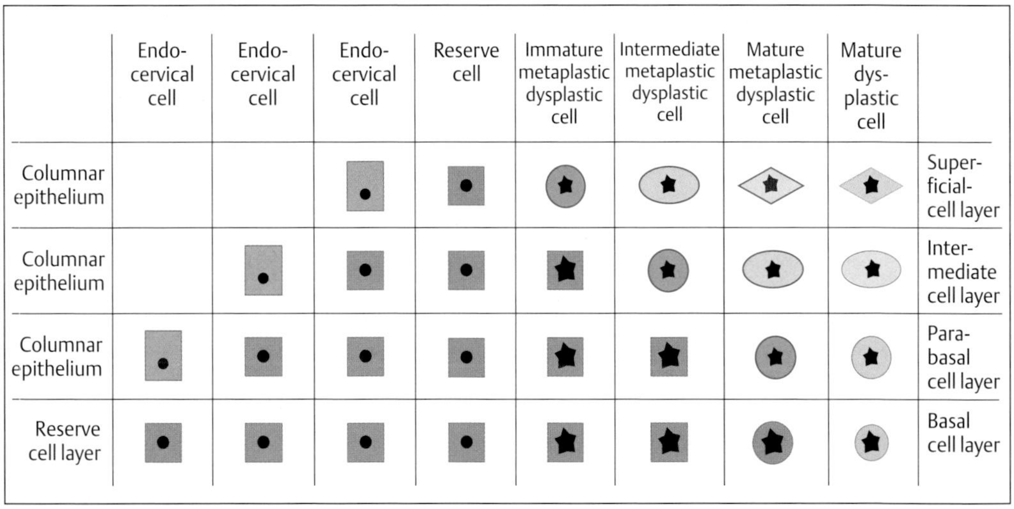

Duration of dysplasia. The precancerous stages may last for various lengths of time. If dysplasia gives rise to a carcinoma, this will be preceded by mild dysplasia for 6 years on average, moderate dysplasia for 4 years, and severe dysplasia for about 1 year (312, 367). If the epithelium already consists of nothing but immature abnormal cells, the lesion has a homogeneous, monotonous structure. This state, called carcinoma in situ, then exists for another period of 5–15 years on average (175, 362). There are, however, indications that these periods show considerable variation (18a). As a rule, it will take at least 6 months for mild dysplasia to develop into severe dysplasia (413a). The small-cell variant of CIN III lesions seems to have a very brief course, since it does not have a dysplastic precursor stage but develops fairly suddenly from an inflammatory state of basal cell hyperplasia. Its frequency of about 20% correlates with the practical experience with cases of CIN III: previous cytological findings have been positive (CIN I, II) in only about 50% of cases and negative in 25%, whereas there were no findings in another 25%. Hence, it can be assumed that at least one quarter of all CIN III lesions develop much faster than previously thought.

Tumor invasion. At some point, the abnormal epithelium no longer expands further toward the surface but begins to produce enzymes known as proteases. These enzymes dissolve the basal membrane, which has so far repre-

	Super-ficial cell	Inter-mediate cell	Para-basal cell	Basal cell	Monomor-phic dys-plastic cell
Super-ficial-cell layer	◆	◯	◯	●	★
Inter-mediate cell layer	◯	◯	◯	●	★
Para-basal cell layer	●	●	●	●	★
Basal cell layer	●	●	●	●	★

Fig. 4.**10 Development of small-cell carcinoma in situ.** When HPV lesions occur in the squamous epithelium during inflammatory basal cell hyperplasia, **abnormal small basal cells** develop. This leads to squamous carcinoma in situ without going through a dysplastic stage first.

	Endo-cervical cell	Endo-cervical cell	Abnormal reserve cell	Abnormal glandular cell
Columnar epithelium	●	●	★	★
Reserve cell layer	●	★	★	★

Fig. 4.**11 Development of adenocarcinoma in situ.** A HPV lesion in the endocervical columnar epithelium initially leads to **abnormal reserve cells** and then to abnormal glandular cells.

sented the last barrier between epithelium and connective tissue. This enables the abnormal cells to infiltrate the stroma; this process is called tumor invasion (285, 407). By definition, the **microinvasive carcinoma** thus formed has a depth of invasion of 3 mm or less and a horizontal spread of 7 mm or less (56, 204). By contrast, a **microcarcinoma** is defined as having an invasive depth of 5 mm or less and a horizontal spread of 7 mm or less (169).

Once there is a **clinically manifest carcinoma**, the tumor cells aggressively attack the lymphatic vessels and blood vessels by releasing enzymes known as collagenases. This facilitates their metastatic spread to the lymph nodes and organs (11). Here, they produce vasoendothelial growth factors, which provide for

their own vascular supply (2). Because of their enormous growth potential, the metastases occupy more and more space at the expense of healthy tissues, thus leading to the death of the patient.

Vaginal carcinoma. Morphogenesis of vaginal carcinoma probably takes place in a similar way as in cervical carcinoma, with the same histological and cytological changes occurring also in the vaginal epithelium, though far less often (see p. 212).

Colposcopy

The precancerous and early stages of cervical carcinoma show characteristic colposcopic changes, especially when the acetic acid test is applied:
▶ Acetowhite epithelium
▶ Punctuation
▶ Mosaic pattern
▶ Abnormal transformation zone
▶ Prominent leukoplakia
▶ Abnormal vascularization
▶ Tumor or ulcer

The clinical symptoms of epithelial cancer development are nonspecific, and there is usually no discomfort involved. During its precancerous stages, however, a developing cervical carcinoma exhibits characteristic manifestations under the colposcope. Here, both the acetic acid test and the iodine test are valuable diagnostic tools.
▶ **Acetowhite epithelium:** Dysplasia developing within the original squamous epithelium consists of nonkeratinizing cells and is therefore completely inapparent in the native condition. It turns temporarily white, however, upon topical application of acetic acid (see p. 29).
▶ **Punctuation:** When the epithelial dysplasia intensifies, the stromal papillae are already visible as red dots before the acetic acid test, but they become more prominent after application of acetic acid (Figs. 4.**12**, 4.**13**) (223).
▶ **Mosaic pattern:** If the dysplasia is severe, the epithelial cones become much broader as they grow toward the stroma. As a result, the laterally compressed stromal papillae appear as narrow red ridges on the surface of the epithelium, where they create a mosaic pattern (Figs. 4.**12**, 4.**14**). Here, too, the acetic acid test intensifies the pattern, whereas the iodine test is negative (405).

4

Fig. 4.**12 Colposcopic findings in case of epithelial dysplasia.** With low-grade dysplasia, the stromal papillae are still well preserved (**punctuation**). With high-grade dysplasia, they become laterally compressed due to the increased pressure, thus appearing on the surface as red crests (**mosaic pattern**).

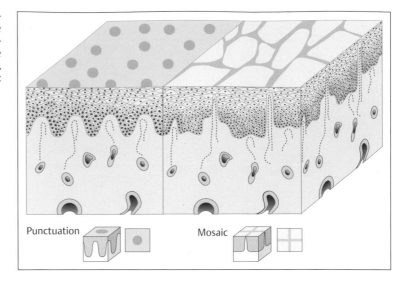

Punctuation

Mosaic

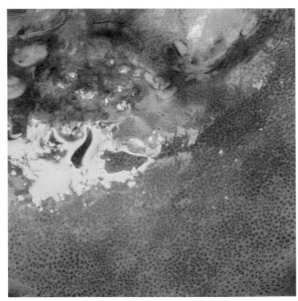

Fig. 4.**13 Area with abnormal punctuation** after application of the acetic acid test. The area below the transformation zone constitutes a CIN I or CIN II lesion.

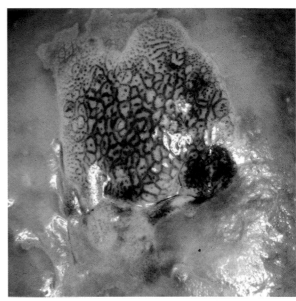

Fig. 4.**14 Area with abnormal mosaic pattern** after application of the acetic acid test. The area in the center of the prominent cervical leukoplakia constitutes a CIN III lesion.

▶ **Abnormal transformation zone:** If a carcinoma develops within the metaplastic epithelium of the transformation zone, the epithelial structure becomes irregular and contains areas of both erythroplakia and leukoplakia. The glandular openings appear particularly prominent and stick out like buttons. This pattern is called an abnormal transformation zone (Fig. 4.**15**) (404) and results in an uneven surface of the epithelium. Application of acetic acid causes distinct acetowhitening (Fig. 4.**16a, b**), whereas the iodine test is negative (54).

▶ **Prominent leukoplakia:** In keratinizing dysplasia, a flaky, prominent leukoplakia already exists before the application of acetic acid. The extent of the leukoplakia does not permit any conclusions regarding the severity of the dysplasia, because the keratinized layer acts as a filter and makes details deep in the epithelium invisible (Fig. 4.**17**) (54).

▶ **Abnormal vascularization:** Under the colposcope, invasive carcinoma is characterized by abnormal vascularization with bizarre ramification and changing caliber of

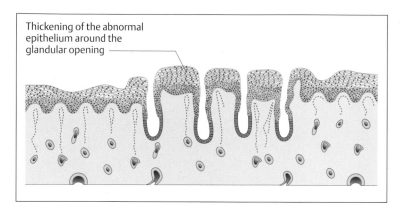

Thickening of the abnormal epithelium around the glandular opening

Fig. 4.**15 Abnormal transformation zone.** Diagram explaining the colposcopic findings. In the glandular part of the cervix, the abnormal epithelium thickens around the glandular openings, thus forming circular protrusions.

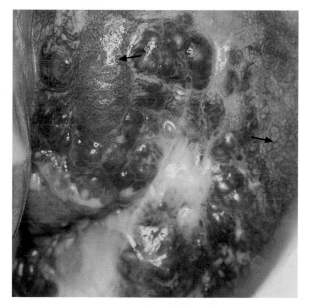

Fig. 4.**16 a, b Abnormal transformation zone.**
a The prominent zone with irregular areas of leukoplakia constitutes a CIN III lesion. Areas with mosaic pattern or punctuation are visible as well (——▶).

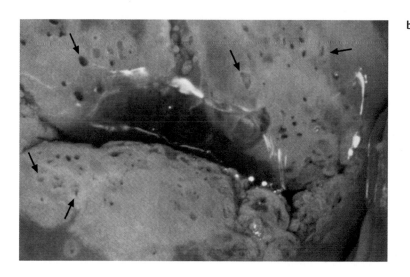

b The prominent zone with circular thickening around the glandular openings (——▶) constitutes a CIN III lesion.

4

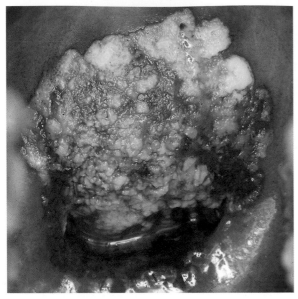

Fig. 4.**17** **Abnormal leukoplakia.** The very prominent, flaky area on the anterior lip of the cervical os constitutes a CIN III lesion.

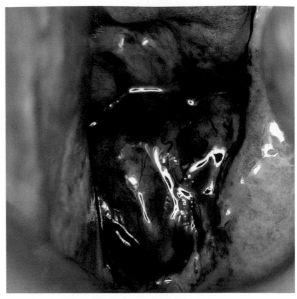

Fig. 4.**18** **Abnormal vascularization.** The wide variation in vascular caliber and the irregular ramifications of blood vessels on the anterior and posterior lips of the cervical os are caused by an invasive cervical carcinoma.

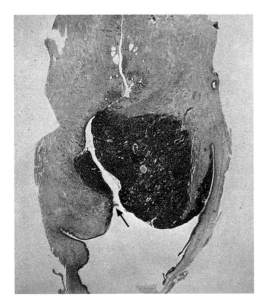

Fig. 4.**19** **Squamous cell carcinoma of the cervix.** Histological gross section. The carcinoma grows at the level of the portio vaginalis while spreading and infiltrating deep into the tissue, thus pushing the cervical canal (⟶) to the left of the photograph. HE stain, actual size.

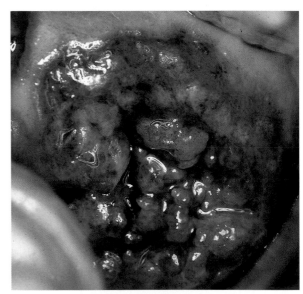

Fig. 4.**20** **Irregular crater-shaped ulcer** due to invasive squamous cell carcinoma of the cervix.

the vessels (Fig. 4.**18**) (337, 372). Furthermore, there may be a tendency for bleeding and the tissue may be fragile (171).

▶ **Tumor or ulcer:** The carcinoma may form an exophythic tumor, infiltrate the stroma in the area of the portio vaginalis (Fig. 4.**19**), or cause a well-defined ulcer due to superficial necrosis (Fig. 4.**20**) (405). The acetic acid test and the iodine test show complementary reactions, thus providing diagnostic clues. Usually, it is only at this late stage of cancer development that the first clinical symptoms appear in the form of irregular vaginal bleeds due to vascular erosion.

Histology

The precursors of cervical cancer are subdivided into four stages: mild, moderate, and severe dysplasia, and carcinoma in situ (preinvasive carcinoma). The international nomenclature does not differentiate between severe dysplasia and carcinoma in situ and defines only three stages: **cervical intraepithelial neoplasia (CIN) grades I–III.** Independently of this subdivision by grade, a distinction is made between nonkeratinizing, metaplastic, keratinizing, small-cell, and glandular dysplasia. The various types of invasive cervical carcinoma are defined as nonkeratinizing (large-cell), keratinizing, and small-cell carcinoma.

Grades of dysplasia. Squamous epithelial carcinogenesis affects primarily the original **nonkeratinizing epithelium**. The histological criteria for the degree of malignancy of the dysplasia depend on the vertical expansion of the abnormal basal cell layer. This layer is not only thicker than in the normal epithelium, but its nuclei are polymorphic, enlarged, and hyperchromatic and exhibit an abnormal chromatin structure and increased mitotic rates (22, 305). The WHO classification distinguishes between the following grades:

▶ **Mild dysplasia (CIN I):** the thickening of the abnormal epithelium is restricted to the lower third of the epithelial layer (Fig. 4.**21 a**).

▶ **Moderate dysplasia (CIN II):** the lower two thirds of the epithelial layer are abnormal (Fig. 4.**21 b**).

▶ **Severe dysplasia (CIN III):** the upper third of the epithelial layer is abnormal as well (Fig. 4.**21 c**) (311, 313).

Koilocytes are detected not only in CIN I and CIN II lesions but sometimes also in the upper cell layers of CIN III lesions, although these usually do not exfoliate (Fig. 4.**22**).

Carcinoma in situ. When there is no further maturation and the epithelium becomes monomorphic, the lesion is called carcinoma in situ. However, its biological significance does not differ from that of a severe dysplasia. Hence, it is usually referred to as CIN III/CIS (Fig. 4.**23**). Severely abnormal squamous epithelium frequently lines the cervical glands (Fig. 4.**24**), although these are normally lined with columnar epithelium. It was originally assumed that the abnormal squamous epithelium grows into the glands to replace the columnar epithelium. However, recent studies suggest that reserve cells underneath the columnar epithelium transform into abnormal basal cells after infection with HPV and then differentiate into squamous epithelial cells (21).

Types of dysplasia. In **nonkeratinizing dysplasia** the epithelium contains abnormal basal cells, whereas **metaplastic dysplasia** is distinguished by the presence of abnormal metaplastic cells (Fig. 4.**25**) (28). In **keratinizing dysplasia**, the different degrees of severity are histologically distinct, but the superficial horny layer prevents exfoliation of the deeper layers of the epithelium, thus making the cytological diagnosis difficult (Fig. 4.**26**) (368). When a carcinoma does not develop via dysplasia but rather via abnormal basal cell hyperplasia, a **small-cell carcinoma in situ** develops relatively suddenly, and this is often referred to as CIN III of the basal cell type (Fig. 4.**27**) (22).

Types of carcinoma. The classic histological nomenclature (309) distinguishes between the following growth types of invasive cervical carcinoma:

▶ **Nonkeratinizing** (large-cell) squamous carcinoma
▶ **Keratinizing** squamous carcinoma
▶ **Small-cell** squamous carcinoma (Figs. 4.**28**– 4.**30**) (306)

4

Fig. 4.**21a–c Dysplasia of the cervical epithelium.**

a Mild dysplasia. The basal cell layer consists of immature cells with enlarged, polymorphic, and hyperchromatic nuclei. The thickening of the basal layer is restricted to the lower third of the epithelium (⟶). HE stain, ×100.

b Moderate dysplasia. The basal cell layer consists of immature cells with enlarged, polymorphic, and hyperchromatic nuclei. The thickening of the basal cell layer occupies the lower two thirds of the epithelium (⟶). HE stain, ×100.

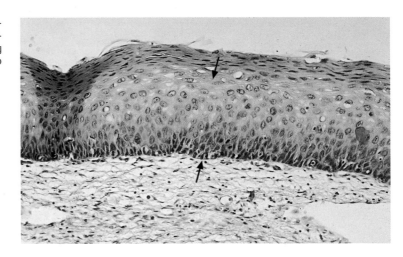

c Severe dysplasia. The basal cell layer consists of immature cells with enlarged, polymorphic, and hyperchromatic nuclei. The thickening of the basal cell layer occupies the total width of the epithelium (⟶). Only at the very surface is there a cell layer with condensed nuclei and signs of keratinization—a condition called pseudomaturation. HE stain, ×100.

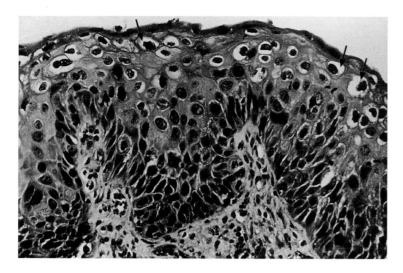

Fig. 4.**22** **Severe, koilocytic dysplasia of the cervical epithelium.** An increased number of koilocytes are found in the upper third of the abnormal epithelium (⟶). HE stain, ×400.

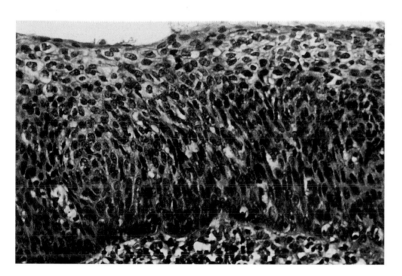

Fig. 4.**23** **Large-cell carcinoma in situ of the cervical epithelium.** The abnormal epithelium consists of immature cells with enlarged, hyperchromatic, but monomorphic nuclei. The thickening of the basal cell layer occupies the total width of the epithelium. The uniform size of the cells gives the epithelium a monomorphic character. HE stain, ×400.

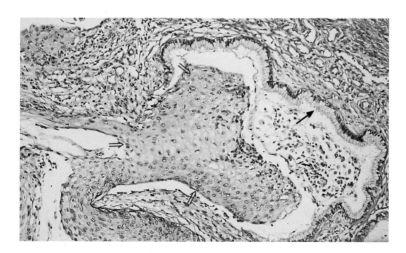

Fig. 4.**24** **Cervical glands lined with severely dysplastic cervical epithelium.** While there is still a normal, single-layered columnar epithelium visible on the right (⟶), it has been replaced by abnormal squamous epithelium on the left (⟹). HE stain, ×100.

Fig. 4.**25** **Moderate metaplastic dysplasia of the cervical epithelium.** Cervical glandular epithelium is still present around the tubular gland (—▶), but at the epithelial surface it has been replaced by abnormal epithelium. A dark line of hyperplastic reserve cells is visible at the base of this epithelium (⟹), and the thickening of the basal cell layer reaches into the middle third of the epithelium. The maturing cells in the upper layer (➡) show an increase in cytoplasm and more defined cell boundaries. HE stain, ×400.

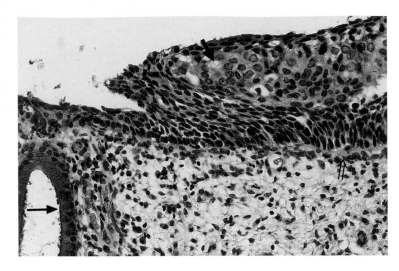

Fig. 4.**26** **Mild keratinizing dysplasia of the cervical epithelium.** The basal cell layer consists of immature cells with enlarged, polymorphic, and hyperchromatic nuclei. The thickening of the basal cell layer remains restricted to the lower third of the epithelium, although individual abnormal cells do reach the surface. Superficially, there is a relatively thick dyskeratotic zone of keratinization (—▶). HE stain, ×200.

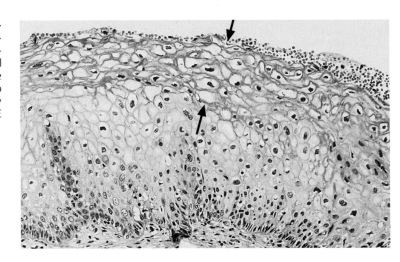

Fig. 4.**27** **Small-cell carcinoma in situ of the cervical epithelium.** The basal cell layer consists of very small immature cells with hyperchromatic, monomorphic nuclei, and it occupies the entire thickness of the epithelium. The uniform size of the cells gives the epithelium a monomorphic character. HE stain, ×200.

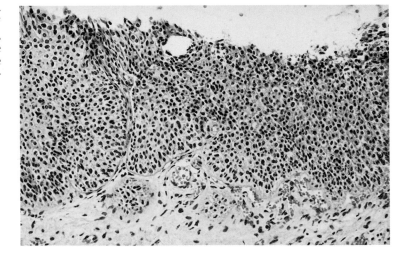

180

Fig. 4.**28** **Large-cell squamous carcinoma of the cervix.** The connective tissue (stroma) exhibits an increase in vascularization and leukocytic infiltration. The basal membrane of the epithelium is interrupted at several locations, and the abnormal epithelium massively invades the stroma (⟶). The variation in cell size gives the abnormal epithelium a polymorphic character. HE stain, ×100.

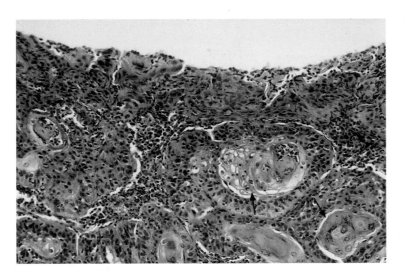

Fig. 4.**29** **Keratinizing squamous carcinoma of the cervix.** Almost all cells show keratinization, either as individual cells or as keratin pearls (⟶). The variation in cell size gives the abnormal epithelium a polymorphic character. HE stain, ×100.

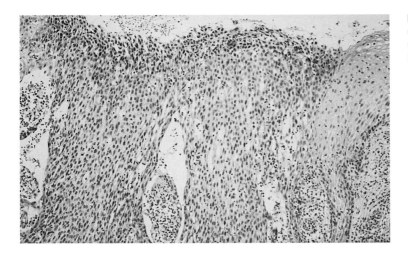

Fig. 4.**30** **Small-cell squamous carcinoma of the cervix.** The abnormal epithelium consists of small cells only and is monomorphic in character. HE stain, ×100.

4

Fig. 4.**31 Anaplastic carcinoma of the cervix.**
This is an undifferentiated small-cell carcinoma.
The abnormal epithelium on the right is well preserved, and its nuclei are large, bright, and monomorphic. However, the nuclei tend to become
necrotic so that the tissue on the left contains only
very small, degenerated, pyknotic nuclei and hyalinized cytoplasm (➞). HE stain, ×400.

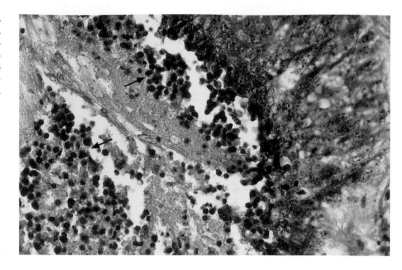

Table 4.**1** WHO classification of malignant tumors
of the cervix (according to R. E. Scully, et al., 1994:
Histological Typing of Female Genital Tract Tumors)

Squamous carcinomas

▶ Keratinizing squamous carcinoma
▶ Nonkeratinizing squamous carcinoma
▶ Verrucous squamous carcinoma
▶ Condylomatous squamous carcinoma
▶ Papillary squamous carcinoma
▶ Reticular squamous carcinoma

Adenocarcinomas

▶ Mucinous adenocarcinoma
▶ Endometrioid adenocarcinoma
▶ Clear cell adenocarcinoma
▶ Serous adenocarcinoma
▶ Mesonephric adenocarcinoma

Other epithelial carcinomas

▶ Adenosquamous carcinoma
▶ Glassy cell carcinoma
▶ Adenoid cystic carcinoma
▶ Adenoid basal carcinoma
▶ Carcinoid carcinoma
▶ Small-cell anaplastic carcinoma (undifferen-
tiated, neuroendocrine)

Sarcomas

▶ Leiomyosarcoma
▶ Stromal sarcoma
▶ Rhapdomyosarcoma
▶ Endometrioid sarcoma
▶ Alveolar sarcoma

Mixed tumors

▶ Adenosarcoma
▶ Mesodermal mixed tumor
▶ Wilms tumor

Mixed cell tumors

▶ Malignant melanoma
▶ Malignant lymphoma
▶ Yolk sac carcinoma

Formation of metastases

The 1994 WHO classification (Table 4.**1**) no
longer distinguishes a small-cell squamous
carcinoma (348). Instead, it lists an **undifferentiated (anaplastic) small-cell carcinoma**,
which is of neuroendocrine origin and corresponds to the oat cell carcinoma of the lung.
This type of carcinoma is very rare and represents only about 1 % of all cervical carcinomas.
Because of the small size of its cells (Fig. 4.**31**),
it is usually not recognized cytologically, thus
leading to early onset of metastasis with high
mortality. Since the small-cell squamous carcinoma mentioned before (see p. 177) is observed much more often, it may not be wise to
abandon the classic nomenclature (309) in
favor of the new WHO classification (348).
Furthermore, the three established variants of
squamous cell carcinoma can be easily verified
in the cytological smear.

Cytology

The great success of cervical cytology in early
cancer detection is based on the fact that abnor-
mally altered cell nuclei can be assigned to different degrees of cytoplasmic maturation. This
makes it possible to distinguish between various
grades of dysplasia and to recognize precursors
of cervical carcinoma very early on.

Nucleus of a normal intermediate cell
- regular, fine chromatin structure
- small chromocenter
- smooth nuclear envelope
- round nucleus

Nucleus of a mildly dysplastic cell
- irregular, coarse chromatin structure
- large chromocenter
- hyperchromatic nuclear envelope
- irregular shape of nucleus

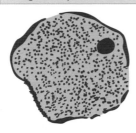

Nucleus of a severely dysplastic cell
- irregular, coarse chromatin structure
- cribriform pattern
- large chromocenter
- hyperchromatic nuclear envelope
- irregular shape of nucleus

Nucleus of a cancerous cell
- irregular, coarse chromatin structure
- cribriform pattern
- nuclear clearing
- nucleolus
- hyperchromatic nuclear envelope
- irregular shape of nucleus

Uncertain cytoplasmic criteria for malignancy
- koilocytosis
- dyskeratosis
- spindle-shaped cells
- cell-in-cell figures (cannibalism)

Fig. 4.**32** **Criteria for malignancy** of squamous epithelia. Nuclear criteria and inconsistent (unreliable) cytoplasmic criteria.

Malignancy Criteria

The morphology of cells exfoliating at the surface of the squamous epithelium makes it possible to determine the degree of severity of the pathological lesion lying underneath. HPV infects exclusively the cell's nucleus, where it stimulates the replication of DNA and chromosomes. Hence, the criteria for malignancy focus primarily on the nucleus. We distinguish between the following features:

▶ Abnormal chromatin structure
▶ Hyperchromatic nuclear envelopes and presence of chromocenters
▶ Hyperchromatic nuclei
▶ Polymorphic nuclei (variable size and shape)
▶ Absolute enlargement of nuclei
▶ Increased nucleocytoplasmic ratio
▶ Presence of nucleoli

Abnormal chromatin structure. The most important criterion for malignancy is an irregular, coarse, and therefore abnormal, chromatin structure. The Papanicolaou method uses he-

matoxylin to stain the nuclear chromatin. The biologically active portion, the **euchromatin**, appears bright whereas the genetically inactive portion during interphase, the **heterochromatin**, appears dark. The heterochromatin particles of normal nuclei are finely granular and evenly distributed (Fig. 4.**32**) (137, 146).

In contrast, abnormal nuclei show an increase in heterochromatin, and this is proportional to the degree of dysplasia. The more severe stages of dysplasia show dark chromatin bands that form bridges between the coarsened granules, thus creating a **cribriform pattern**—which is nevertheless still fairly regular (116). In invasive carcinoma, however, the nuclei exhibit irregular distribution of the chromatin due to the clumping of heterochromatin. Large bright areas within the chromatin pattern represent active parts of the nucleus and are referred to as **nuclear clearings** (44).

A coarse chromatin structure needs to be distinguished from degenerative nuclear changes (81). Karyorrhexis is characterized by

183

disintegration of the chromatin bridges described above, leaving behind chromatin fragments (see pp. 57 ff.). This is accompanied by gaps in the nuclear envelope (26).

Hyperchromatic nuclear envelopes and chromocenters. Heterochromatin adhering to the inner nuclear membrane results in a **hyperchromatic nuclear envelope** of irregular thickness, and clumping of heterochromatin inside the nucleus results in the formation of **chromocenters** (138). During degenerative and inflammatory processes, on the other hand, the nuclear membrane usually remains smooth despite being thickened, and chromocenters are less prominent (see pp. 57 ff. and 109 ff.).

Hyperchromatic nuclei. This additional sign of malignancy is due to an increased amount of **heterochromatin** (4). In highly malignant cells with rapid cell division and short interphases, however, **euchromatin** often exceeds heterochromatin, and the nuclear stain is only faint. Inflammation may also result in hyperchromatic nuclei, and this is sometimes so pronounced that it becomes difficult to distinguish between inflammation and malignancy (368).

Polymorphic nuclei. Another common criterion for malignancy is that the nuclei vary in size and shape, which is called nuclear polymorphism.

Differences in nuclear size result from uncoordinated DNA synthesis; this sign of malignancy is called **anisonucleosis** (64).

Changes in nuclear shape include loss of roundness, indentations, and lobes. These irregularities result from an increased metabolic exchange between the nucleus and the cytoplasm. The increased metabolism requires an enlarged nuclear surface, which is achieved by folds and indentations of the nuclear membrane. Extreme cytoplasmic invagination creates ring-shaped nuclei.

However, nuclear changes are not necessarily a sign of malignancy as they also occur with degeneration and inflammation.

Absolute nuclear enlargement. This is an important criterion, but not very specific because it is also observed with inflammation (see pp. 109 ff.). In dysplasia, the nuclear area is about three to four times larger than in normal superficial cells, but it is only about twice the size in carcinomas (306). Cells with enlarged nuclei should only be evaluated as malignant when other signs of malignancy are present as well.

Increased nucleocytoplasmic ratio. The ratio of nuclear volume to cytoplasmic volume indicates malignancy if there is a greater shift in favor of the nucleus than is the case with benign processes.

Presence of nucleoli. Another unreliable criterion for malignancy is the formation of nucleoli, a feature also associated with many benign changes, particularly during cell regeneration (see pp. 123, 138 ff.) (394). Although nucleoli are essentially absent from dysplastic cells and even from carcinoma in situ, they are regularly found in invasive carcinomas. Hence, their detection in a carcinoma in situ indicates the start of invasion (206). The presence of nucleoli signifies an increase in metabolic activity and protein synthesis in the cell. Since formation of a nucleolus requires a relatively long interphase, nucleoli are not detected in very rapidly growing, anaplastic carcinomas (72).

Reliability of the criteria. None of the criteria for malignancy (Table 4.2) is absolutely reliable, since they occasionally also occur with certain benign processes. The more criteria are detected simultaneously in the cytological smear, the more likely they are to indicate the existence of a malignant lesion. The differential diagnosis between benign and malignant nuclear changes may be very difficult in some cases.

If the cells in the preparation are well preserved and the critical features are very prominent, the diagnosis is relatively simple. If inflammatory or degenerative processes influence the cytological image and the signs are poorly developed, the diagnosis becomes considerably more difficult or even impossible. On the other hand, inflammation and degeneration may cause cellular and nuclear changes so severe that they mimic the criteria for malignancy, thus giving rise to a false positive diagnosis.

Table 4.**2** Cytological criteria for malignancy expressed by the nuclei of cervical lesions. Most criteria are also found in almost all benign lesions, with the exception of abnormal chromatin

	Abnormal chromatin	Presence of nucleoli	Polymorphic nuclei	Hyperchromatic nuclei	Enlarged nuclei
Squeezed cells					+
Fixation error					+
Cytolysis					+
Degeneration					+
Radiotherapy					+
Chemotherapy					+
HPV infection				+	+
Inflammation		+	+	+	+
Metaplasia			+	+	+
Regeneration		+	+	+	+
Dysplasia	+		+	+	+
Carcinoma	+	+	+	+	+

Classification Criteria

Corresponding to the histologically defined **grades of dysplasia** that lead to **invasive carcinoma**, we can cytologically distinguish the following precancerous stages based on the cytoplasmic maturation of dysplastic cells (see also Fig. 4.7, p. 171):

▶ mild dysplasia
▶ moderate dysplasia
▶ severe dysplasia
▶ squamous carcinoma in situ
▶ invasive squamous carcinoma

Additional **facultative cellular phenomena** include:

▶ Signs of HPV infection
▶ Cell-in-cell configurations
▶ Spindle-shaped cells
▶ Gland-shaped cell clusters

With respect to the **morphological origin**, we distinguish:

▶ Nonkeratinizing dysplasia
▶ Metaplastic dysplasia
▶ Keratinizing dysplasia
▶ Small-cell dysplasia
▶ Radiation-induced dysplasia
▶ Glandular dysplasia (see p. 218 ff.)

Grades of dysplasia. The term **dyskaryosis**, which was introduced by Papanicolaou and is partly still in use today, is used when nuclei in the cytological smear meet the criteria for malignancy described earlier. The morphology of the cytoplasm surrounding these nuclei provides the decisive clue to the origin of the dyskaryotic cells, thus permitting classification of the lesion.

A dyskaryotic cell is a squamous epithelial cell of a certain degree of maturity—or a columnar epithelial cell—that contains an abnormal nucleus (Fig. 4.**33**) (278). Table 4.**3** presents data obtained by measuring cytoplasm and nuclei of abnormal squamous cells associated with various degrees of malignancy. The data reveal that the size of the cells diminishes with increasing malignancy. The size of the nuclei, on the other hand, remains relatively constant, although the nuclei are larger in dysplasia than in carcinoma. Abnormal cells exfoliate as small clusters of cells and sometimes as single cells. In most cases, dysplasia affects squamous epithelia that mature normally and, therefore, leads to **nonkeratinizing dysplasia**.

Mild dysplasia. This involves superficial or intermediate (mature) cells that show dyskaryosis (Fig. 4.**34 a–c**).

Moderate dysplasia. This involves parabasal (intermediate) cells and that show dyskaryosis (Fig. 4.**35 a–c**) (310, 413).

Severe dysplasia. This involves basal cells that show dyskaryosis (Fig. 4.**36 a, b**).

Squamous carcinoma in situ. Monomorphic immature cells showing dyskaryosis and ap-

Fig. 4.**33 Criteria for classification** of abnormal squamous cells. The relative size of the nucleus increases with increasing degrees of malignancy.

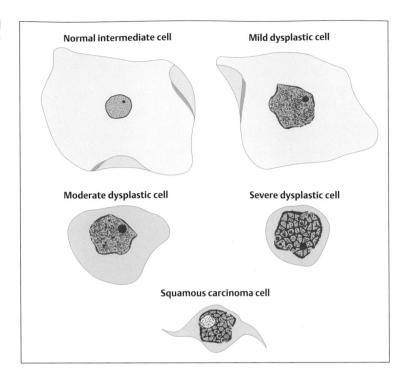

Table 4.**3**   Sizes of cells and nuclei found in dysplasia, carcinoma in situ, and invasive carcinoma of the cervix (according to Reagan et al., 1957)

	Normal findings	Dysplasia	Carcinoma in situ	Invasive carcinoma
Number of cases examined	50	100	100	100
Number of cells measured	2 500	5 000	10 000	5 000
Cell diameter (µm)	44.93 ± 4.28	36.81 ± 5.59	20.85 ± 3.16	16.85 ± 2.94
Cell area (µm²)	1604.2 ± 312.25	1087.6 ± 311.0	352.51 ± 115.78	222.49 ± 82.81
Nuclear diameter (µm)	6.75 ± 1.20	14.52 ± 1.57	11.67 ± 1.65	9.78 ± 1.59
Nuclear area (µm²)	36.51 ± 13.31	167.20 + 38.20	109.38 ± 33.04	77.04 ± 26.57

pearing in single-file arrangement ("Indian files") indicate squamous carcinoma in situ (Figs. 4.**37 a–c**, 4.**38**) (304).

Facultative cellular phenomena. In addition to dysplastic cells, there are occasionally other morphological abnormalities. Koilocytes or other indirect signs of **HPV-induced changes** are often detected (Figs. 4.**39 a, b**, 4.**40 a, b**) (373). The lower the grade of dysplasia, the more common these are. They are almost never associated with CIN III lesions and invasive carcinomas. The detection of cytological signs of HPV, however, is in no way correlated with the microbiological detection of the pathogen; the latter is several-fold more efficient (393).

A sign frequently observed during the entire process of carcinogenesis is the **cell-in-cell configuration** (Figs. 4.**41 a, b**, 4.**42 a, b**) (81). Here, individual abnormal squamous cells either have incorporated leukocytes (phagocytosis) or have engulfed other squamous epithelial cells (cannibalism). The foreign cellular elements usually become encapsulated in a cytoplasmic vacuole.

All precancerous stages may occasionally be associated with **abnormal spindle-shaped cells** appearing in the smear. These are the result of a dysplastic surface effect (Figs. 4.**43**, 4.**44 a, b**, 4.**45 a, b**) (281).

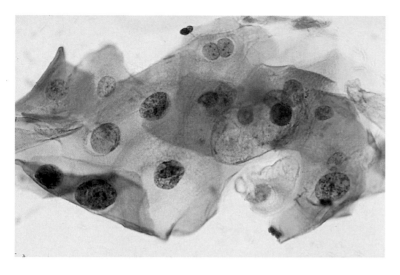

Fig. 4.**34 a–c** **Mature dysplastic cells** associated with **mild dysplasia** of the cervix.
a Superficial and intermediate cells with enlarged, hyperchromatic nuclei and irregular chromatin structure. ×630.

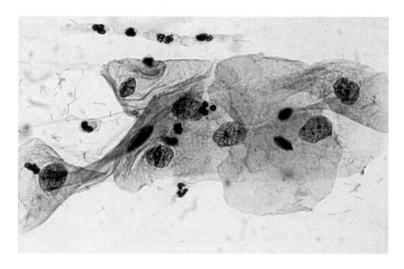

b Superficial and intermediate cells with enlarged, hyperchromatic nuclei and irregular chromatin structure. ×630.

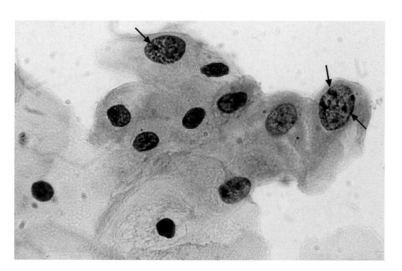

c The cells at the top show enlarged, hyperchromatic nuclei with irregular chromatin pattern, hyperchromatic nuclear envelope, and formation of chromocenters (⟶). Normal superficial cells are visible at the bottom. ×1000.

4

Fig. 4.**35 a–c   Intermediate dysplastic cells** associated with **moderate dysplasia** of the cervix.

a Small intermediate cells with enlarged, hyperchromatic nuclei and irregular chromatin structure. ×630.

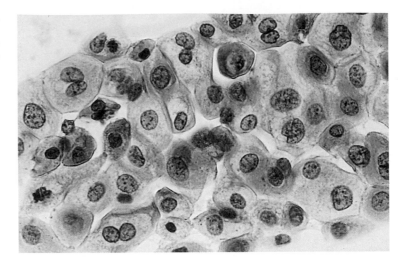

b Small intermediate and parabasal cells with enlarged, abnormal, and partly condensed nuclei. Normal superficial and intermediate cells are visible nearby. ×400.

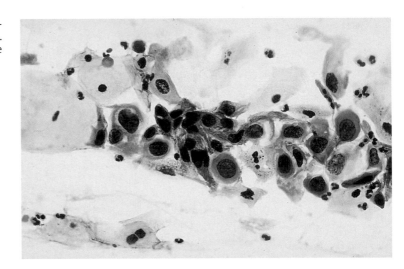

c Small intermediate cells with enlarged, hyperchromatic nuclei and irregular chromatin structure. Normal intermediate cells are also visible (⟶). ×630.

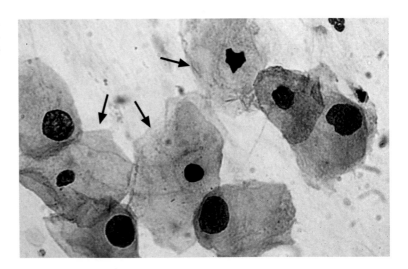

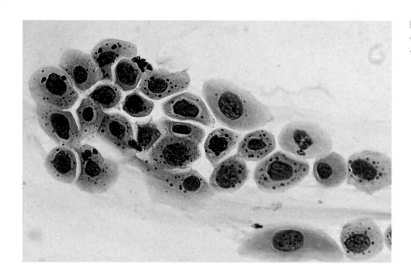

Fig. 4.**36 a, b Immature dysplastic cells** associated with **severe dysplasia** of the cervix.
a Parabasal cells with abnormal, polymorphic nuclei. The cytoplasm contains incorporated
granules of blood pigment (hemochromogen).
×630.

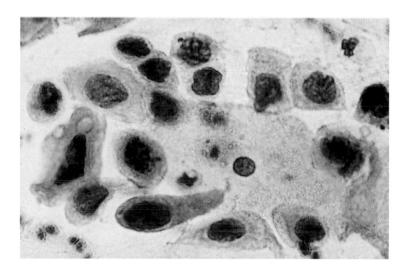

b Parabasal epithelial cells with polymorphic, abnormal, and partly condensed nuclei. A normal
intermediate cell is visible in the center. ×1000.

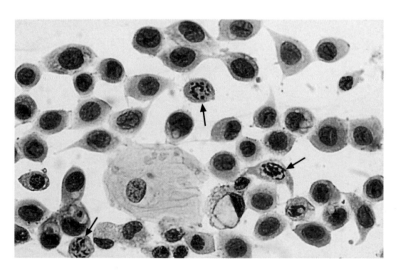

Fig. 4.**37 a–c Squamous carcinoma in situ cells.**
a Fairly uniform abnormal basal cells with irregular, coarse chromatin structure. Some nuclei
show karyorrhexis (⟶). A small normal intermediate cell is visible at the bottom. ×630.

4

b Abnormal basal cells of uniform morphology. They all show signs of malignancy. ×1000.

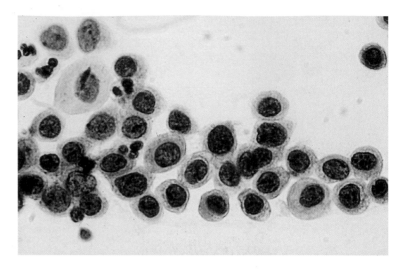

c Characteristic striplike arrangement of abnormal basal cells with partly active, partly degenerated nuclei. Normal superficial cells are visible nearby. ×400.

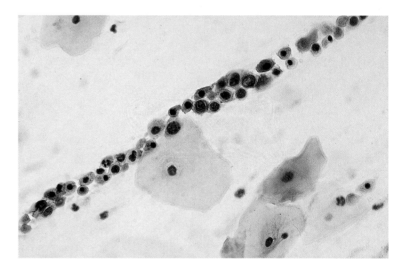

Fig. 4.**38 Monomorphic dyskaryosis** associated with **carcinoma in situ** of the cervix. Abnormal basal cells in single file (→), showing irregular chromatin structure and chromocenters. Two normal superficial cells are visible on the left. ×1000.

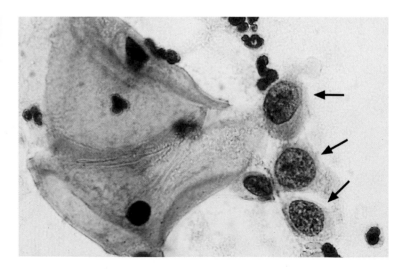

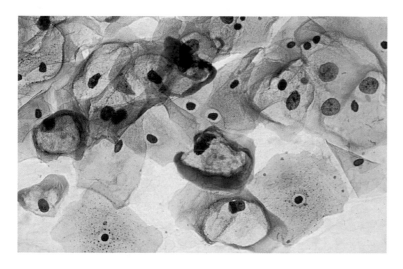

Fig. 4.**39 a, b Koilocytic superficial cells** associated with **mild dysplasia** of the cervix.
a The nuclei are partly abnormal. Single normal superficial cells with pyknotic nuclei are also visible. ×400.

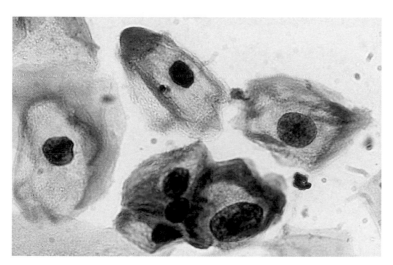

b The nuclei are partly abnormal, and some cells contain two nuclei. ×630.

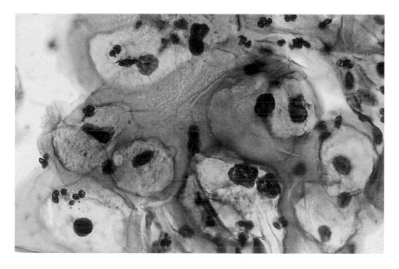

Fig. 4.**40 a Koilocytotic intermediate cells** associated with **moderate dysplasia** of the cervix.
a The nuclei are abnormal. ×1000.

4

b The nucleus of the large koilocyte is abnormal. ×1000.

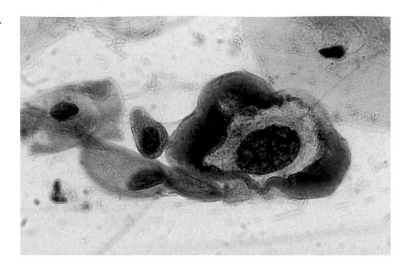

Fig. 4.**41 a, b Cell-in-cell configuration.**

a Associated with **mild dysplasia** of the cervix. A dysplastic intermediate cell (⟶) contains nuclear components in several vacuoles. Mature dyskaryotic cells are visible on the right. ×500.

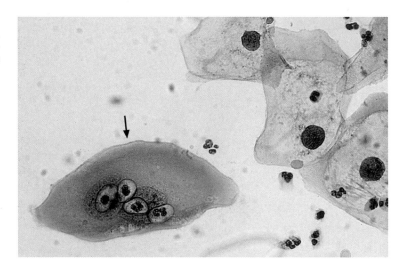

b Associated with **severe dysplasia** of the cervix. Cannibalism of an abnormal basal cell is shown in the center (⟶). ×1000.

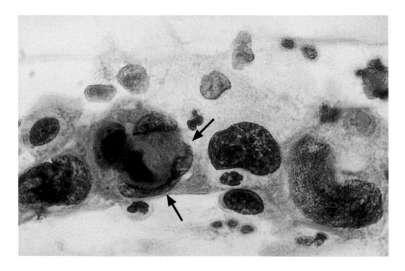

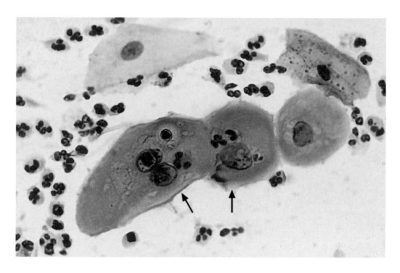

Fig. 4.**42 a, b   Cell-in-cell configuration**

a Associated with **mild metaplastic** dysplasia. Two dysplastic cells (⟶) with phagocytosed leukocytes. Intermediate cells are visible at the top. ×630.

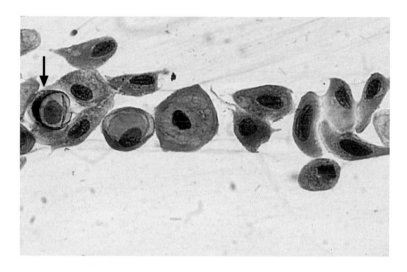

b Associated with **moderate metaplastic dysplasia**. Bird-eye configuration on the left (⟶). ×630.

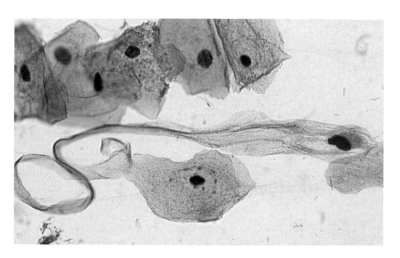

Fig. 4.**43   An abnormal spindle cell** associated with **mild dysplasia** of the cervix. The nucleus is enlarged and hyperchromatic. Normal superficial and intermediate cells are visible nearby. ×1000.

4

Fig. 4.**44 a A cluster of abnormal spindle cells** associated with **moderate dysplasia** of the cervix. The nuclei are hyperchromatic and condensed. Normal intermediate cells are visible nearby. ×250.

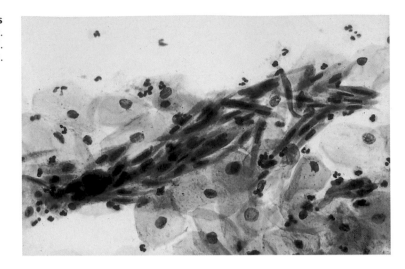

Fig. 4.**44 b Small abnormal spindle cells** associated with **severe dysplasia** of the cervix. The nuclei are enlarged and condensed, and isolated immature dyskaryotic cells are visible at the top (—→). ×400.

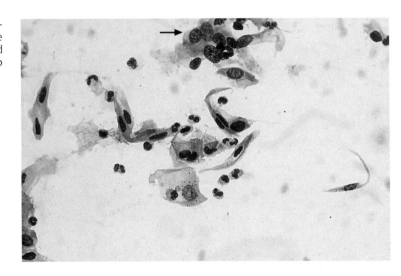

Fig. 4.**45 a, b Abnormal spindle cells** associated **with carcinoma in situ** of the cervix.
a The nuclei are enlarged and condensed, and the cytoplasm stains relatively faint. ×400.

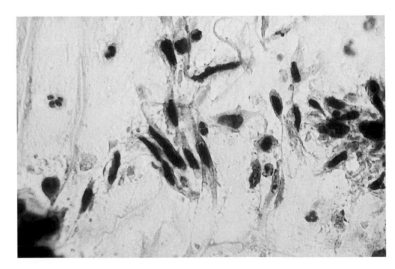

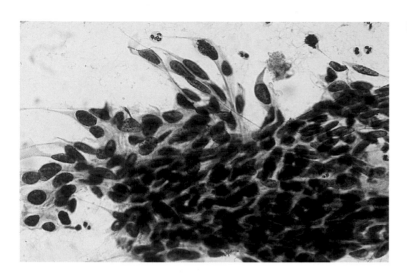

When the cervical glands are lined with highly abnormal cells, compact **gland-shaped cell clusters** are commonly found in the smear (Fig. 4.**46 a, b**), and these are often difficult to distinguish from clusters of endocervical columnar cells (203). The irregular arrangement of the hyperchromatic, enlarged nuclei, as well as the three-dimensional arrangement of the cells, are typical signs of malignancy.

Types of dysplasia. According to their morphological origin, we distinguish various types of dysplasia.

▶ **Metaplastic dysplasia:** Dyskaryotic changes affecting metaplastic cells are classified as **mild, moderate**, or **severe metaplastic dysplasia** depending on the appearance of the cells (Figs. 4.**47 a, b**, 4.**48 a, b**, 4.**49**) (355). The cytoplasm of these cells is often vacuolated and shows the characteristic color and structure of metaplastic cells, and cell-in-cell configurations are frequently observed (see Fig. 4.**42 a, b**). When of metaplastic origin, carcinomas in situ often exhibit severely vacuolated, foamy cytoplasm and indistinct cell borders (Fig. 4.**50 a, b**).

▶ **Keratinizing dysplasia:** Here, the subdivision into different degrees of maturity is difficult (Figs. 4.**51**–4.**54**) (368), because keratinization prevents the exfoliation of the deeper, immature cell layers. Only absolute enlargement and hyperchromatism of the nuclei provide clues to the degree of severity of such a lesion (Fig. 4.**55**).

▶ **Small-cell dysplasia:** Small-cell CIN III develops directly from abnormal basal cell hyperplasia; it therefore has no dysplastic precursors (281). The abnormal basal cells are very small—often not much larger than lymphocytes—and are mostly evenly distributed over the smear (Fig. 4.**56**). The cells have a strong tendency to dissociate, i. e., they appear isolated rather than in sheets or clusters. This feature is characteristic for small-cell squamous carcinomas and hampers their diagnostic identification. As a rule, small abnormal cells can only be identified at high magnification, and it is difficult to distinguish them from normal basal cells, particularly when the latter show nuclear degeneration (Figs. 4.**57 a, b**, 4.**58 a, b**).

▶ **Radiation-induced dysplasia:** Postradiation dysplasia develops as a result of gynecological radiation therapy. It exhibits abnormal nuclear changes similar to nonkeratinizing dysplasia (183). Based on the medical history of the patient and on detectable cytoplasmic radiation effects, however, it is possible to establish a tentative diagnosis (Figs. 4.**59**, 4.**60 a–c**).

4

Fig. 4.**46 a, b A gland-shaped cluster of dysplastic cells** associated with **carcinoma in situ** of the cervix. These clusters represent cells lining the cervical glands.

a A small, disorganized (three-dimensional) cell cluster with hyperchromatic abnormal nuclei. Superficial cells are visible nearby. ×400.

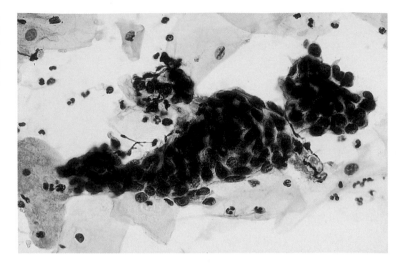

b A large, disorganized (three-dimensional) cell cluster that resembles a cluster of degenerated endocervical glandular cells. ×200.

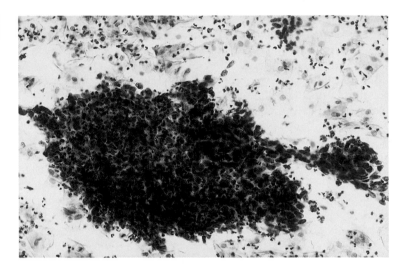

Fig. 4.**47 a, b Immature metaplastic dysplastic cells** associated with **moderate metaplastic dysplasia.**
a Partly still intact, partly degenerated nuclei and occasional cytoplasmic vacuolation (⟶). ×400.

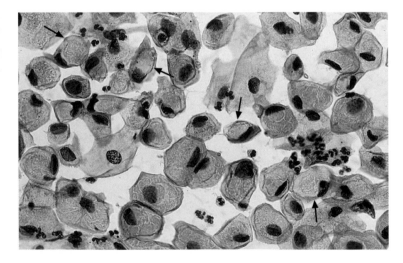

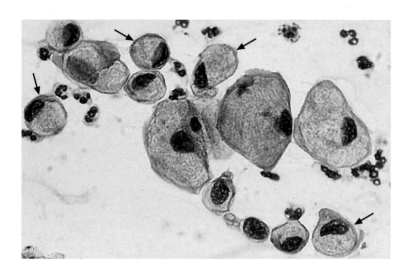

b Abnormal nuclei and vacuolated, foamy cytoplasm. The nuclei are partly pushed to the cell margin, thus forming signet-ring cells (⟶). ×630.

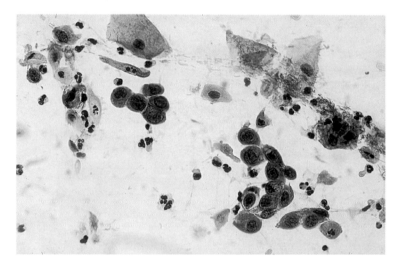

Fig. 4.**48 a, b** **Immature metaplastic dysplastic cells** associated with **severe metaplastic dysplasia.**

a Irregularly structured and hyperchromatic nuclei. Normal intermediate cells are visible at the top. ×400.

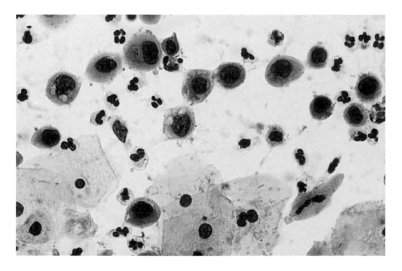

b Abnormal nuclei and cytoplasmic vacuolation. Normal superficial cells are visible at the bottom. ×630.

Fig. 4.**49** **Severe metaplastic dysplastic cells** associated with **carcinoma in situ**. Loose cluster of cells with irregular, hyperchromatic chromatin structure, formation of chromocenters, and cytoplasmic vacuolation. ×630.

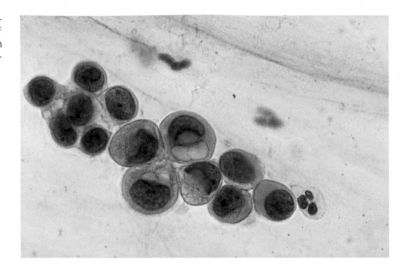

Fig. 4.**50 a, b** **Metaplastic squamous carcinoma in situ cells** associated with **carcinoma in situ** of the basal cell type.
a The nuclei show considerable irregularities, and the foamy cytoplasm is often without distinct borders. Some cell-in-cell configurations are visible (⟶). ×400.

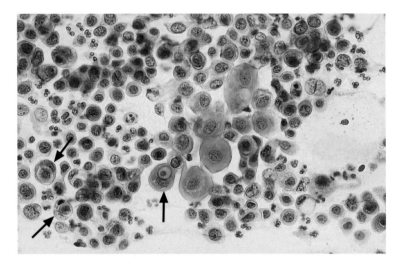

b The nuclei show all signs of malignancy, and the cytoplasm is foamy and without distinct borders. A normal superficial cell and a normal parabasal cell are visible at the top. ×630.

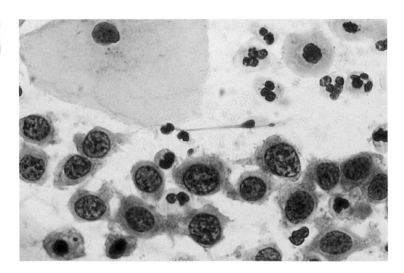

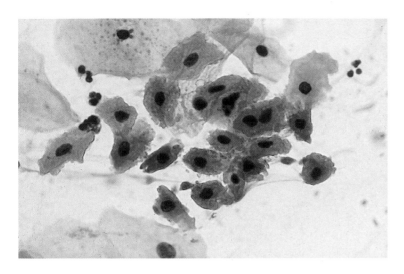

Fig. 4.**51 a, b Keratinizing dysplastic cells** associated with **mild keratinizing dysplasia** of the cervix.
a The nuclei are slightly enlarged and condensed. Normal superficial cells are seen nearby. ×630.

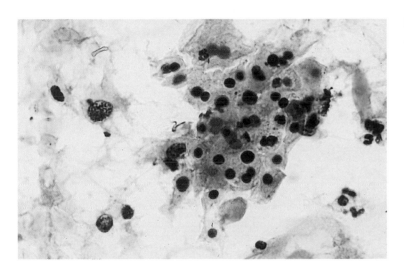

b A loose cell cluster with clearly enlarged, hyperchromatic, and condensed nuclei. ×630.

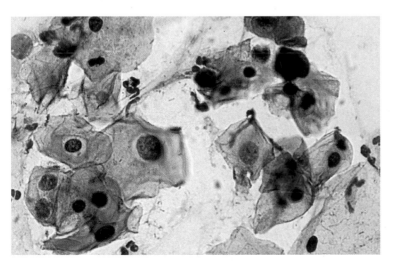

Fig. 4.**52 a, b Keratinizing dysplastic cells** associated with **moderate keratinizing dysplasia** of the cervix.
a The nuclei are partly still intact, partly degenerated, and they clearly vary in size. Normal intermediate cells are seen nearby. ×630.

b Abnormal, much enlarged nuclei with irregular chromatin structure. ×1000.

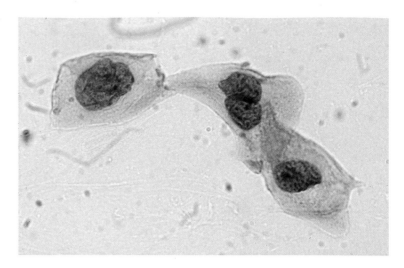

Fig. 4.**53 a, b** **Keratinizing dysplastic cells** associated with **severe keratinizing dysplasia** of the cervix.

a The nuclei are mostly still intact, showing extreme polymorphism and irregular chromatin pattern. Some vacuolated cells and cell-in-cell configurations are visible (⟶). ×400.

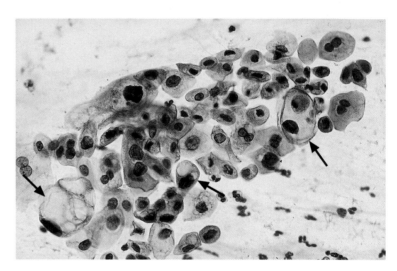

b The nuclei are mostly degenerated, condensed, and hyperchromatic. The cytoplasm stains mostly light yellow. A normal intermediate cell is visible on the lower right. ×400.

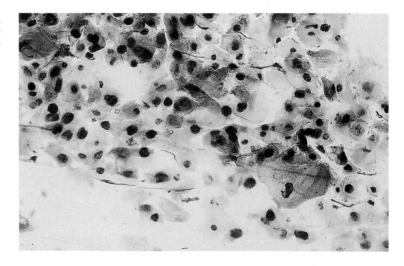

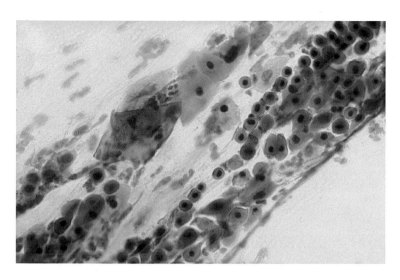

Fig. 4.**54 a, b   Squamous keratinizing carcinoma in situ cells.**

a The small cells are arranged in a striplike formation. Their condensed, hyperchromatic nuclei show only moderate enlargement. Some superficial cells are also visible. ×250.

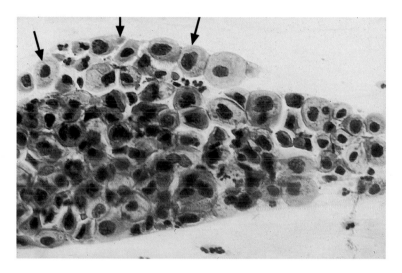

b A loose cluster of cells with much enlarged nuclei. Some nonkeratinized immature dysplastic cells are visible at the top (———▶). ×400.

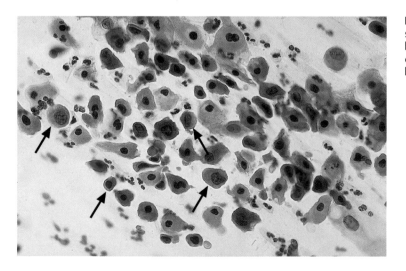

Fig. 4.**55   Squamous keratinizing carcinoma in situ cells.** The degenerated, condensed nuclei are hardly enlarged. Several nonkeratinized immature dysplastic cells are visible (———▶), thus indicating a lesion of high degree of malignancy. ×400.

4

Fig. 4.**56 Small dysplastic cells** associated with **severe small-cell dysplasia** of the cervix. Diffusely arranged, abnormal basal cells with partly still intact, partly degenerated nuclei. Some normal superficial cells are visible in between. ×400.

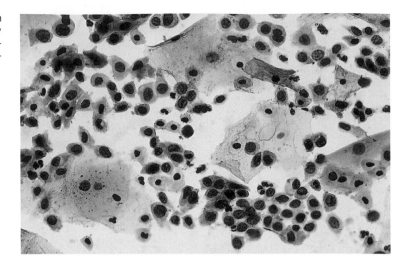

Fig. 4.**57 a, b Small dysplastic cells** associated with **severe small-cell dysplasia** of the cervix. Basal cells with hyperchromatic, condensed nuclei.
a Groups of cells with well-defined cytoplasmic borders (→). Normal superficial cells are visible nearby. ×400.

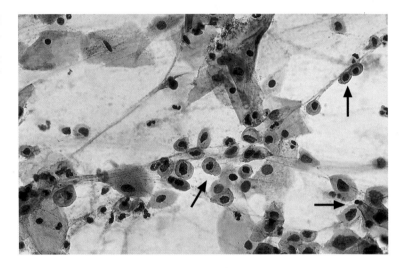

b A group of cells (→) with more or less well-defined cytoplasmic borders, surrounded by partial cytolysis of intermediate cells. ×630.

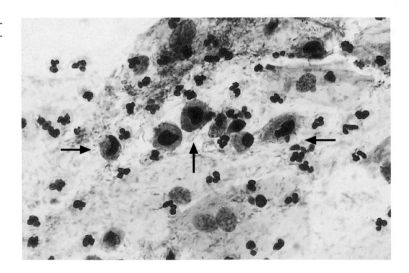

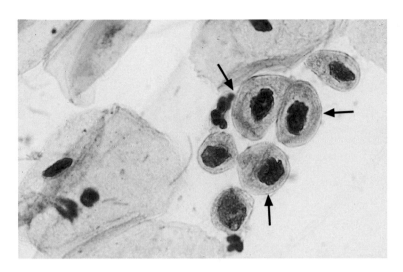

Fig. 4.**58 a, b Small squamous carcinoma in situ.**

a A group of cells (→) with hyperchromatic nuclei that show signs of degeneration. Superficial cells are seen nearby. ×1000.

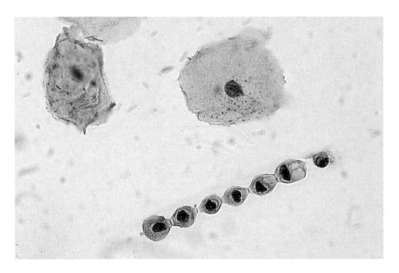

b Cells in single-file arrangement. The nuclei of these basal cells are hyperchromatic and condensed. Two normal intermediate cells are visible at the top. ×630.

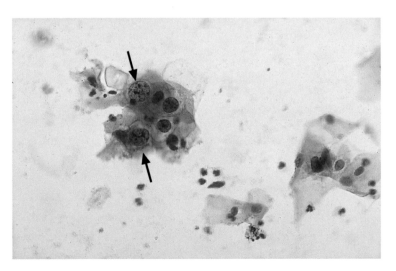

Fig. 4.**59 Radiation-induced mild dysplasia.** The nuclei are hyperchromatic with irregular chromatin pattern (→). Typical radiation effects are not seen in this area of the smear. ×250.

Fig. 4.**60 a–c Radiation-induced moderate dysplasia.**

a The abnormal nuclei partly show signs of degeneration. The cytoplasm exhibits pseudo-eosinophilia and amphophilia, and the cell borders are partly blurred. ×400.

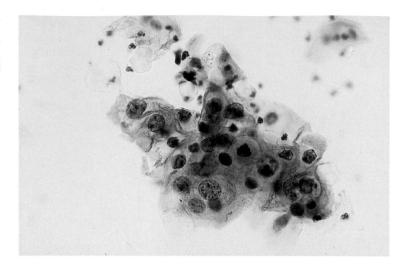

b The nuclei are polymorphic, hyperchromatic, and mostly condensed. The cells vary in size, are binuclear, and partly have indistinct borders. ×1000.

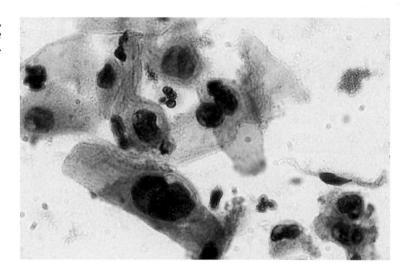

c The nuclei appear hyperchromatic and are mostly condensed. Radiation-induced changes include binuclear cells, variable cell size, and vacuolated cytoplasm reminiscent of HPV-induced vacuolation (⟶) (×630).

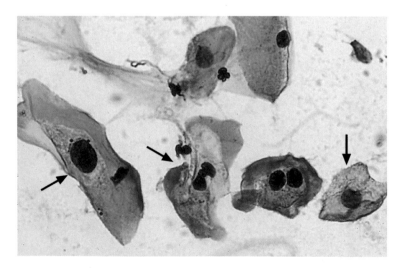

Table 4.**4** Sizes of cells and nuclei in nonkeratinizing, keratinizing, and small-cell squamous carcinomas of the cervix (according to Reagan et al., 1957)

	Nonkeratinizing squamous carcinoma	Keratinizing squamous carcinoma	Small-cell squamous carcinoma
Cell diameter (µm) Cell area (µm²)	18.06 ± 9.38 256 ± 69	18.7 ± 11.66 275 ± 107	14.68 ± 6.86 169 ± 37
Nuclear diameter (µm) Nuclear area (µm²)	10.58 ± 6.18 88 ± 30	9.9 ± 5.98 77 ± 28	9.1 ± 4.06 65 ± 13

Table 4.**5** Shift in the frequency of cervical carcinoma types (according to Reagan and Ng, 1974)

Time period	Number of cases	Nonkeratinizing squamous carcinoma		Keratinizing squamous carcinoma		Small-cell squamous carcinoma	
		Absolute	%	Absolute	%	Absolute	%
1941–1945	221	85	38.5	95	43.0	41	18.5
1946–1950	206	90	43.7	76	35.9	40	19.4
1951–1955	161	82	50.9	61	37.9	18	11.2
1956–1960	115	80	69.6	26	22.6	9	7.8
1961–1965	101	70	69.3	24	23.8	7	6.9
1966–1970	93	67	72.0	20	21.5	6	6.5
1941–1970	897	474	52.8	302	33.7	121	13.5

Invasive Carcinoma

Classification. The classic subdivision into nonkeratinizing (large-cell), keratinizing, and small-cell squamous carcinoma is abandoned in the 1994 WHO nomenclature, and the small-cell squamous carcinoma is no longer considered as an independent entity (348). Nevertheless, the former division still seems to make sense for practical purposes, since the three classic types of carcinoma are frequently reflected by the cells found in the smears (306). A cytomorphological grading into three categories based on the degree of tumor differentiation has also been proposed, but has not gained acceptance (58).

Sizes of cells and nuclei. Table 4.**4** presents data obtained by measuring the cells and nuclei found in the three classic types of cervical carcinoma. The data reveal that the cells are about the same size in large-cell (nonkeratinizing) and keratinizing carcinomas, whereas the cells of small-cell carcinomas are clearly smaller. The nuclear diameters are similar in all three growth types.

Shift in carcinoma type frequencies. A still unexplained phenomenon is the distinct shift in the frequencies of various types of carcinoma

between 1941 and 1970 (Table 4.**5**) (309). During this period, the keratinizing type and the small-cell type declined in favor of the large-cell type. Today, we find that only about 7 % of carcinomas are of the keratinizing type and 20 % are of the small-cell type (see Fig. 4.**8**, p. 172).

Squamous carcinoma cells. The presence of nucleoli within a lesion with severe dysplasia provides reason to suspect microinvasion (Fig. 4.**61 a, b**) (255). This assumption is supported by the simultaneous polymorphism of cytoplasmic features, such as variable cell shape, presence of spindle cells, and abnormal keratinization (Fig. 4.**62 a, b**). These cells are the typical sign of invasive squamous cell carcinoma (413).

Large-cell squamous carcinoma. The most common type of squamous carcinoma originates from either nonkeratinizing or metaplastic dysplasia. The mean cell diameter is 18 µm, and the mean nuclear diameter is 11 µm (216). The nuclei are highly hyperchromatic with an abnormal chromatin structure (salt and pepper pattern); the intensity of these features surpasses that seen with dysplasia. **Macronucleoli** are frequently detected; they differ from chromocenters by their mild eosinophilia (Figs. 4.**63**, 4.**64**). Unlike the situation with dysplasia, the cell clusters exhibit an irregular, three-dimensional arrangement of cells, and there is no polarization of nuclei. Cell-in-cell configurations are common (Figs. 4.**65**, 4.**66**).

The tumor cells often show signs of degeneration, including naked nuclei. The background of the preparation indicates **tumor diathesis**. The degenerative processes within the tumor are a result of its rapid growth—the tissue is no longer adequately supplied with blood vessels (119). Disintegration of the tissue gives rise to a fine granular protein precipitation, also called detritus. Unlike the precipitation resulting from lactobacillus-induced cytolysis, which contains only cytoplasmic components, it contains also lysed nuclei, connective tissue cells, and blood cells (Fig. 4.**67**).

A dirty background of the cell image is frequently considered as a typical sign of malignancy. Nevertheless, it is not safe to rely on this criterion. Dirty cell images may also be associated with benign processes, whereas a squamous carcinoma may be present even when the smear image is absolutely clean or shows only lactobacillus-induced cytolysis (see Figs. 4.**69 c**, 4.**70**).

Keratinizing squamous carcinoma. Here, the mean cell diameter is 19 μm, and the mean nuclear diameter is 10 μm (36). The nuclei are often condensed, and they stain dark to black without having a recognizable chromatin structure. Characteristic signs are severe polymorphism of the predominantly keratinized tumor cells, presence of **spindle cells**, and a wide range in cell size (Fig. 4.**68 a–c**). Nucleoli are rarely observed, but keratin pearls are found occasionally. It is sometimes difficult to distinguish this type of carcinoma from keratinizing dysplasia.

Small-cell squamous carcinoma. Here, the mean cell diameter is 15 μm, and the mean nuclear diameter is 9 μm (306). The nuclei have an abnormal chromatin pattern or are condensed. Cellular polymorphism is less pronounced than in the two other types of squamous carcinoma, and the nucleocytoplasmic ratio is clearly shifted in favor of the nucleus. Nucleoli are as common as in large-cell squamous carcinoma. Because of the small size of the cells, differentiating them from endometrial and nonepithelial cells is difficult (Fig. 4.**69 a–c**). The cells of this neuroendocrine tumor are hardly larger than lymphocytes and are often overlooked by the cytologist because of the relatively uniform chromatin structure of their nuclei and the absence of nucleoli (Fig. 4.**70**) (348). This type of tumor is very rare and extremely aggressive.

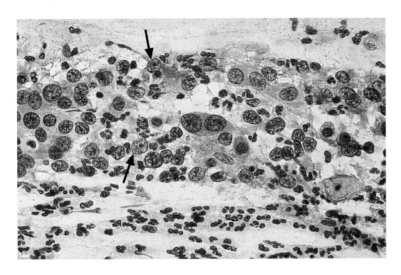

Fig. 4.**61 a, b Microinvasive squamous carcinoma of the cervix.**
a Polymorphic dysplastic cells with nucleoli (—►). Lactobacillus induced cytolysis is seen in the background. ×400.

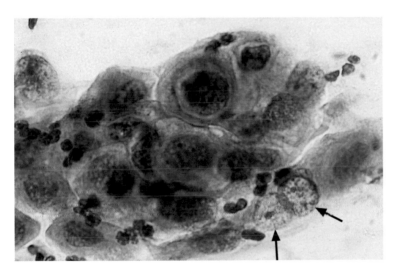

b Irregular cluster of cells. The abnormal nuclei are still of uniform size, although they overlap and may already contain nucleoli (—►). ×1000.

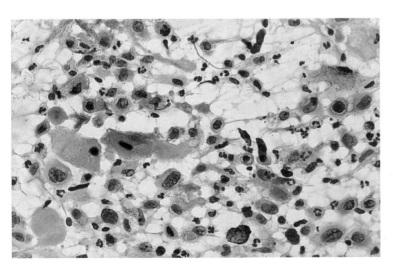

Fig. 4.**62 a, b Squamous carinomas cells.** Cells and nuclei vary in size, and individual cells tend to keratinize.
a The abnormal nuclei are mostly still intact, but some are degenerated. ×400.

4

b The dirty background of the preparation indicates **tumor diathesis**. ×630.

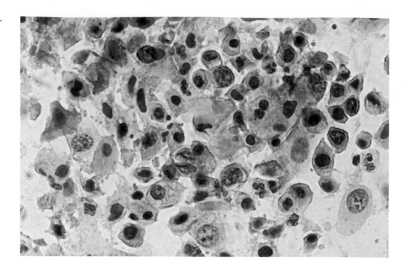

Fig. 4.**63 Large-cell squamous carcinoma of the cervix.** A loose cluster of abnormal cells. The cytoplasm has indistinct borders, and the nuclei clearly contain nucleoli. Cell-in-cell configurations are also visible (→). ×630.

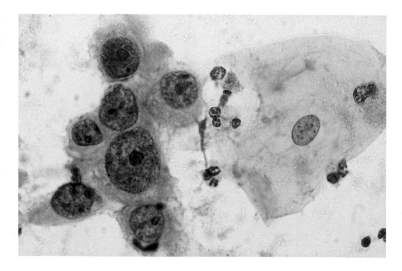

Fig. 4.**64 Large-cell squamous carcinoma of the cervix.** A loose cluster of cells with macronucleoli and indistinct cell borders. A normal superficial cell is visible on the right. ×1000.

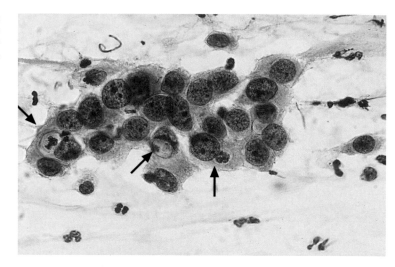

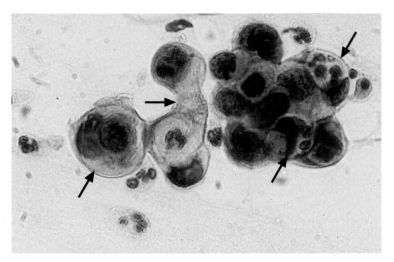

Fig. 4.**65 Large-cell squamous carcinoma of the cervix.** A cluster of abnormal cells in an irregular, three-dimensional arrangement. Several cell-in-cell configurations resulting from phagocytosis and cannibalism are visible (⟶). The cells have distinct cytoplasmic borders. ×1000.

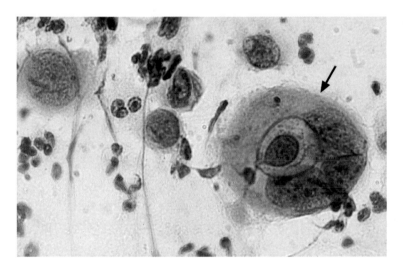

Fig. 4.**66 Large-cell squamous carcinoma of the cervix.** A multinuclear tumor cell showing cannibalism (⟶). The naked nuclei of some tumor cells are visible nearby. The background of the preparation indicates inflammation. ×1000.

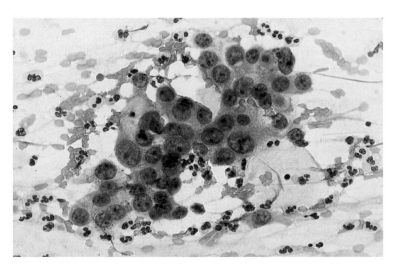

Fig. 4.**67 Large-cell squamous carcinoma of the cervix.** A loose cluster of cells showing pseudo-eosinophilia due to the admixture of blood. The nuclei contain macronucleoli, and the cytoplasm has indistinct borders. The background of the preparation indicates inflammation and **tumor diathesis**. ×400.

Fig. 4.**68 a–c Keratinizing squamous carcinoma of the cervix.**

a Keratinized, abnormal spindle cells. The nuclei are condensed and pyknotic. ×250.

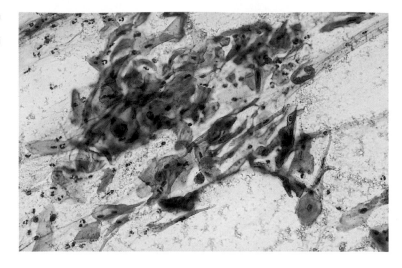

b Polymorphic spindle cells with partly still intact, but mostly condensed and pyknotic nuclei. The background of the preparation indicates tumor diathesis. ×400.

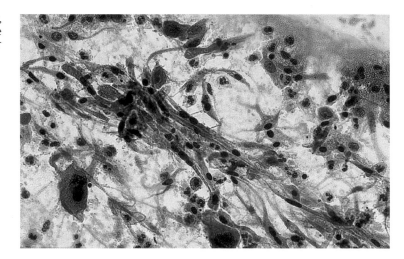

c Polymorphic, only partly keratinized spindle cells. The nuclei are pyknotic or show karyorrhexis (⟶), and the cells vary in size. ×790.

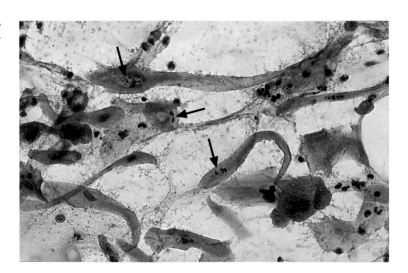

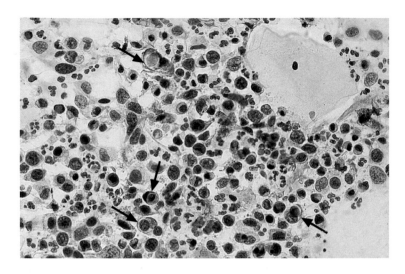

Fig. 4.**69** **Small-cell squamous carcinoma of the cervix.**

a Diffuse arrangement of small, abnormal basal cells of histiocytic appearance. The nuclei are partly still intact, partly degenerated. Several cell-in-cell configurations are present (—➤). The cytoplasm has mostly indistinct borders. A normal superficial cell is visible on the upper right. ×630.

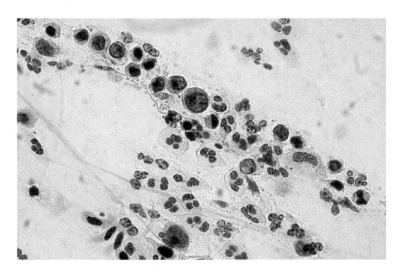

b Abnormal basal cells of histiocytic appearance in a loose striplike arrangement. The nuclei are partly still intact, partly degenerated, and the cytoplasm has indistinct borders. The background of the preparation indicates inflammation. ×400.

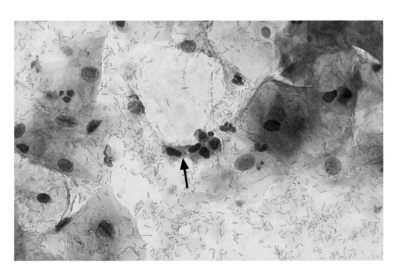

c A small group of abnormal basal cells (—➤). The nuclei are hyperchromatic and condensed. Several normal superficial cells and lactobacilli are seen in the vicinity. ×500.

4

Fig. 4.**70** **Small-cell anaplastic carcinoma of the cervix.** The small tumor cells have the size of lymphocytes and show a diffuse, dissociated arrangement (⟶). The nuclei are hyperchromatic and condensed. Two normal superficial cells are visible on the left. The background of the preparation shows lactobacillus-induced cytoloysis. ×400.

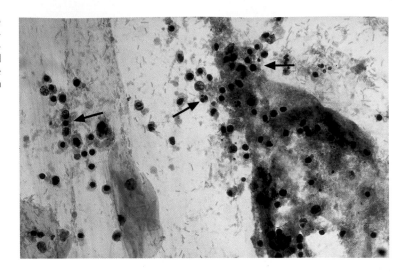

Malignant Tumors of the Vagina

Primary vaginal carcinoma. This is an extremely rare carcinoma with an incidence of 0.005‰ (208, 265, 325). Its morphogenesis is similar to that of cervical carcinoma (213).

Secondary vaginal carcinoma. Far more common are precancerous lesions in the blind vaginal pouch following hysterectomy, which may finally develop into secondary vaginal carcinoma (242). Although the frequency of vaginal preneoplasia in the blind vaginal pouch is about one third of that of cervical preneoplasia, the frequency of secondary vaginal carcinoma is said to equal that of cervical carcinoma (242). Interestingly, in case of precancerosis of the bind vaginal pouch, dyskeratocytes are often detected in the cytological smear (242). Vaginal dysplasia is also called **vaginal intraepithelial neoplasia (VAIN).**

Primary vaginal adenocarcinoma. In very rare cases, a primary vaginal adenocarcinoma may develop from vaginal adenosis in women whose mothers had been treated with diethylstilbestrol during the pregnancy (132) (see also p. 125).

Therapy

Treatment of severe precursors of cervical carcinoma usually consists of conization, whereas treatment of invasive carcinoma is mostly by extended hysterectomy and, if necessary, subsequent radiation.

Therapeutic strategies. Great hopes have been placed on **prophylactic vaccination,** i.e., inoculation with HPV (387). Vaccines are already available against HPV types 6, 11, 16, and 18, and clinical tests have been promising (131a). The vaccines consist of virus-like particles (VLP) assembled from a synthetically produced structural protein of the virus (capsid protein L1), which stimulates antibody production (131b).

It is advisable to vaccinate before a person becomes sexually active, namely, during puberty. It remains to be seen whether the forthcoming vaccination campaign will bring the desired results. For the time being, there are some doubts that individuals who have so far not participated in the screening for early cancer detection will be more willing to undergo HPV vaccination. On the other hand, those individuals who did participate will possibly continue to rely on the affordable preventive examination, rather than undergo a relatively expensive vaccination. Furthermore, it is also uncertain whether parents will have their children vaccinated against a sexually transmitted disease, particularly since we currently do not have reliable data about the duration of the protective immunity; the preventive examination will therefore still be necessary. Finally, it should be taken into account that the vaccination targets only a few viral types and suppression of these types might promote the development of more virulent mutants of the virus (similar to the situation with HIV).

For treatment of cervical carcinoma and its precursors, **invasive methods** are almost exclusively considered. In view of the relatively high rate of spontaneous regression of dysplastic lesions at this stage of carcinogenesis, therapeutic caution and the utmost restraint are advised. The individual needs of the patient and, particularly, her age and her desire to have children, should have priority over schematic treatment strategies.

It is not possible to determine a clearly defined point in time at which surgical assessment and treatment of an existing dysplastic lesion become necessary. Even the availability of the currently much favored **HPV hybridization test** for identifying potentially aggressive types of the virus, such as types 16 and 18, does not change this view (172). The mere presence of such a high-risk virus does not justify a surgical intervention in the absence of any morphological correlates. On the other hand, one can hardly do without such assessment if there is a cytological reason to suspect carcinoma in situ, even if the test for high-risk viruses is negative.

Even the establishment of **dysplasia outpatient clinics**—where colposcopic, cytological, and histological punch biopsy data are collected and combined to establish a diagnosis—will hardly improve this strategic concept (224). There is no guarantee that excision of only a few millimeters, taken from a colposcopically suspicious region, will identify the maximum degree of an existing epithelial abnormality. Furthermore, only a few highly specialized centers can provide detailed colposcopic diagnostics. Colposcopy is unsuitable for broad application in gynecological practice because the sound clinical knowledge required for such examinations is often missing. In most cases, therefore, any decision regarding the strategic concept for treating epithelial dysplasia depends on the interaction between the gynecologist and the cytology laboratory.

Mild and moderate dysplasia. On account of the long time course typical of these lesions, cautious waiting for months and even years may be justified as long as examinations are carried out at regular intervals—usually every 3 months (164). However, both patient and physician may be impatient for immediate diagnostic assessment. If the lesion lasts for many years, HPV analysis may be helpful in making the decision to extend these intervals to about 6 months if the infection is due to a low-risk type of the virus.

Severe dysplasia. Early assessment should be attempted when severe dysplasia is suspected. If possible, surgical clean-up should be done by conization in order to achieve a definite treatment together with the diagnostic assessment (47). Subsequent monitoring by means of cytological smears should follow at intervals of 3 months for at least 1 year, thus making sure that a potential persistence or relapse of the lesion is recognized in time.

Severe dysplasia during pregnancy. Severe dysplasia detected during pregnancy does not need to be treated right away, but it should be monitored cytologically at monthly intervals until delivery (343). The lesions often heal at the time of delivery or during the postpartum period as a result of the traumatic cervical changes and the regenerative processes that follow (355). On the other hand, immediate conization becomes necessary when microinvasion is suspected, since invasive cervical carcinoma may rapidly expand during this period, depending on the topographic location.

Invasive cervical carcinoma. Removal of the uterus, i.e., hysterectomy, is obligatory when invasive cervical carcinoma is discovered. This therapy may be sufficient as long as there is only microinvasive growth.

In the presence of a manifest carcinoma, extensive stage-dependent surgery is required, which includes the removal of parametrial connective tissue, vaginal cuff, and the regional lymph nodes (**radical hysterectomy**, or **Wertheim operation**) (87). Any involvement of metastatic lymph nodes requires subsequent radiotherapy.

In advanced cervical carcinoma, surgery may no longer be possible. Primary radiotherapy is necessary in such cases (90).

Even after hysterectomy, cytological smears from the blind vaginal pouch are still required for monitoring. As already mentioned before, there is an increased risk of dysplastic relapses in this region (Fig. 4.**71 a, b**) (24). Severe dysplasia in this region needs to be removed by means of excision, laser treatment, or radiotherapy.

4

Fig. 4.71 An irregular pale lesion at the apex of the blind vaginal pouch after hysterectomy.
a The discrete leukoplakia with poorly defined borders (⟶) corresponds to a CIN III lesion.

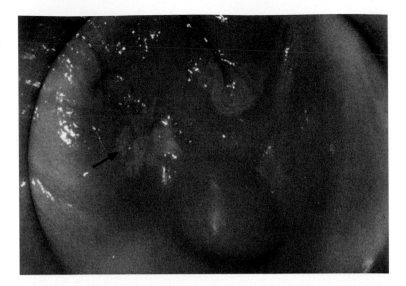

b The same patient, showing an iodine-negative focus after application of iodine.

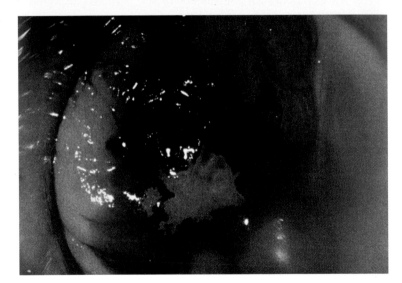

Endocervix

Epidemiology

Endocervical adenocarcinoma is about 10 times less frequent than squamous carcinoma. However, an increase has been observed in recent years, and the reason for this is not known. Glandular precursors of adenocarcinoma have also been described, although these are rare.

Incidence. The incidence of adenocarcinoma of the endocervix is about one tenth of that of squamous carcinoma (0.01–0.02‰) (106). Over the last 30 years, an increase by a factor of 3–4 from the original value has been recorded, and it is not clear whether this is due to a true increase in incidence or to an increased rate of detection (417).

This type of tumor is more difficult to detect by means of cervical cytology than squamous carcinoma. The reasons include, first of all, the low tendency of the endocervical epithelium to exfoliate and, second, the hidden location of endocervical lesions. Third, the monomorphic cell population of the endocervical glandular epithelium does not permit assignment of nuclear atypia to different degrees of epithelial maturation, and precursors are therefore difficult to recognize (261).

The occurrence of endocervical adenocarcinoma—unlike that of squamous carcinoma—is not clearly age-dependent. This type of cancer occurs from childhood to old age (95).

Precursors. Precancerous stages of endocervical adenocarcinoma are rarely detected by histological and cytological means, although it is

	Endo-cervical cell	Endo-cervical cell	Abnormal reserve cell	Abnormal glandular cell
Columnar epithelium				
Reserve cell layer				

Fig. 4.**72 Development of adenocarcinoma in situ.** A HPV lesion in the endocervical columnar epithelium initially leads to abnormal reserve cells and then to abnormal glandular cells.

assumed that about 14% of all cervical carcinomas have a glandular component (328, 96). On the other hand, about 5% of glandular lesions simultaneously contain areas of squamous epithelial atypia. Among the precursors of endocervical adenocarcinoma we distinguish **glandular dysplasia** and **adenocarcinoma in situ of the endocervix** (39). The latency periods of glandular precancerous lesions seem to resemble those of squamous dysplasia. The average age is reported to be 35 years for glandular dysplasia, 40 years for adenocarcinoma in situ, and 50 years for invasive adenocarcinoma, although there is a wide range in age distribution (403).

Morphogenesis

Like squamous carcinoma, adenocarcinoma of the endocervix is probably caused by HPV. Although development of this tumor is likely to go through intermediate stages that last for many years, these stages are only rarely detected. In detail, we distinguish:
▶ Abnormal reserve cell hyperplasia
▶ Glandular dysplasia
▶ Adenocarcinoma in situ
▶ Invasive adenocarcinoma

Abnormal reserve cell hyperplasia. As is the case with squamous carcinoma, HPV infection is thought to be responsible for the development of endothelial adenocarcinoma, with HPV type 18 probably playing an important role (109). The virus infects the reserve cells, which lie beneath the endocervical columnar epithelium and are still able to divide, and induces an increase in cell proliferation. However, this leads to a qualitative change (malignant transformation) of the cells (Fig. 4.**72**), unlike the situation with squamous epithelium where the virus causes a quantitative change (thickening) of the basal cell layer.

Abnormal reserve cell hyperplasia is histologically difficult to distinguish from reactive reserve cell hyperplasia (354). The signs of malignant transformation include enlarged and hyperchromatic nuclei, presence of nucleoli, and a reduction in mucus production (104). Because of the increased space occupied by the nuclei within the single-layered epithelium, the nuclei give way in different directions, thus leading to apparent stratification (pseudostratification). This is associated with an increase in mitotic figures detectable in the epithelium.

Glandular dysplasia and adenocarcinoma in situ. Whereas the morphology of the glands still reveals a normal structure in case of glandular dysplasia, the adenocarinoma in situ exhibits an increase in deformation of tubular glands, papillary evagination within the glandular lumina, and direct contact of glands with one another without any connective tissue between them (back-to-back configuration) (317).

Invasive adenocarcinoma. When the abnormal epithelium is no longer able to expand by proliferation and deformation of glands, it breaks through the basal membrane in the form of interconnected, compact glandular formations and infiltrates the surrounding stroma. Here, it penetrates into lymph and blood vessels and starts to metastasize, thus meeting the criteria of an invasive adenocarcinoma.

Colposcopy

Abnormal glandular epithellum is difficult to diagnose by colposcopy.

Invasive adenocarcinoma and its precursors have an irregular surface and show intense redness within the glandular area (Fig. 4.**73a**), although this is often difficult to see because the carcinoma is usually located inside the endocervix. The clinical symptoms are not very characteristic; occasionally there is irregular vaginal bleeding. Because of the hidden location in the endocervix, the carcinoma is often recognized very late and may have expanded locally for a long period of time. The growth of a barrel-shaped carcinoma often causes the entire cervix to enlarge (Fig. 4.**73b**) (189).

Fig. 4.**73 a, b Adenocarcinoma of the endocervix.**

a Glandular villi and focal hemorrhage in the center.

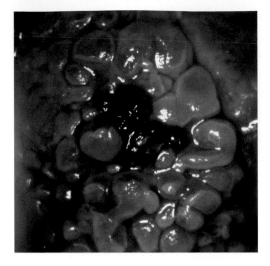

b The tumor expands deep into the endocervix and in the direction of the uterine corpus, thus forming a "barrel-shaped carcinoma." The ectocervix and the cervical canal (→) are not affected. Gross section, HE stain, actual size.

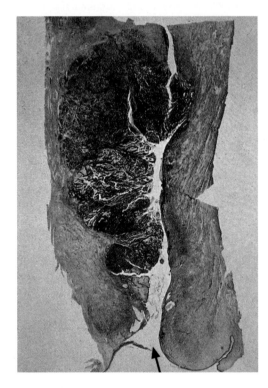

Histology

Precancerous and malignant changes of the endocervical glandular epithelium are histologically characterized by the following changes:

▶ Enlarged and polymorphic nuclei
▶ Hyperchromatic nuclei and abnormal chromatin structure
▶ Increased mitotic rate and presence of nucleoli
▶ Apparent stratification (pseudostratification)
▶ Decrease in mucus production
▶ Back-to-back configuration of glands
▶ Intraluminar epithelial budding
▶ Proliferation and coiling of glands
▶ Deep invasion into the stroma

Glandular dysplasia. The carcinogenesis of endocervical glandular epithelium has received increased attention in recent years. Different degrees of glandular dysplasia have been diagnosed. The criteria used for classification include the presence of enlarged, hyperchromatic, and polymorphic nuclei, intensity of hyperchromatism, presence of nucleoli, mitotic rate, mucus production, and glandular structure (9, 10). The abnormal epithelial changes may affect single glands (Fig. 4.**74**), portions of glands, or even many groups of glands.

Adenocarcinoma in situ. At this stage, the proliferation of abnormal glands is particularly intense, thus causing the glands to stand back-to-back without the connective tissue bridges normally present between them (Fig. 4.**75a**). The glands show increased coiling, thus creating a serrated pattern in histological sections, and also evagination into the glandular lumina by epithelial budding (Fig. 4.**75 b**). The nuclei are enlarged and polymorphic with an abnormal chromatin structure, and nucleoli are detected in large numbers.

Invasive adenocarcinoma. The invasive adenocarcinoma is histologically easier to recognize than its precursors because the ratio of glandular epithelium to stroma is considerably disturbed. The deep penetration of abnormal glands into the stroma, their confluence into groups with back-to-back configuration, and their irregular acinotubular structures are clear morphologic signs of invasion.

We distinguish different growth types, which essentially differ from one another in mucus production, cell size, and glandular structure (154).

▶ The less differentiated the carcinoma, the larger the nuclei and their nucleoli. The most common type is **endocervical adenocarcinoma** (Fig. 4.**76**). Its cells resemble those of normal cervical glands and, therefore, often have a vacuolated cytoplasm.
▶ There is also **endometroid adenocarcinoma**, the cells of which resemble endometrial cells, and several other variants, such as **clear cell adenocarcinoma** and **intestinal adenocarcinoma** of the endocervix (195).

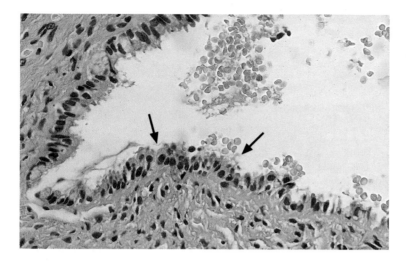

Fig. 4.**74** **Glandular dysplasia of the endocervical glandular epithelium.** The columnar epithelium shows enlarged, hyperchromatic, and polymorphic nuclei. The enlarged nuclei cause pseudostratification in some areas (——►). HE stain, ×200.

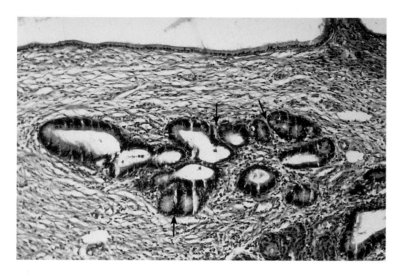

Fig. 4.**75 a, b** **Adenocarcinoma in situ of the endocervical glandular epithelium.**

a The glandular lumina are numerous and of variable size. Many of them exist back-to-back (——►) and exhibit epithelial budding. The columnar epithelium shows enlarged, hyperchromatic, and polymorphic nuclei. The enlarged nuclei cause pseudostratification in some areas. HE stain, ×200.

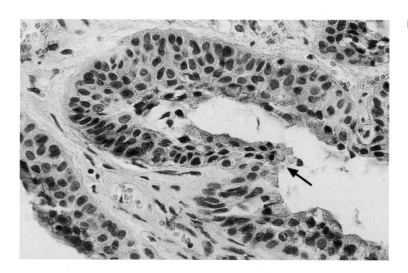

b The abnormal columnar epithelium shows pseudostratification due to nuclear enlargement. It also exhibits intraluminar epithelial budding (——►). HE stain, ×400.

Cytology

Precancerous and malignant changes of the endocervical glandular epithelium show the following cytological features:

▶ Enlarged and polymorphic nuclei
▶ Hyperchromatic nuclei and abnormal chromatin structure
▶ Presence of nucleoli
▶ Loss of polarization of cell clusters
▶ Three-dimensional cell clusters
▶ "Bird feather" or "fireworks" structure of cell clusters
▶ Vacuolated cytoplasm and cell-in-cell configurations

Criteria for malignancy. The cytological interpretation of precancerous and malignant glandular lesions is much more difficult than it is with lesions of the stratified squamous epithelium, where different degrees of cellular maturity make it possible to assign dyskaryotic nuclei to different degrees of epithelial dysplasia.

Unlike squamous epithelium, glandular epithelium is always single-layered, even at various stages of malignancy. All of its cells have the same degree of maturity, thus representing a **monomorphic cell population**. Because of the morphology of the columnar epithelium, there are only criteria for malignancy and none for classification. Furthermore, columnar epithelial cells have a lower tendency to exfoliate and a lower resistance to environmental factors than squamous epithelial cells. For these reasons, precursors of endocervical adenoma are particularly difficult to recognize cytologically.

Glandular dysplasia. Columnar epithelial cells derived from glandular dysplasia exfoliate either as single cells (Fig. 4.**77a**) or, more often, in clusters (Fig. 4.**77b**), and they show depolarization of their nuclei (268). Furthermore, the nuclei are usually enlarged and hyperchromatic. Abnormal glandular cells may occasionally show metaplastic features, such as more distinct cell borders, vacuolation, and cell-in-cell configuration (Fig. 4.**78a, b**).

Adenocarcinoma in situ. Here, the depolarization of nuclei is even more pronounced, and three-dimensional clusters in which the nuclei can no longer be observed in the same optical plane are more common. The nuclei are increasingly polymorphic in shape and size, and the chromatin structure is more coarse but still rather uniform (Fig. 4.**79a–d**). The cytoplasm

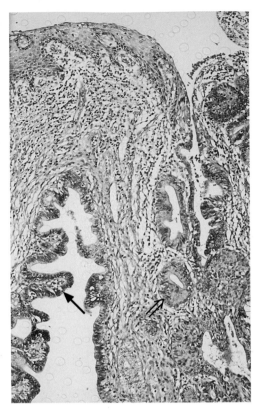

Fig. 4.**76 Adenocarcinoma of the endocervix (endocervical type).** The surface epithelium exhibits metaplastic transformation. The cervical glands are lined with an abnormal epithelium. The epithelium shows pseudostratification due to nuclear enlargement and secretory vacuoles; it also shows intraluminar epithelial budding (⟶). Some of the glands break through the basal membrane and invade the stroma (⟹). HE stain, ×100.

is often foamy and has indistinct borders. Once degeneration sets in, the cytoplasm may be completely absent, and only naked nuclei remain.

Invasive adenocarcinoma. Invasive adenocarcinoma of the endocervix is characterized by an increase in exfoliated cells. Cell clusters are even more irregular than those derived from adenocarcinoma in situ (13), and those with "bird feather" or "fireworks" patterns occur more frequently. Clusters of abnormal glandular cells appear like grapes attached to a delicate scaffold of connective tissue when viewed from the side (Fig. 4.**80a**), and seem to spread from the center to the periphery when viewed from above (Fig. 4.**80b**). The cytoplasm of abnormal glandular cells, unlike that of squamous epithelial cells, has indistinct borders and a loose, foamy structure (105). The nuclei are clearly enlarged. They are polymor-

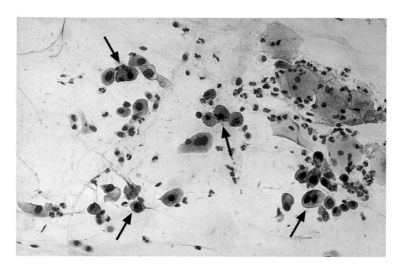

Fig. 4.**77 a, b Glandular dysplasia of the endocervix.**
a Dysplastic endocervical cells in dissociated arrangement. The cytoplasm resembles that of
metaplastic cells (⟶). Normal superficial cells
are visible on the upper right. ×200.

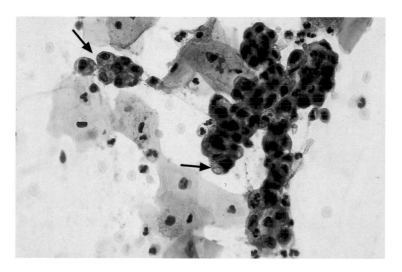

b A compact cluster of dysplastic endocervical
cells in irregular, three-dimensional arrangement. The cytoplasm is partly vacuolated (⟶).
Normal superficial cells are visible in the vicinity.
×400.

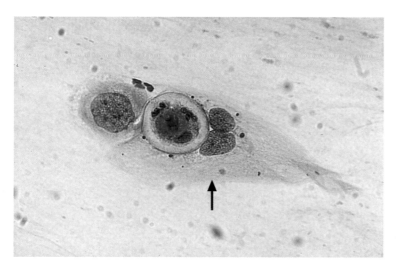

Fig. 4.**78 a, b Cell-in-cell configuration associated with glandular dysplasia.**
a Next to the cell-in-cell configuration lies a binuclear intermediate cell (⟶). ×630.

4

b Next to the cell-in-cell configuration lie one intact and one pseudoeosinophilic, degenerated abnormal glandular cell (→). ×630.

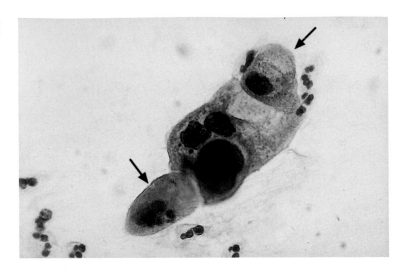

Fig. 4.79 a–d   Adenocarcinoma in situ of the endocervix.

a A compact irregular cluster of abnormal glandular cells. The nuclei are hyperchromatic, but the chromatin structure is regular and nucleoli are absent. Normal superficial cells are visible nearby. ×500.

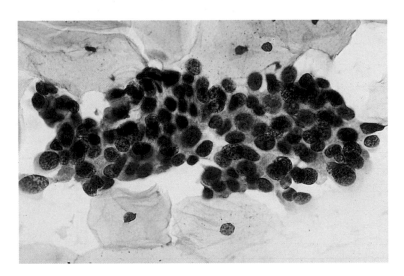

b A loose cluster of abnormal glandular cells with a slightly irregular, coarse chromatin structure and indistinct cytoplasmic borders. Normal superficial cells are visible nearby. ×630.

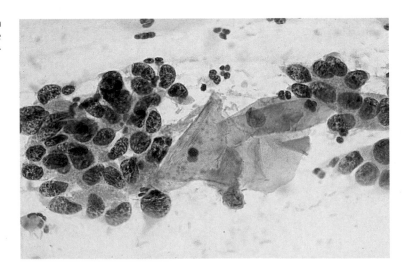

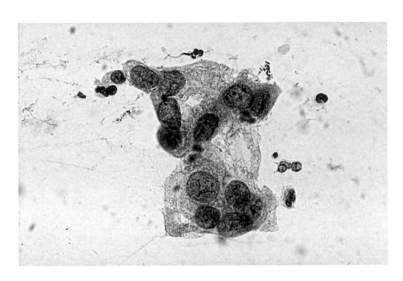

c A loose cluster of abnormal glandular cells with coarse, but regular chromatin structure and foamy cytoplasm. ×630.

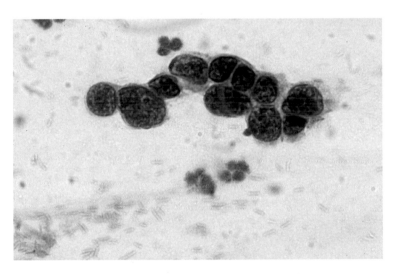

d A cluster of abnormal glandular cells. The hyperchromatic nuclei vary in shape and show a coarse, irregular chromatin structure. ×1000.

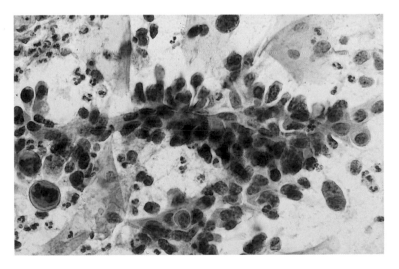

Fig. 4.**80** **Adenocarcinoma of the endocervix.**
a A loose cluster of abnormal glandular cells in lateral view. The cells have a tendency to dissociate, thus creating a structure resembling a bird feather. The nuclei vary in size and are hyperchromatic. The cytoplasm is sparse and has indistinct borders. Normal superficial cells are visible nearby. ×250.

b A loose, irregular cluster of abnormal glandular cells viewed from above. The nuclei are of uniform size and exhibit a coarse, irregular chromatin structure (nuclear clearings). The cytoplasm is sparse and has indistinct borders. ×500.

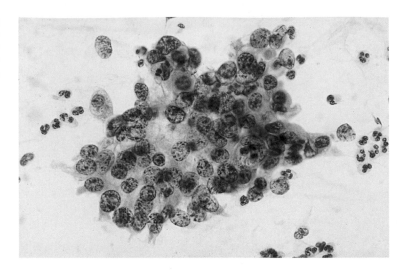

Fig. 4.81 Highly differentiated adenocarcinoma of the endocervix.

a The mostly isolated cells resemble normal endocervical cells and contain basal nuclei, but they have an irregular chromatin structure. Single normal superficial and intermediate cells are visible in the center. The background of the preparation is dirty. ×400.

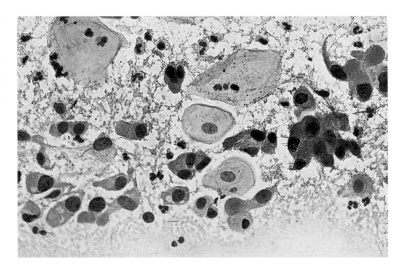

b Multinuclear, syncytial sheet. The nuclei have a relatively regular structure, and some of them contain nucleoli. Plenty of erythrocytes are also visible. ×630.

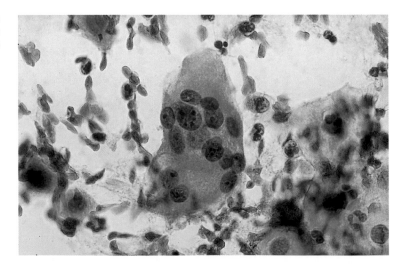

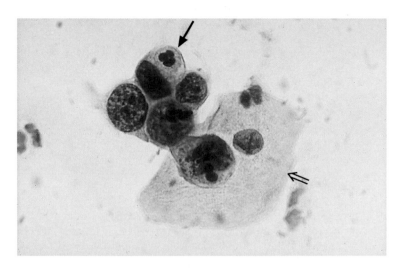

Fig. 4.**82** **Poorly differentiated adenocarcinoma of the endocervix.** A small cluster of abnormal glandular cells with much enlarged nuclei, irregular and coarse chromatin structure, and chromocenters. A cell-in-cell configuration is also present (→). A normal intermediate cell is visible nearby (⇒). ×1000.

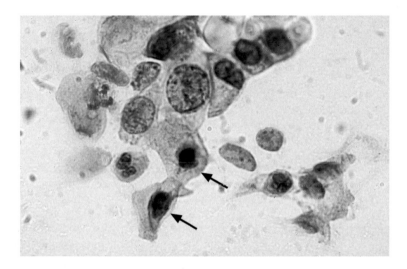

Fig. 4.**83** **Squamous epithelial component of an adenocarcinoma of the endocervix.** In addition to intact abnormal glandular cells, there are two dyskeratotic squamous epithelial cells with condensed nuclei (→). ×1000.

phic, have a coarse but mostly uniform chromatin structure, and often contain nucleoli.

Types of carcinoma. Structure and size of cells and nuclei are clearly determined by the degree of differentiation of the carcinoma type. **Well-differentiated carcinomas** have cells that are rich in cytoplasm and are sometimes multinuclear. The cells are similar in size to normal endocervical glandular cells, and they often have basal nuclei (Fig. 4.**81 a, b**). These features are not found in **poorly differentiated carcinoma**, where the nuclei are clearly enlarged (Fig. 4.**82**) (14). **Endometroid endocervical adenocarcinoma** is characterized by small cells that very much resemble normal endometrial cells. In rare cases, endocervical adenocarcinoma may also have a **squamous epithelial component** (Fig. 4.**83**).

Therapy

Treatment of adenocarcinoma of the endocervix is essentially like that of squamous carcinoma.

Treatment of endocervical adenocarcinoma corresponds to that of squamous carcinoma because its localization is almost the same. As glandular carcinomas are usually recognized later than squamous carcinomas, prognosis is less favorable and surgical treatment tends to be more radical (97) (see also p. 212 f.).

4

Endometrium

Epidemiology

Today, uterine corpus carcinoma is slightly more common than cervical carcinoma. In contrast to cervical carcinoma, it has not been possible to reduce its absolute frequency (prevalence) by the introduction of preventive cancer screening. Certain lesions are obligatory precancerous stages.

Incidence. The incidence of uterine corpus carcinoma is currently 0.2–0.3‰, slightly higher than that of cervical carcinoma (182). At the beginning of preventive cancer screening about 50 years ago, cervical carcinoma was still about 15 times more frequent than uterine corpus carcinoma. It is obvious from this development that the reduction in the frequency of uterine corpus carcinoma has been far less pronounced during this period than that of cervical carcinoma.

Early detection. There is no program aimed at early detection of adenocarcinoma of the uterine corpus comparable to the cervical carcinoma program. Access to the endometrium is much more difficult than to the cervix, and direct smears from the uterine cavity are therefore not possible within the context of preventive examinations. As a result, uterus corpus carcinoma and its precursors are detected more or less accidentally when routine cervical smears contain suspect endometrial glandular cells (120).

Unlike the situation with cervical carcinoma, the precursors of uterine corpus carcinoma are rarely detected; hence, their development into a carcinoma can hardly be prevented. Diagnostic attempts regarding carcinoma of the uterine corpus focus therefore less on the detection of precursors and more on the detection of **early stages** (182). Uterine corpus carcinoma is usually a disease of postmenopausal women and has its peak around age 60 (402). It rarely occurs before this age, but then occurs at a fairly constant frequency until old age.

Morphogenesis

It is not known what causes the development of uterine corpus carcinoma, and no reliable data are available on the duration of carcinogenesis. Hormonal factors have been discussed as possible causes. Adenomatous and atypical hyperplasia, as well as adenocarcinoma in situ, are regarded as precursors.

Risk factors. Chronic estrogen exposure that is not opposed by gestagen has been discussed as a risk factor. So far, however, there has been no evidence for this. This type of carcinoma occurs predominantly during a phase of life when, normally, estrogen production has ceased and the endometrium is atrophic (329). Other potential risk factors may include obesity, hypertension, and diabetes mellitus. In the final analysis, the causes for development of uterine corpus carcinoma are still unknown (347).

Endometrial hyperplasia. The initial step in the course of carcinogenesis is probably an abnormal change of endometrial reserve cells, followed by increased proliferation of endometrial glandular cells. Not much is known about latency periods, but it is assumed that **adenomatous (complex) hyperplasia** and **atypical adenomatous (complex atypical) hyperplasia** of the endometrium are precursors of endometrial cancer (67, 190).

Endometrial hyperplasia is associated with enlargement of the nuclei, which then move in a vertical direction because they occupy more space. This leads to apparent stratification (pseudostratification) of the epithelium. At the same time, the proliferation of glands causes deformation and coiling of the glandular tubes, while papillary protrusions into the glandular lumina create a serrated pattern in histological sections. Due to increased glandular proliferation, the stroma between the glands is reduced, and the glands often come into direct contact (back-to-back configuration). Such a quantitative increase in glands causes a cushion-like thickening of the endometrium and may exist for a long period of time.

Invasive carcinoma. Finally, the glands break through the basal membrane and infiltrate the surrounding stroma. The invasive carcinoma thus created later penetrates the muscles of the uterine wall, where lymph and blood vessels may become eroded and metastasis may develop (303). Clinical symptoms in the form of irregular vaginal bleeding appear only at a relatively late stage (94). If the carcinoma expands toward the cervix of the uterus, metastasis is expected to occur much earlier, since the blood supply is better in the cervix than in the body of the uterus.

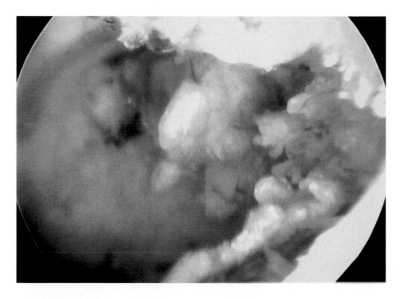

Fig. 4.**84** **Uterine corpus carcinoma.** Hysteroscopic image. The carcinoma causes irregularly protrusions of the endometrium, as shown on the right.

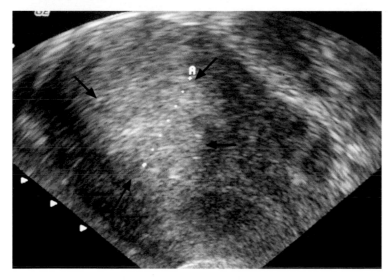

Fig. 4.**85** **Uterine corpus carcinoma.** Ultrasonography. The carcinoma causes the endometrium to appear as an irregular, highly reflective area ("snowstorm" appearance) of 10 mm in diameter (⟶).

Endoscopy and Sonography

By using an endoscopic technique known as **hysteroscopy**, it is possible to evaluate the uterine mucosa. If the lining of the uterine cavity is thickened or irregular, one should always suspect a carcinoma, especially when the tissue bleeds upon contact (Fig. 4.**84**) (234).

An endometrial thickness of about 1 mm in the ultrasound image is not suspect, but a well-developed, irregular mucosa should prompt further investigation (Fig. 4.**85**) (113). A negative sonographic finding is more convincing than a positive one, since the rate of false positives is higher than that of false negatives.

An uterine corpus carcinoma is even more hidden from view than an adenocarcinoma of the endocervix. It can therefore grow for a long time into the uterine cavity, thus forming a cushion-like thickening of the mucosa (Fig. 4.**86**).

Histology

Precancerous and malignant changes of the endometrium show the following histological features:

▶ Enlarged and polymorphic nuclei
▶ Hyperchromatic nuclei and abnormal chromatin structure
▶ Increased mitotic rate and presence of nucleoli
▶ Apparent stratification (pseudostratification)
▶ Back-to-black configuration of glands
▶ Intraluminar epithelial budding
▶ Proliferation and coiling of glands
▶ Deep invasion into the stroma

4

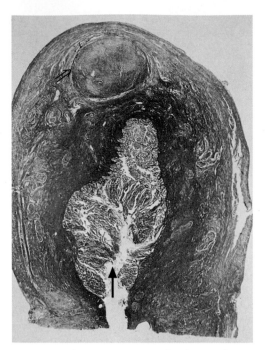

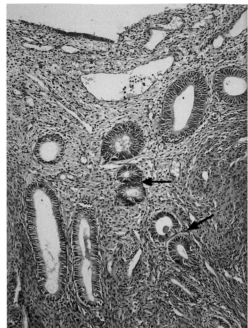

Fig. 4.86 Adenocarcinoma of the endometrium. The tumor causes a severe, cushion-like thickening of the uterine mucosa (→) but does not yet show much infiltration into the myometrium. A round, smooth-surfaced intramural myoma is visible at the top (⇒). Gross section, HE stain, actual size.

Fig. 4.87 Adenomatous (complex) endometrial hyperplasia. The glandular lumina are numerous and of variable size. Many glands have back-to-back contact (→) and exhibit intraluminar epithelial buddings. The columnar epithelium shows enlarged, hyperchromatic, and polymorphic nuclei that give rise to pseudostratification. HE stain, ×100.

Precursors. We distinguish the following precursors of the uterine corpus carcinoma: **adenomatous (complex) hyperplasia**, **atypical adenomatous (complex atypical) hyperplasia**, and **carcinoma in situ**. Diagnostic clues are derived from the presence of enlarged, hyperchromatic, and polymorphic nuclei, the degree of hyperchromatism, the mitotic rate, and the presence of nucleoli. Further criteria are the intensity of abnormal gland formation—glandular proliferation, coiling, back-to-back configuration, and intraluminar epithelial budding—and the extent of pseudostratification (Fig. 4.87) (256).

Invasive carcinoma. We distinguish between different growth types with regard to the degree of differentiation on the one hand, and the presence of squamous epithelial structures on the other (348):

▶ The more the abnormal glands resemble normal endometrial glands, the more differentiated the carcinoma, corresponding to grades 1 and 2 of the Broders index (41). More than 90% of all uterine corpus carcinomas are of small-cell growth type and exhibit a **high degree of differentiation** (Fig. 4.88).

▶ Glandular structures are no longer recognizable in **dedifferentiated carcinoma**, and solid epithelial cones dominate, corresponding to grades 3 and 4 of the Broders index (290). This large-cell growth type slightly resembles a poorly differentiated squamous epithelial carcinoma (Fig. 4.89).

▶ The pluripotent endometrial reserve cells are not only able to generate endometrial glandular cells but also squamous epithelial cells, which may even have a tendency for keratinization (308). This is the case in about one third of all uterine corpus carcinomas (Fig. 4.90). If the squamous epithelial cells exhibit benign characteristics, the carcinoma is called **adenoacanthoma**, whereas dyskaryotic changes indicate an **adenosquamous carcinoma**.

Sarcoma. Apart from carcinomas, sarcomas of the uterine corpus do occasionally occur, e.g., **stromal sarcoma** or **leiomyosarcoma**. The latter develops from smooth muscle and is histologically characterized by spindle cells and large nuclei with nucleoli (Fig. 4.91). Mesodermal tumors also occur, although they are very rare.

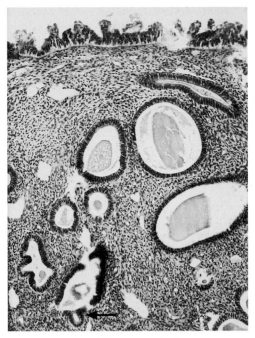

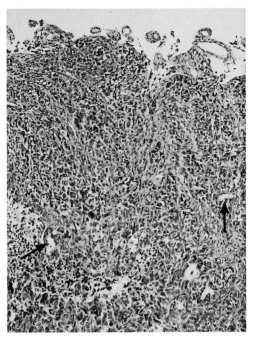

Fig. 4.**88** **Highly differentiated adenocarcinoma of the uterine corpus.** The glandular lumina are numerous and of variable size. Many glands have back-to-back contact and show intraluminar epithelial buddings. Nuclear enlargement gives rise to pseudostratification. In addition, some glands have broken through the basal membrane and now invade the stroma (⟶). HE stain, ×100.

Fig. 4.**89** **Dedifferentiated adenocarcinoma of the uterine corpus.** The abnormal glandular epithelium consists of relatively large cells with hyperchromatic, polymorphic nuclei. It forms no or only few glands (⟶). HE stain, ×100.

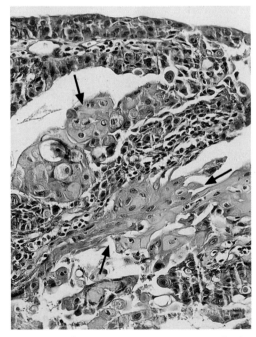

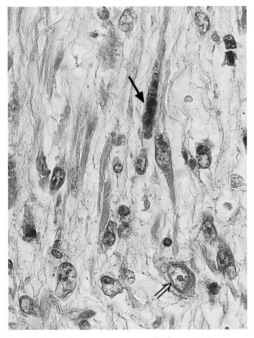

Fig. 4.**90** **Adenosquamous carcinoma of the uterine corpus.** The abnormal epithelium partly has the glandular structure of a columnar epithelium but partly also a squamous epithelial structure (⟶). HE stain, ×200.

Fig. 4.**91** **Leiomyosarcoma of the uterine corpus.** The abnormal smooth muscle cells are partly differentiated and spindle-shaped (⟶) and partly undifferentiated with sparse cytoplasm, large nuclei, and nucleoli (⟹). HE stain, ×500.

4

Cytology

Precancerous and malignant changes of the endometrium show the following cytological features:

▶ Enlarged and polymorphic nuclei
▶ Hyperchromatic nuclei and abnormal chromatin structure
▶ Presence of nucleoli
▶ Loss of polarization of cell clusters
▶ Three-dimensional cell clusters
▶ Vacuolated cytoplasm
▶ Atypical nonkeratinized or keratinized squamous cells

Direct sampling. Although the direct removal of cells from the uterine cavity promises a hit rate of about 90%, this method is associated with such high costs in material and expenses that its routine application in the context of screening is not possible (302). The method is also painful for the patient. If necessary, curettage followed by histological diagnosis seems to make more sense.

Cell population. The cytodiagnosis of endometrial carcinoma and its precursors is therefore usually restricted to abnormal endometrial glandular cells appearing in the cervical smear (177). Interpretation of these cells is more difficult than with abnormal squamous cells. The tendency of the glandular epithelium to exfoliate is very low, and—unlike squamous epithelium—the glandular epithelium represents a monomorphic epithelial population that is lacking different degrees of maturation.

Precursors. For the above reason, precursors are rarely detected by cytological means. Additional unfavorable factors include the small size of the endometrial cells and their poor state of preservation due to the large distance they travel between the sites of exfoliation and sample collection (257). The cytological distinction of endometrial carcinoma from its precursors, e. g., **adenomatous hyperplasia**, is rarely possible (256). The extent of nuclear atypia and the degree of vacuolation are very similar in both cases (Fig. 4.**92 a, b**).

Invasive carcinoma. The background of the preparation nevertheless indicates inflammation more frequently in case of invasive carcinoma than in case of its precursors. The smears contain more leukocytes, histiocytes, debris, protein precipitate, and erythrocytes.

▶ **Highly differentiated abnormal endometrial cells** hardly differ cytologically from normal endometrial cells (Fig. 4.**93 a–c**) (308). They may occur solitary or in clusters. The cytoplasm is loose with indistinct borders. The abnormal nuclei are usually slightly larger than those of normal endometrial cells, the chromatin shows more irregularities, and nucleoli are found more often.

▶ The cells of **moderately differentiated carcinoma** often occur in clusters and show vacuolation of the cytoplasm (Fig. 4.**94 a, b**) (53, 152). Cannibalism and phagocytosis of granulocytes are observed more frequently (Fig. 4.**94 c, d**) (134). Certain types of differentiation are associated with an increase in cytoplasm and the formation of glandular structures, such as acinar or syncytial cell clusters (Fig. 4.**94 e, f**). Abnormal differentiation may also be accompanied by concentrically laminated foci of calcification, called psammoma bodies (Fig. 4.**94 g**) (127).

▶ The cells of **poorly differentiated endometrial carcinoma** are characterized by an increase in nuclear size and macronucleoli (Figs. 4.**95 a, b**, 4.**96 a**) (31). The abnormal cells are often solitary and evenly distributed across the smear. As is the case with ovarian carcinoma, cell clusters occasionally form spherical bodies with an external line of retraction (Fig. 4.**96 b**).

Other malignant tumors. When horny flakes, parakeratocytes, or dyskeratocytes are detected in addition to abnormal glandular cells, one should suspect **adenoacanthoma** or **adenosquamous carcinoma** (Fig. 4.**97**) (348). Cytological smears from the rarely occurring **leiomyosarcoma** contain spindle-shaped cells with foamy cytoplasm, considerably enlarged nuclei, and macronucleoli (Fig. 4.**98**) (125). The abnormal cells of **endometrial stromal sarcoma** are unusually small (178). **Mesodermal tumors**, also called müllerian mixed tumors, contain both a carcinomatous and a sarcomatous component. The exfoliated cells are poorly differentiated, and the abnormal cells include spindle-shaped as well as glandular cells (178).

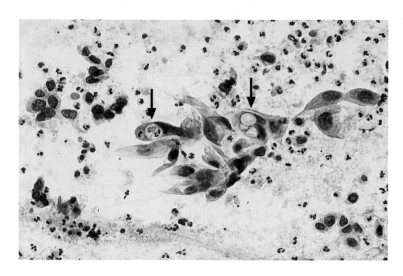

Fig. 4.92 a, b Adenomatous hyperplasia of the endometrium.

a A loose cluster of abnormal glandular cells. The cells are columnar to spindle-shaped, and some have a vacuolated cytoplasm (→). Most nuclei are active, but some have degenerated. The background of the preparation indicates inflammation. ×320.

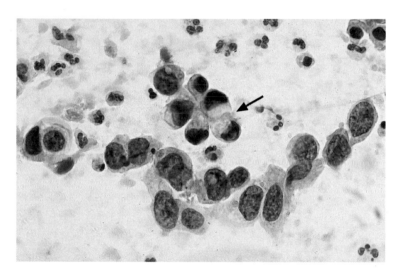

b A band-shaped cluster of abnormal glandular cells. The cells have considerably enlarged, hyperchromatic nuclei with coarse chromatin structure. The sparse cytoplasm has indistinct borders and is partly vacuolated (→). ×630.

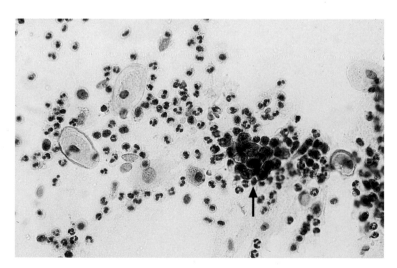

Fig. 4.93 a–c Highly differentiated uterine corpus carcinoma.

a A cluster of relatively small cells (→) with hyperchromatic nuclei; these cells are difficult to distinguish from normal endometrial cells. The background of the preparation indicates inflammation, and single normal parabasal cells are also visible. ×400.

b A group of small cells (→) with hyperchromatic nuclei; these cells are difficult to distinguish from normal endometrial cells. The background of the preparation indicates inflammation, and normal parabasal cells are also visible. ×400.

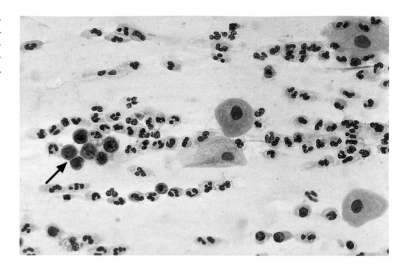

c A regularly structured, loose cluster of small, slightly hyperchromatic glandular cells (center); these cells are difficult to distinguish from normal endometrial cells. Normal intermediate and superficial cells are visible nearby. ×400.

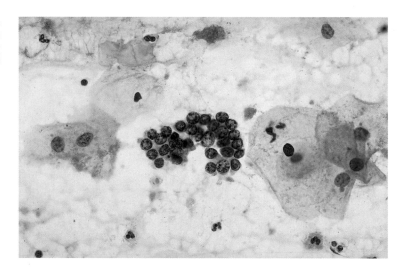

Fig. 4.**94 a–g Moderately differentiated uterine corpus carcinoma.**

a A loose cluster of irregular and severely vacuolated cells. The nuclei are slightly enlarged, have a relatively regular chromatin structure, and contain small nucleoli. They are pushed to the cell margin by large vacuoles (→). The background of the preparation indicates inflammation. ×250.

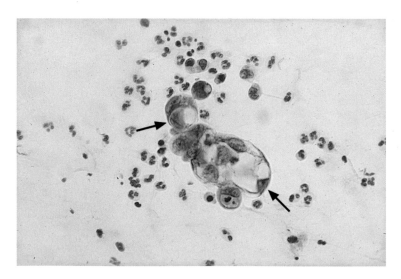

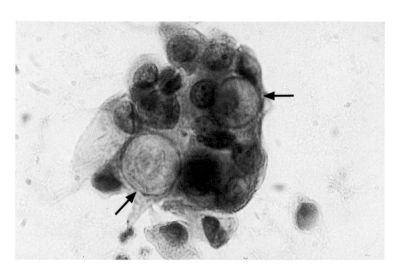

b An irregular, three-dimensional cluster of cells, containing isolated large vacuoles (—▶). The intact nuclei have a regular chromatin structure, and partly contain small nucleoli. ×1000.

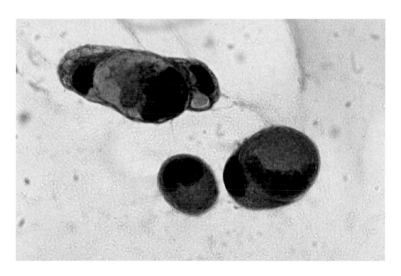

c Solitary abnormal glandular cells with well-defined cytoplasmic borders. They have hyperchromatic, condensed nuclei that are often pushed to the cell margin by vacuoles. A cell-in-cell configuration is visible on the right. ×1000.

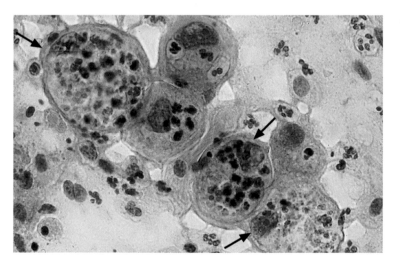

d The background of the preparation indicates inflammation. The abnormal glandular cells have phagocytosed masses of leukocytes, thus pushing the nuclei to the cell margin (signet-ring cells) (—▶). The nuclei have a relatively regular chromatin structure and contain nucleoli. The cells strongly resemble normal macrophages. ×630.

e A loose, acinus-shaped cluster of columnar, partly vacuolated glandular cells with basal nuclei that are slightly enlarged, but have a relatively regular chromatin structure and contain nucleoli. ×790.

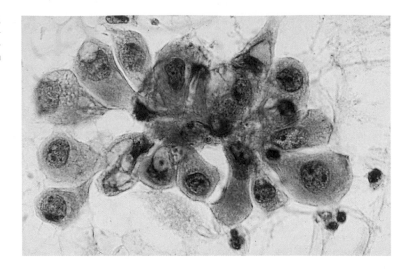

f A syncytial, sharply outlined cell cluster; cells and nuclei both vary in size. The nuclei have an irregular chromatin structure and contain macronucleoli. Normal superficial cells and red blood cells are visible in the vicinity. ×400.

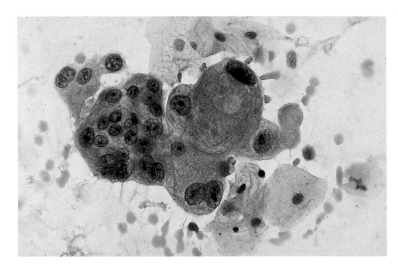

g A papillary cluster of degenerated cells, showing a calcified structure in the center (psammoma body). Plenty of erythrocytes are present in the vicinity. ×630.

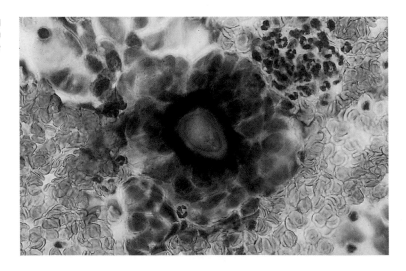

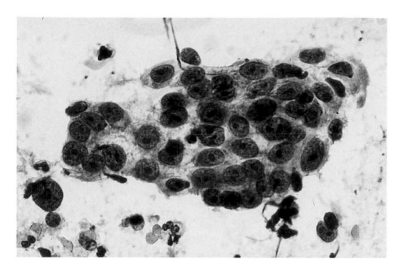

Fig. 4.**95 a, b   Poorly differentiated uterine corpus carcinoma.**

a A loose cell cluster. The severely enlarged and hyperchromatic nuclei contain macronucleoli. The cytoplasm is sparse and shows indistinct borders. ×630.

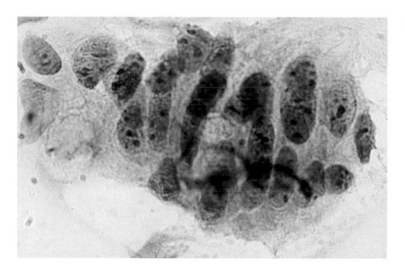

b A loose cluster of severely abnormal glandular cells. The enlarged nuclei have an irregular chromatin structure and contain numerous macronucleoli. ×1000.

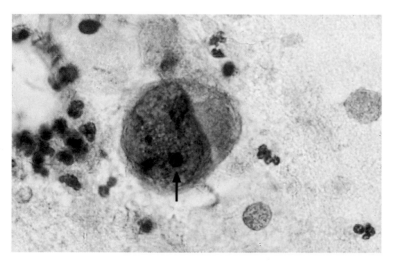

Fig. 4.**96 a, b   Dedifferentiated uterine corpus carcinoma.**

a An abnormal glandular cell with extremely enlarged nucleus, relatively regular chromatin structure, and a maronucleolus (⟶). The cytoplasm is sparse. The background of the preparation indicates tumor diathesis. ×1000.

4

b A three-dimensional cluster of degenerated, irregular cells with severely enlarged nuclei and an external line of retraction. A mitotic figure is visible on the left (——►). Plenty of erythrocytes surround the cluster. x1000.

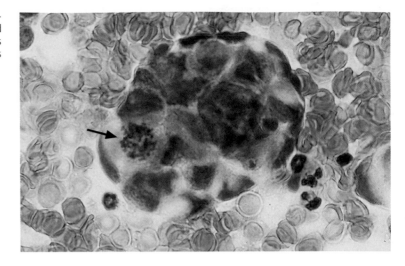

Fig. 4.**97 Adenoacanthoma of the endometrium.** Partly keratinizing, dyskaryotic squamous epithelial structures are visible in the center. The background of the preparation indicates inflammation, and solitary abnormal glandular cells are also present. ×630.

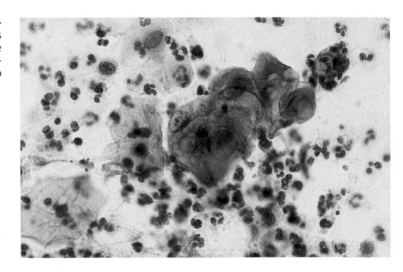

Fig. 4.**98 Leiomyosarcoma of the uterine corpus.** A multinuclear giant cell with foamy cytoplasm, indistinct borders, and cell-in-cell configuration (——►). The enlarged nuclei of this abnormal spindle-shaped cell show an irregular chromatin structure and macronucleoli. The background of the preparation indicates inflammation. ×630.

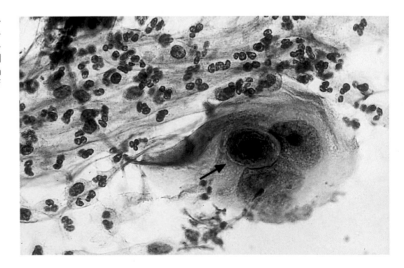

Therapy

The diagnosis of uterine corpus carcinoma should be confirmed by curettage before treatment. The definitive therapy is hysterectomy with extirpation of the adnexa, and this is often supplemented by postoperative radiotherapy.

As the diagnostic procedures essentially focus on the detection of early—though already invasive—uterine corpus carcinoma, any further approach is determined by relevant clinical symptoms. These include mainly bleeding during or after the menopause, sonographic evidence of endometrial thickening, or distinctive cytological features during or after the menopause.

The presence of such symptoms almost always calls for curettage in which material from the endocervix and from the uterine cavity is collected separately (fractional curettage), in order to assess a possible expansion of the carcinoma across the cervical region (427) (see also p. 343).

If the curettage yields evidence for **precancerosis** or for **uterine corpus carcinoma that is restricted to the uterine cavity**, hysterectomy is the therapy of choice. Because of the potential for metastatic spreading of the carcinoma into the ovaries and the upper region of the vagina, the operation usually includes bilateral extirpation of the adnexa and removal of the portion of the vaginal part that borders on the portio vaginalis, called the vaginal cuff (140). If the histological evaluation reveals **deep infiltration of the myometrium** or **metastatic invasion of the ovaries**, hysterectomy is followed by postoperative radiotherapy or, if necessary, even by chemotherapy (92).

If **cervical involvement** of the carcinoma is already detected during curettage, an extended operation is carried out, corresponding to that used for advanced squamous carcinoma of the cervix (radical hysterectomy, or Wertheim operation) (326). Radiotherapy and/or chemotherapy usually completes the treatment.

Even after complete treatment, continued clinical and cytological monitoring is required during aftercare because there is an increased risk that **relapses** occur in the blind vaginal pouch (70).

Ovary

Epidemiology

Ovarian carcinoma is slightly more common than cervical carcinoma and slightly less common than uterine corpus carcinoma. Its prevalence has been unchanged for many years because preventive screening has not been very successful. Certain lesions are considered precancerous.

Ovarian carcinoma has an incidence of 0.15–0.25‰, intermediate between cervical carcinoma and uterine corpus carcinoma (235). Preventive screening essentially focuses on clinical palpation and ultrasonography, which occasionally detect precursors but mostly only invasive carcinomas (420). This tumor has the highest mortality rate of all genital carcinomas (69). Sometimes abnormal glandular cells are observed in the cytological smear more or less by accident, and they lead one to suspect ovarian carcinoma. The age distribution of ovarian carcinoma peaks at age 50 (63).

Morphogenesis

Ovarian carcinoma probably develops within cysts of the germinal epithelium and then spreads relatively rapidly to other organs.

Ovarian carcinoma and its possible precursors probably develop from cysts of the germinal epithelium. These cysts persist over a long period of time and occur preferentially after the menopause (420). Little is known about their cause. The malignant glandular epithelium gradually replaces the germinal epithelium and finally infiltrates the stromal tissue underneath. As the ovary does not have a peritoneal surface epithelium, abnormal cells soon exfoliate into the abdominal cavity, leading to early onset of metastasis. Ovarian carcinoma has an unfavorable prognosis (205).

Endoscopy and Sonography

Ultrasonographic and endoscopic methods are widely used when ovarian carcinoma is suspected.

4

Fig. 4.**99** **Ovarian carcinoma.** Laparoscopic image of a carcinoma of the left ovary, showing an irregular, tuberous surface and focal hemorrhages. A biopsy forceps is seen on the left, and the uterine fundus is partly visible on the right (—►).

Fig. 4.**100** **Ovarian carcinoma.** Sonographic image of an ovarian carcinoma of about 50 mm in diameter (—►), showing irregular solid (highly reflective) and cystic (echo-free) portions.

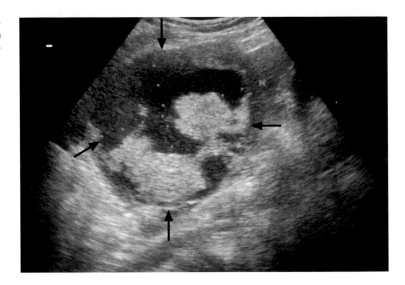

Large irregular and multilocular cysts detected by laparoscopy always lead one to suspect ovarian carcinoma (Fig. 4.**99**) (263). Such a suspicion mostly arises during ultrasound examination (Fig. 4.**100**). Clinical symptoms are rare, though patients may occasionally report nonspecific pain in the lower abdomen. For this reason, timely detection of early stages is an exception (Fig. 4.**101**) (188).

Histology

Ovarian carcinoma is usually of glandular growth type, although there are numerous morphological variants.

Ovarian cystoma occasionally has the character of a borderline tumor, with abnormal cellular changes being detectable in the wall of the cyst (see also p. 152 ff.).

With respect to invasive ovarian carcinoma, we distinguish between **serous** and **mucinous cystadenocarcinoma** (65). As the mucinous type exhibits a higher degree of differentiation, its prognosis is twice as good as that of the serous type. Highly differentiated epithelial ovarian carcinoma represents the most common type with about 70%, showing either a tubular or a papillary growth pattern (Fig. 4.**102**), whereas poorly differentiated ovarian carcinoma has a solid structure (81). **Germ cell tumors** are rare; they occur predominantly during adolescence. With 20%, the ovary represent a relatively common site for the manifestation of **extraovarian malignomas**, such as mammary, gastric, or renal carcinomas (374).

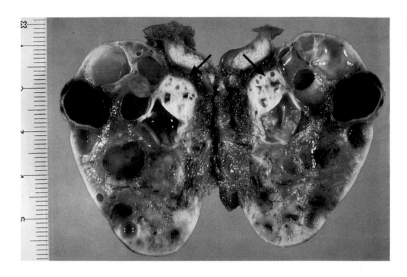

Fig. 4.**101 Multicystic ovarian tumor.** A gross section of the ovary showing a malignant tumor (white tissue) (→).

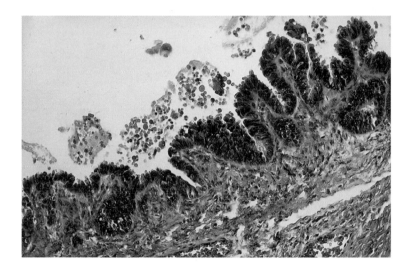

Fig. 4.**102 Serous papillary cystadenocarcinoma of the ovary.** The abnormal epithelium shows numerous epithelial protrusions. It consists of differentiated cells with large hyperchromatic, and polymorphic nuclei. The enlarged nuclei cause apparent stratification (pseudostratification) of the epithelium. HE stain, ×100.

Cytology

Tumor cells are frequently detected in the puncture fluid of malignant ovarian cysts. In rare cases, tumor cells end up in the cytological cervical smear and thus provide diagnostic clues based on their special morphology.

If a malignant cystic ovarian tumor is discovered and a **puncture specimen** is taken, tumor cells in the cyst fluid may be detected cytologically. They usually form spherical, three-dimensional clusters with an external line of retraction.

Occasionally, **exfoliated** malignant cells are caught by the fimbriae of the uterine tube and thus reach the uterine cavity, the cervix, and finally the vagina (69). The cytological preparation then shows three-dimensional clusters of abnormal glandular cells arranged in a rosette structure, while the background is often clean, i. e., free of inflammatory cells (Fig. 4.**103a, b**) (308). Because they travel a long distance from the ovary to the vagina, the cell clusters are often spherical and show an external line of retraction (Fig. 4.**103c**) (112).

Occasionally, a cell contains a mucous vacuole of various size which pushes the nucleus to the cell margin, thus creating signet-ring cells (Fig. 4.**103d**). Phagocytosis and cannibalism are common. The nuclei vary in size, are often hyperchromatic, and contain prominent nucleoli (Fig. 4.**103e**). Sometimes, concentrically laminated foci of calcification (psammoma bodies) are detected within a cell cluster (Fig. 4.**104**) (29).

Fig. 4.103 a–d   Ovarian carcinoma.

a   A well preserved, loose cluster of naked nuclei in the **cervical smear**. The nuclei are similar in size and hyperchromatic, show a relatively regular chromatin structure, and contain macronucleoli. Normal superficial cells are visible nearby. ×630.

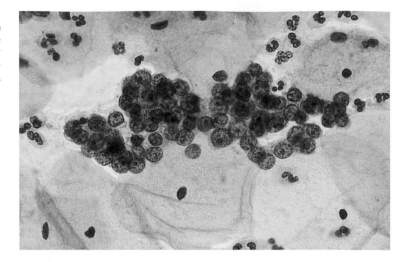

b   A well-preserved, loose cell cluster in the **cervical smear**. The nuclei have a relatively uniform structure and contain macronucleoli. The cytoplasm is sparse. Normal superficial cells are visible nearby. ×630.

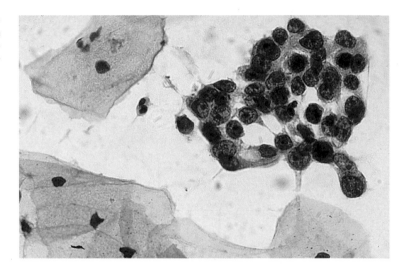

c   An irregular, three-dimensional cell cluster in the **puncture fluid** showing severely enlarged nuclei, macronucleoli, and an external line of retraction (⟶). ×790.

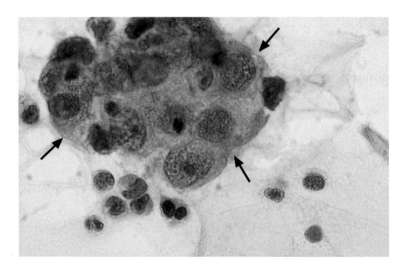

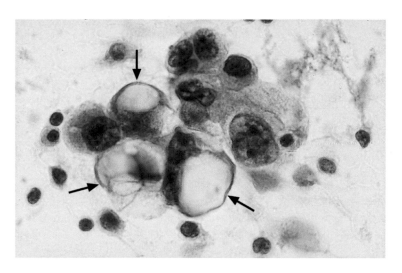

d A loose, irregular cell cluster in the **puncture fluid** showing severely enlarged, polymorphic nuclei, and macronucleoli. The signet-ring cells are the result of cytoplasmic vacuolization (⟶). ×790.

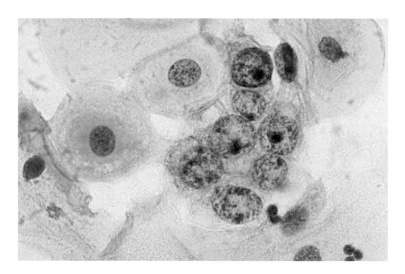

e **Papillary cystadenocarcinoma of the ovary.** A loose cell cluster in the **cervical smear** showing severely enlarged nuclei, coarse irregular chromatin structure, and macronucleoli. Normal parabasal and intermediate cells are visible in the background. ×1000.

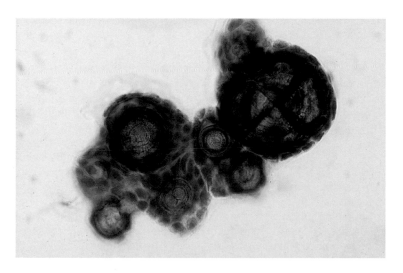

Fig. 4.**104** **Papillary ovarian cystoma.** A degenerated cell cluster in the **puncture fluid** showing several focal structures of calcification (psammoma bodies). ×400

4

Therapy

Every cyst or enlargement of the ovary that is more than 3–4 cm in diameter should be regarded as potentially malignant and therefore needs to be assessed (332). Suspicion of ovarian carcinoma may arise during palpation in the context of preventive screening or, more frequently, during ultrasound examination that is done either routinely or because of pain in the lower abdomen. In young women, cysts are monitored at regular intervals of 3 months. In older women, or if the cysts persist, cysts should be punctured or removed during laparoscopy (389) (see also p. 344).

The content of the cyst is subjected to cytological and hormonal analysis (see also p. 155). If the result is suspect, extirpation of the adnexa is required. If suspicious glandular cells are found in the cytological smear, diagnostic confirmation is required even when the sonographic findings are negative. A reliable cytological differentiation between endometrial and ovarian carcinoma is rarely possible; hence, curettage and laparoscopy are usually done during the same session (286).

Treatment of ovarian carcinoma mainly consists of surgical removal of the uterus and both ovaries, since metastases often occur in the endometrium and in the contralateral ovary (420). Chemotherapy is an important part of the treatment because ovarian carcinoma responds poorly to radiotherapy. Chemotherapy is often used preoperatively to achieve a reduction in tumor size and thus create better conditions for the operation.

Vulva

Epidemiology

Carcinoma of the vulva occurs about 20 times less frequently than other genital carcinomas. However, its precursors are currently detected almost 10 times more often than invasive vulvar carcinoma.

Incidence. The frequency of vulvar carcinoma is 0.01‰ and has been fairly constant during the last 50 years (150). By contrast, an increase in the frequency of carcinoma precursors has been reported during the last 10 years, presumably because HPV infection in the general population has increased and the rate of diagnostic detection has improved (157). Reliable data about the incidence of precancerous stages are currently not available (269, 270).

Early detection. The latency periods between precancerous stages and invasive carcinoma are about 10 years longer for the vulva than for the uterine cervix. As the vulvar carcinoma has attracted more attention in recent years, its precursors—particularly high-grade dysplasia—are now detected much earlier and are removed before an invasive carcinoma can develop. Although invasive vulvar carcinoma occurs almost exclusively after the menopause, with the frequency peaking around age 60, its precursors often occur during sexual maturity with the frequency peaking around age 40 (81). In half of the women who show such a precursor, a cervical precancerous lesion is present at the same time.

Morphogenesis

Carcinoma of the vulva develops either from an HPV infection or from dystrophy of the epithelium. As in the case of cervical carcinoma, carcinogenesis is a long-term process and mostly, but not always, involves several precancerous stages. These precursors are called dysplasia, carcinoma in situ, or vulvar intraepithelial neoplasia (VIN); they are subdivided according to their chronological development as follows:
▶ Mild dysplasia (VIN I)
▶ Moderate dysplasia (VIN II)
▶ Severe dysplasia (VIN III)
▶ Carcinoma in situ (VIN III)
▶ Invasive carcinoma

Development. Two different factors seem to be equally responsible for the morphogenesis of vulvar carcinoma:
▶ **HPV infection:** As with cervical carcinoma, certain types of the virus—particularly types 16 and 18—first lead to abnormal basal cell hyperplasia and finally, via increasing degrees of dysplasia, to invasive vulvar carcinoma. This course seems to take place preferentially in young women (114).
▶ **Vulvar dystrophy:** This condition presumably has a hormonal origin. Here, chronic inflammatory stimulation in the subepithelial stroma causes reactive acanthotic epithelial hyperplasia. Although maturation of the epithelium is still mostly normal, cellular dysplasia in the basal layer develops more frequently. This

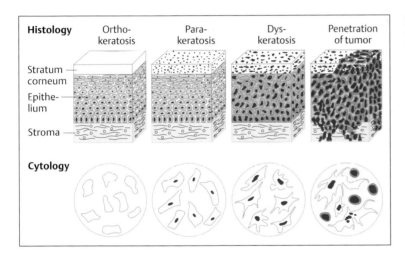

Cytology	Ortho-keratosis / para-keratosis	Mild dyskeratosis		Moderate dyskeratosis		Severe dyskeratosis
Horny layer						
Superficial cell layer						
Intermediate cell layer						
Intermediate cell layer						
Parabasal cell layer						
Basal cell layer						
Histology	Normal	Mild dys-plasia	Moderate dys-plasia	Severe dys-plasia	Carci-noma in situ	Invasive carcinoma

Fig. 4.**105 Stages of epithelial carcinogenesis** in the keratinizing squamous epithelium. The dysplastic basal cell layer thickens with increasing degrees of dysplasia, as is the case with nonkeratinizing squamous epithelium. Normally, the **horny layer** prevents the exfoliation of immature cells, but cells with mild, moderate, and severe dyskeratosis do exfoliate depending on the severity of the disease.

Histology	Ortho-keratosis	Para-keratosis	Dys-keratosis	Penetration of tumor
Stratum corneum				
Epithe-lium				
Stroma				
Cytology				

Fig. 4.**106 Morphology of the horny layer.** Diagram illustrating the histological and cytological findings associated with different types of keratotic lesions of the vulva.

or directly. This development is observed especially in elderly women (89).

Grades of dysplasia. The stages of epithelial carcinogenesis are classified as mild, moderate, and severe. They basically correspond to those of cervical carcinoma, although the vulvar epidermis is a keratinizing squamous epithelium and, therefore, always keratinizes at the surface (Fig. 4.**105**). Benign processes are associated with orthokeratotic or parakeratotic keratinization, whereas dysplasia and carcinoma show dyskeratotic keratinization with abnormal nuclei detectable in the horny layer (Fig. 4.**106**) (201).

Invasive carcinoma. The carcinoma penetrates the basal membrane and infiltrates the stroma. The invasion occurs by means of the immature abnormal cells of the epithelium, and the keratinizing cells form keratin pearls. The latter are characteristic for keratinizing squamous carcinoma (42). Despite invasion of the stroma, about half of all carcinomas show a complete dyskeratotic layer, the nuclei of which sometimes seem to be reduced in size and number; this is called pseudoorthokeratosis. In all other cases, immature anaplastic tumor cells penetrate the dyskeratotic zone and reach the surface (247). Once the increased metabolic needs in this region are no longer met, necrosis and inflammatory reactions occur, thus causing chronic ulcers. The invasion of lymph and blood vessels finally leads to metastatic spreading of the tumor.

4

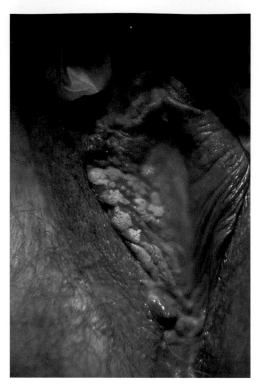

Fig. 4.**107**   **Carcinoma in situ.** Prominent diffuse (flaky) leukoplakia on the inside of the right labium minus.

Fig. 4.**108**   **Carcinoma in situ.** The lesion is located in the perianal region and partly affects the rectal mucosa. It shows prominent pigmentation with distinct borders.

Macroscopy

Vulvar carcinoma and its precursors usually impress macroscopically as prominent leukoplakia, occasionally also as erythroplakia or pigmented areas. Advanced carcinomas may manifest themselves as tumors or ulcers. Both the acetic acid test and the toluidine blue test are valuable diagnostic tools.

Dysplasia induced by HPV. Dysplastic lesions developing as the result of HPV infection usually show sharply outlined, prominent leukoplakia, erythroplakia, pigmentation, or a mixture of these (Figs. 4.**107**, 4.**108**) (157). The acetic acid test and the toluidine blue test make these lesions stand out more clearly.

Vulvar dystrophy. Clinically, dysplasia associated with dystrophy differs little from benign dystrophy. The changes are mostly diffuse and spread over large skin areas (Fig. 4.**109**) (168). Occasionally, the toluidine blue test reacts with skin areas in which parakeratotic or dyskeratotic keratinization is particularly severe; this is regarded as an indicator for potentially abnormal development (Fig. 4.**110**). However, the method is only semiquantitative,

with benign lesions yielding less intense staining than malignant ones (89).

Invasive carcinoma. The carcinoma spreads within the skin, forming either an exophythic tumor or an endophythic ulcer (Fig. 4.**111 a, b**). As long as the abnormal lesion has no horny layer, colposcopy often reveals irregular blood vessels (Fig. 4.**111 c**). Carcinomas developing as the result of HPV infection usually have well-defined borders. By contrast, those developing from dystrophy are surrounded by numerous, still benign lesions that appear mostly in the form of leukoplakia (201). Carcinogenesis in such cases is diffuse and multicentric (Fig. 4.**112**). The clinical symptoms of vulvar carcinoma and its precursors are nonspecific. Itching, burning, pain, or occasional spot bleeding may occur.

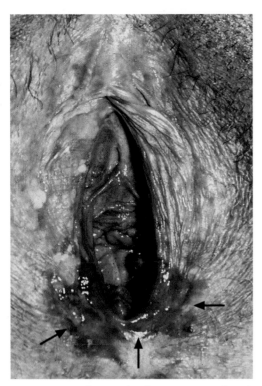

Fig. 4.**109** **Moderate dysplasia** associated with diffuse vulvar dystrophy (diffuse leukoplakia). Circumscribed erythroplakia on the posterior commissure and the fossa of the vaginal vestibule (→).

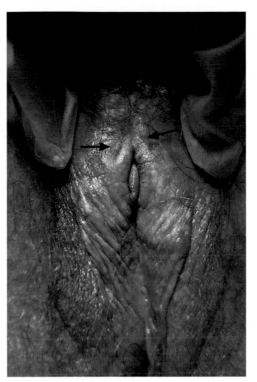

Fig. 4.**110** **Mild dysplasia** associated with vulvar dystrophy. The discrete leukoplakia in the area of the anterior commissure shows a weak positive reaction with toluidine blue (→).

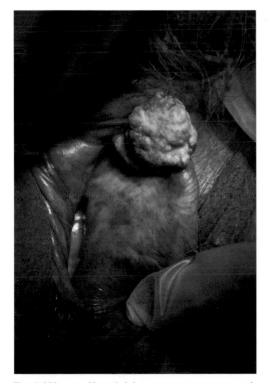

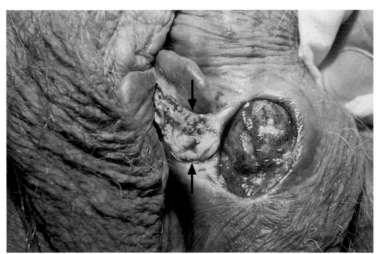

b An oval ulcer on the labium majus is spreading to the labium minus in the form of prominent leukoplakia (→).

Fig. 4.**111 a–c Keratinizing squamous carcinoma.**
a An exophytic tumor grows on the inside of the left labium minus.

4

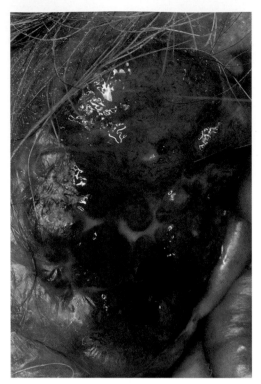

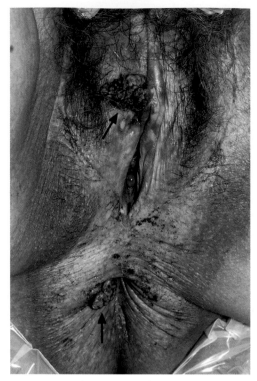

c A partly ulcerating tumor with erythroplakia in the region of the right labium majus. Irregular blood vessels are visible in the upper half, and a circumscribed hemorrhage (⟶) is visible in the lower half. The vaginal orifice and the labia minora are seen on the lower right.

Fig. 4.**112 Multicentric vulvar carcinoma** associated with diffuse dystrophy. Prominent tumors are present on the right labium majus and in the perianal region (⟶). The toluidine blue test is positive.

Histology

In German-speaking countries, the histological classification divides the precursors of vulvar carcinoma into mild, moderate, and severe dysplasia and carcinoma in situ; the latter is often referred to as Bowen disease. By contrast, the international nomenclature distinguishes **vulvar intraepithelial neoplasia (VIN) grades I–III.** Invasive vulvar carcinoma is subdivided into highly and moderately differentiated growth types as well as undifferentiated carcinoma.

Grades of dysplasia. As with cervical dysplasia, we histologically distinguish mild, moderate, and severe dysplasia depending on whether abnormal basal cell hyperplasia occupies the lower, middle, or upper third of the epithelial thickness (Fig. 4.**113**) (157). However, the term "epithelial thickness" defines only the nonkeratinized squamous epithelial layers. Below the zone of keratinization, the abnormal nuclei show all the typical criteria for malig-

nancy, such as polymorphic, enlarged, and hyperchromatic nuclei, abnormal chromatin pattern, and increased mitotic rate. In the keratin layer, the nuclei lose the chromatin pattern due to degeneration, and they finally condense.

Carcinoma in situ. With high grades of dysplasia, especially carcinoma in situ, there is an increasing tendency for accelerated keratinization of individual cells within the epithelium lying underneath, which is not yet keratinized but has widened due to abnormal basal cell hyperplasia. As these deep-lying dyskeratocytes round off, shrink, and dissociate from the rest of the epithelium (acantholysis), they often appear as round, double-contoured bodies ("corps ronds," Fig. 4.**114**) (415). This lesion is still referred to as **Bowen disease** or **Bowen's precancerous dermatosis**—although, according to the international nomenclature, this term should no longer be used officially.

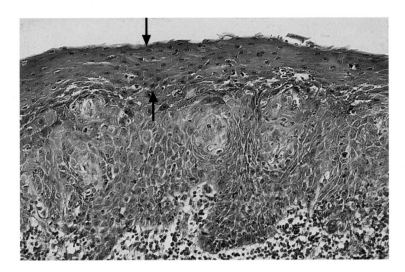

Fig. 4.**113** **Mild vulvar dysplasia.** The basal cell layer consists of immature cells with enlarged, polymorphic, and hyperchromatic nuclei and shows a slight thickening that is restricted to the lower third of the epithelial thickness. The superficial horny layer exhibits **mild dyskeratosis** (⟶). HE stain, ×100.

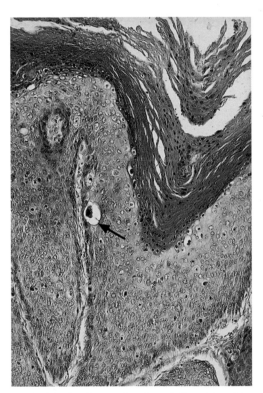

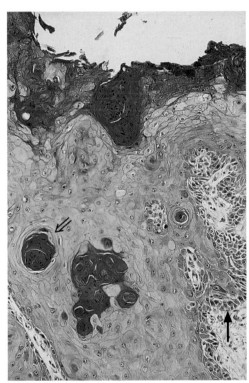

Fig. 4.**114** **Carcinoma in situ of the vulva.** The basal cell layer consists of immature cells with enlarged, hyperchromatic but monomorphic nuclei and shows a thickening that occupies the entire (nonkeratinized) epithelium. The uniform size of the cells gives the epithelium a monomorphic appearance, which is only interrupted by single keratinizing cells that undergo acantholysis, thus creating holes in the epithelium ("corps ronds") (⟶). The horny layer shows hyperkeratosis and **pronounced dyskeratosis**. HE stain, ×100.

Fig. 4.**115 a, b** **Keratinizing squamous carcinoma of the vulva.**

a The abnormal epithelium has two components. The undifferentiated portion on the right (⟶) consists of nonkeratinized cells with large, nucleoli-containing nuclei; these are the invading cells. The highly differentiated portion consists of keratinized cells, which either exist as single cells or form keratin pearls (⟹). The horny layer exhibits hyperkeratosis and **pronounced dyskeratosis**. HE stain, ×100.

b The undifferentiated portion of the tumor consist of nonkeratinized cells with large, nucleoli-containing nuclei. It penetrates the dyskeratotic layer and mixes with the highly differentiated portion of the tumor that consists of keratinized cells. HE stain, ×100.

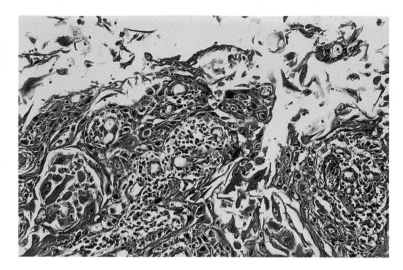

Invasive carcinoma. Here, keratin pearls are detected within the nonkeratinized epithelium, and there is an increase in immature, nonkeratinized cells that contain nucleoli (Fig. 4.**115a**). According to Broders (42), the degree of tumor cell maturity depends on the quantitative portion of keratinized epithelium. Highly differentiated carcinomas have a very high keratin content, but the dedifferentiated growth types are almost completely free of keratin. About half of all carcinomas show penetration of the superficial horny layer by the undifferentiated tumor portion (Fig. 4.**115b**) (247).

Cytology

The cytodiagnosis of vulvar carcinoma and its precursors distinguishes:

▶ Mild dyskeratosis (with mild and moderate dysplasia)
▶ Moderate dyskeratosis (with severe dysplasia and carcinoma in situ)
▶ Severe dyskeratosis (with invasive carcinoma)
▶ Anaplastic tumor cells (with invasive penetrating carcinoma)

Criteria for malignancy. The cytological evaluation of precancerous and malignant changes of the vulva is more problematic than that of the nonkeratinized squamous epithelium of the portio vaginalis because the cell population is monomorphic, like that of the glandular epithelium of the endocervix and the endometrium (245). An assignment of abnormal nuclei to different degrees of epithelial maturity is therefore not possible. Furthermore, the diagnosis is more difficult because of the lower tendency of keratinized squamous epithelium to exfoliate as compared to vaginal epithelium, and also because nonkeratinized squamous epithelial cells from other locations may be admixed. When comparing the sizes of cells and nuclei with those of nonkeratinizing epithelium, there is an increase in cell size with increasing malignancy of the epithelium, unlike the situation with vaginal epithelium (Fig. 4.**116**) (281).

Mild dyskeratosis. Cells typical for **mild and moderate vulvar dysplasias** are classified as mildly dyskeratotic (247). They are slightly enlarged horny cells with polymorphic, slightly enlarged nuclei; hence, there is no clear shift in the nucleocytoplasmic ratio (Figs. 4.**117a, b**, 4.**118a, b**). For this reason, the above conditions are easily confused with benign lesions.

Moderate dyskeratosis. In the presence of **severe dysplasia** and **carcinoma in situ**, moderately dyskeratotic cells exfoliate. The cells are clearly smaller but contain enlarged nuclei; hence, the nucleocytoplasmic ratio has shifted in favor of the nucleus (Figs. 4.**119a, b**, 4.**120a, b**). As the nuclei are always degenerate, their chromatin structure cannot be evaluated, although the condensed nuclei are clearly hyperchromatic. Based on these criteria, high-degree vulvar dysplasia is relatively easy to recognize cytologically.

Severe dyskeratosis. In the presence of **invasive carcinoma**, severely dyskeratotic cells exfoliate. The cells are much enlarged, and their nuclei are smaller than normal. The nucleocytoplasmic ratio has therefore shifted in favor of

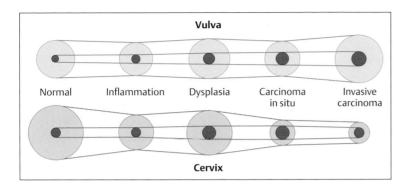

Fig. 4.**116 Sizes of cells and nuclei.** Planimetry of cells and nuclei of vulvar and cervical squamous epithelia associated with various lesions. (Data for the cervix, according to S.F. Patten.)

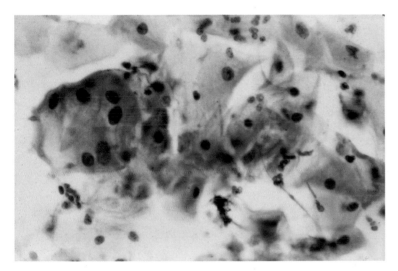

Fig. 4.**117 a, b Mild dyskeratosis associated with VIN I.**
a Cells with slightly polymorphic cytoplasm and slightly enlarged, condensed nuclei. The nuclei vary in size. ×400.

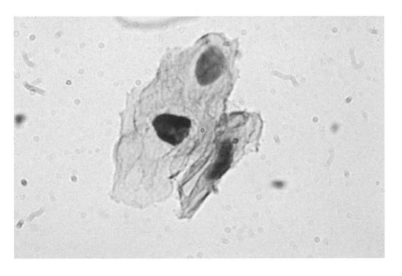

b Cells with plenty of cytoplasm and moderately enlarged, degenerated nuclei. ×1000.

Fig. 4.**118 a, b   Mild dyskeratosis associated with VIN II.**

a The nuclei vary in size. They are hyperchromatic, enlarged, and condensed (——➤). Normal horny flakes and intermediate cells are visible nearby. ×630.

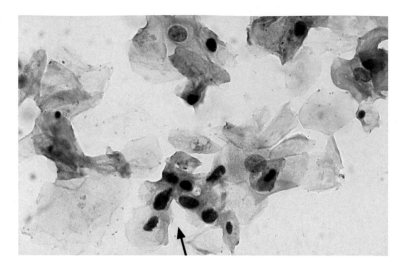

b Hyperchromatic, enlarged, and condensed nuclei (——➤). Normal superficial cells are visible on the right. ×1000.

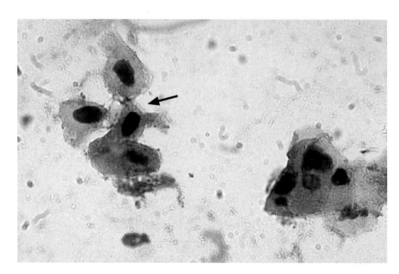

Fig. 4.**119 a, b   Moderate dyskeratosis associated with VIN III.**

a Cells with relatively little cytoplasm and clearly enlarged, hyperchromatic, and condensed nuclei (——➤). Normal superficial cells are visible nearby. ×250.

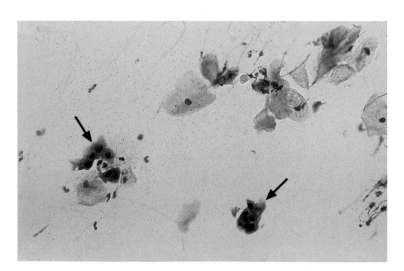

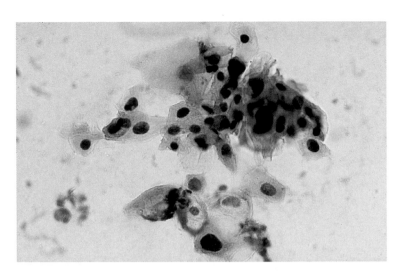

b A loose cluster of cells with relatively little cytoplasm and enlarged, hyperchromatic, condensed nuclei. ×630.

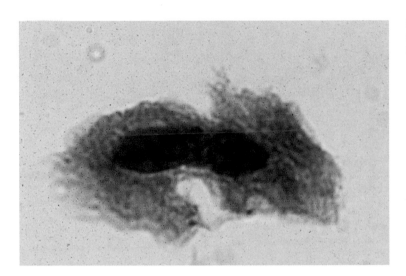

Fig. 4.**120 a, b   Moderate dyskeratosis associated with VIN III.**
a A small cell with polymorphic cytoplasm and enlarged, hyperchromatic, condensed nucleus. ×1000.

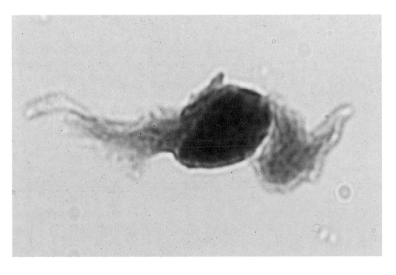

b A small cell with polymorphic, spindle-shaped cytoplasm and enlarged, but degenerated nucleus. ×1000.

4

Fig. 4.**121 a, b Severe dyskeratosis associated with invasive carcinoma.**
a Enlarged cells that vary in size and have polymorphic cytoplasm. Their nuclei are small and either deteriorating or pyknotic (——►). ×250.

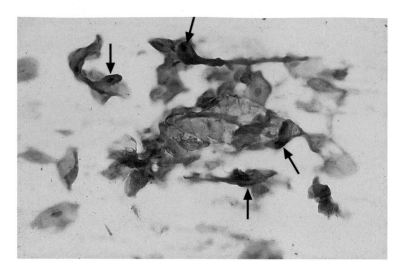

the cytoplasm (Figs. 4.**121 a, b**, 4.**122 a, b**). The wide range in cell size and nuclear size, and the high degree of cytoplasmic polymorphism are important criteria for malignancy (Fig. 4.**123**). Denucleated polymorphic keratin flakes represent pseudoorthokeratosis. Cytological recognition of vulvar carcinoma requires lots of experience in morphological evaluation. When the sizes of cells and nuclei of vulvar carcinoma are compared with those of keratinizing cervical carcinoma, the cells of the former are about four to five times larger, whereas the nuclei are five times smaller.

Anaplastic tumor cells. In addition to severe dyskeratotic cells, about half of all vulvar carcinomas contain anaplastic (undifferentiated) tumor cells that exhibit all typical criteria for malignancy. They show absolute and relative nuclear enlargement and nuclear polymorphism, they have hyperchromatic nuclei with an abnormal chromatin structure, and they contain macronucleoli (Figs. 4.**124 a–c**, 4.**125**). The cytoplasm is usually amphophilic, cyanophilic, or chromophobic. Unlike the resistant horn cells, these cells are sensitive to environmental factors because they are immature. As a result, they often show signs of nuclear and cytoplasmic degeneration, and naked nuclei are common.

Other malignant vulvar lesions. Paget disease is a rare tumor. It presumably develops from apocrine glands, grows in characteristic nests, and—like **basalioma** or **basal cell carcinoma**—destroys the tissue locally, although it hardly metastasizes (192). The cytological smear

shows typical large, bright cells with nuclei that contain nucleoli.

Malignant melanoma of the vulva is also rare, but it has a distinctly unfavorable prognosis (229). Cytological smears from melanomas occasionally contain anaplastic tumor cells that store melanin pigment granules (384).

Therapy

Treatment of the precursors of vulvar carcinoma is limited to surgical excision or destruction by means of laser, and treatment of early invasive carcinoma is restricted to partial vulvectomy. Advanced carcinoma calls for classic total vulvectomy and, if necessary, subsequent radiotherapy.

Precancerous vulvar lesions. If the changes are small and circumscribed, they are excised during local anesthesia (see also p. 344 ff.). In case of larger foci or diffuse affection of the organ, laser vaporization is currently the method of choice and is preferable to lengthy, often painful local attempts at treatment with interferon or fluorouracil (136, 421).

Considering the long latency periods preceding the possible invasion of a precancerous lesion, the attitude of waiting for several years is usually justified. Nevertheless, the situation needs to be closely monitored by clinical examination and cytological smears at intervals of 3 months. If relapses occur, further biopsies and laser treatment become necessary. Radical measures, especially vulvectomy, should be

b Much-enlarged cells that vary in size and have polymorphic cytoplasm. The nuclei also vary in size. ×250.

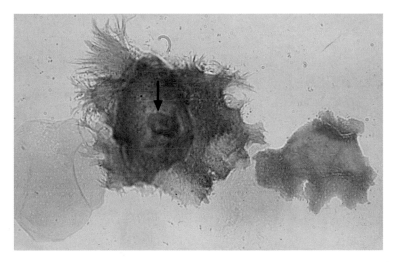

Fig. 4.**122 a, b Severe dyskeratosis associated with invasive carcinoma.**
a A very large cell with polymorphic cytoplasm and slightly enlarged, deteriorating nucleus (⟶). A denucleated dyskeratotic (pseudo-orthokeratotic) cell is seen nearby. ×250.

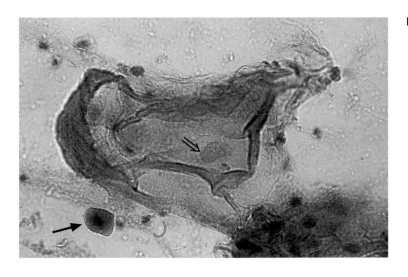

b An extremely large cell with polymorphic cytoplasm. (For size comparison, see the parakeratotic cell at the lower left (⟶).) The central nucleus is slightly enlarged and deteriorating (⟹). ×160.

4

Fig. 4.123 Parakeratotic and dyskeratotic cells of various vulvar lesions. Planimetry for comparing the size and shape of cytoplasm and nuclei.

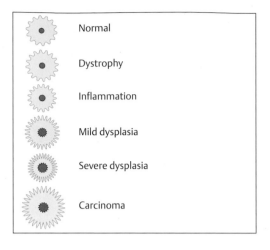

Normal

Dystrophy

Inflammation

Mild dysplasia

Severe dysplasia

Carcinoma

Fig. 4.124 a–c Severe dyskeratosis and anaplastic tumor cells associated with penetrating vulvar carcinoma.

a The severely dyskeratotic cells are enlarged, vary in size, and have degenerated nuclei (→). The anaplastic tumor cells are small, basophilic, and round, and their abnormal nuclei are still intact (⇒). ×400.

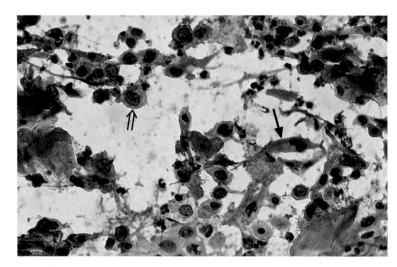

b The severely dyskeratotic cells vary in size, have polymorphic cytoplasm, and contain degenerated nuclei (→). The anaplastic tumor cells are relatively isomorphic, basophilic, and round, and their intact, abnormal nuclei contain nucleoli (⇒). ×400.

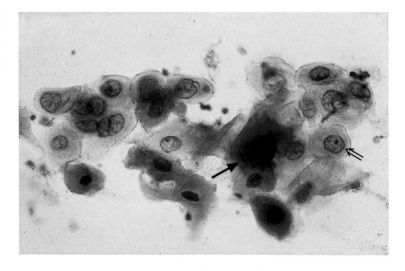

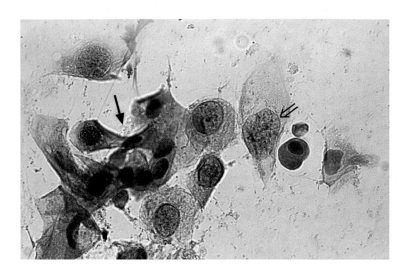

c The severely dyskeratotic cells vary in size, have polymorphic cytoplasm, and contain degenerated nuclei (⟶). The anaplastic tumor cells have much enlarged nuclei with a coarse, irregular chromatin structure (⟹). ×400.

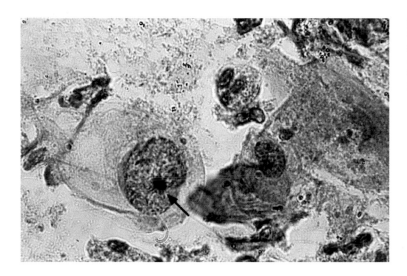

Fig. 4.**125 Anaplastic tumor cells associated with penetrating vulvar carcinoma.** The basophilic cytoplasm has indistinct borders. The nuclei are enlarged, show a coarse, irregular chromatin structure, and contain macronucleoli (⟶). ×1000.

avoided as long as possible, and the region of the clitoris, in particular, should be preserved (339).

Vulvar carcinoma. Surgical treatment depends on the depth of the invasion and the size of the tumor (27). If the depth of invasion is no more than 1 mm and the tumor diameter no more than 2 mm, generous excision is usually sufficient—as long as special circumstances, such as unfavorable location or severe dedifferentiation, do not call for other measures. Such measures include vulvectomy (excision of one half to two thirds of the vulva), together with

extirpation of the inguinal lymph nodes, if necessary (421).

Once the carcinoma has advanced, radical vulvectomy with resection of the inguinal and femoral lymph nodes, and possibly also the pelvic lymph nodes, can no longer be avoided. The value of radiotherapy is controversial, especially because local relapses are common and very difficult to treat on previously radiated and damaged skin (91). Long-term monitoring at intervals of 3 months is essential after the surgery.

Extragenital Manifestations

Tumors Spreading from Adjacent Organs

Rectal carcinoma. Rectal carcinoma has a far higher incidence than urinary bladder carcinoma, and vaginal invasion with fistula formation occurs more often from the rectum than from the bladder (178). Rectal carcinomas are often first diagnosed as a result of such an event. The smear image usually shows a dirty background with plenty of debris and leukocytes (see Fig. 5.**60 b**, p. 300).

Vaginal invasion by rectal carcinoma is often associated with the presence of abnormal glandular cell clusters in the smear. The vacuolated cytoplasm occasionally gives rise to a **signet-ring cell**, in which the nucleus is flatted and pushed to the cell margin by a large secretory vacuole.

Urinary bladder carcinoma. If the invading carcinoma originates from the bladder, abnormal epithelial cells derived from the urothelium (also called **transitional epithelium**) are detected in the smear. They exhibit morphological features of both squamous epithelium and glandular epithelium (Fig. 4.**126**).

Metastatic Tumors of the Genital Tract

Distant metastases in the cervix occur most often in patients with **mammary carcinoma**. They usually appear in the cytological smear as clusters of abnormal glandular cells that do not allow for a clear assignment (178) (see Fig. 5.**96 b**, p. 326). Metastases derived from **ovarian carcinoma**, **urinary bladder carcinoma**, **renal carcinoma**, or **malignant melanoma** (see Fig. 5.**32 b**, p. 280) are rare (225). Their smear images correspond to the morphology of the primary tumors.

Occasionally, **malignant lymphoma** also metastasizes into the cervix. Tumor cells appearing in the smear are usually very small, isomorphic, and diffusely distributed. The cells resemble lymphocytes, although their chromatin structure is coarser and nucleoli are often detected (178).

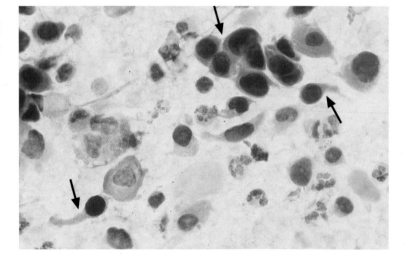

Fig. 4.**126** **Vaginal invasion by urinary bladder carcinoma.** The abnormal urothelial cells, occuring either single or in clusters, partly show a columnar cytoplasm and hyperchromatic, conclused nuclei (**→**). ×630.

Differential
Cytology

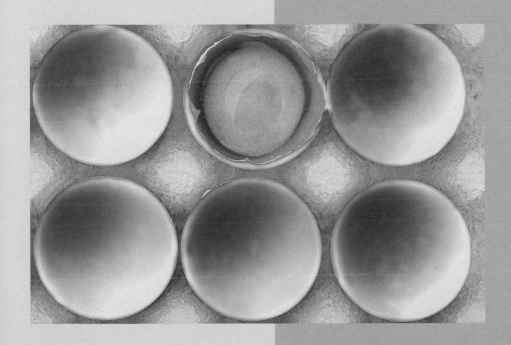

5

5 Differential Cytology

Visual Analysis of the Cell Image

The previous chapters have provided a systematic classification of cellular changes into normal, benign, and malignant categories. In day-to-day practice, however, it is often difficult to verify this classification in the gynecological smear, either because the morphological images are not typical enough, or degenerative, reactive, or proliferative changes exist side by side in the same smear. This creates problems with the identification and interpretation of atypical cells.

Whereas systematic cytology is concerned with the differences between normal, benign, and malignant morphologies, differential cytology focuses on their common features. Because of external and internal influences, benign and malignant lesions may yield cell images that resemble each other, thus leading to misinterpretation of such look alikes.

Changes responsible for problems of classification are subdivided into **squamous epithelial, glandular epithelial,** and **small-cell lesions**, although the transitions between them are fluid. Signs of dedifferentiation within the squamous epithelium include, for example, the presence of vacuoles and a loose, foamy cytoplasm with indistinct cell borders, whereas a loss of these features is observed with dedifferentiation in the columnar epithelium. It is not possible to specify a particular cell size that would indicate where large-cell differentiation ends and small-cell differentiation begins. Both these growth patterns often exist side by side. The diagnostic problem with small-cell lesions begins with a cell diameter of less than 14 μm.

Degenerative and reactive processes affect almost all cells of the smear, thus causing diffuse changes in the **overall image of cells**. However, neoplastic lesions preferentially manifest themselves in changes of single cells (or groups of cells) observed in the **single cell image**. The factors influencing the overall image and the specific morphology of single cell images may be so serious that benign cells appear abnormal whereas malignant cells do not raise any suspicion. Hence, the correct classification of such cells depends on the cytologist's visual acuity, subjective evaluation, and morphological experience.

The interpretation of cellular changes is based on the detection of cytoplasmic and nuclear structures in the smear. Cytoplasmic features provide clues to the origin of a cell, and nuclear analysis helps with the assessment of its integrity. In order to classify the size and structure of problematic nuclei correctly, it is helpful to use clearly identifiable **reference nuclei** for orientation. These are nuclei of cells regularly present in a smear, such as intermediate cells or granulocytes. Technical, physiological, or inflammatory factors may alter both nuclei and cytoplasm to such an extent that only a few morphological criteria remain that make it possible to reach any conclusion about the origin of the cells and their degree of malignancy. The changes may be so extensive that an assignment is no longer possible.

The **visual analysis** of photographic images can be extremely difficult even for an experienced morphologist. Cell images that are clearly recognized in the cytological preparation are often not reproduced in enlarged sections of a microphotograph; first, because any reference to the overall image of cells is missing and, second, because it is not possible to vary the optical magnification and the plane of view. For this reason, the use of photographs, slides, or digitized images is restricted to the purposes of continuing education in cytology. Hence, participation in cytological workshops, where a rich variation of certain topics is presented directly at the microscope and in a practice-oriented manner, cannot be replaced by written teaching material. On the other hand, a textbook, an atlas, or a slide quiz does have the advantage of presenting specific problems in a clear arrangement of text and illustrations, thus providing food for thought with respect to differential diagnosis.

Fig. 5.**1 a** **Normal intermediate cells.** Slightly swollen, active nuclei an d smooth nuclear envelopes. (Differential diagnosis, **DD:** mild dysplasia.) ×630.

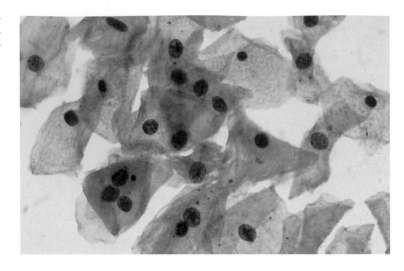

Fig. 5.**1 b** **Mild dysplasia.** Irregular chromatin structure and irregular nuclear envelopes (**DD:** normal intermediate cells.) ×400.

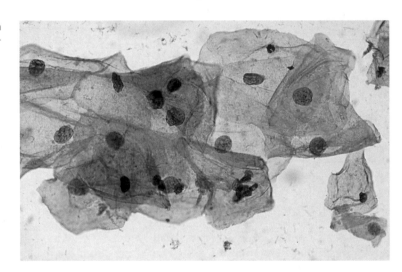

Benign Changes and How to Distinguish them from Malignancies

Squamous Epithelial Lesions ____

The most common problems in differential cytology concern the squamous epithelium. The development of cervical carcinoma is correlated with high endemic HPV infection, which affects a large section of the population, mainly during sexual maturity. Furthermore, inflammatory genital lesions predominate during this period of life and may yield similar cell images.

Portio and Vagina

Physiological Factors

Cytolysis caused by lactobacilli not only lyses the cytoplasm but often results in the edematous swelling of the nucleus, thus making it appear several times larger than it is normally. Such nuclei still have a regular, fine granular chromatin structure, and the nuclear membrane remains smooth. Nevertheless, there are often problems with distinguishing these changes from dysplastic changes (Figs. 5.**1 a, b**, 5.**2 a, b**). When dysplastic cells undergo cytolysis, it is often impossible to identify the epithelial layer from which they originate (Fig. 5.**3**). Superimposed cells may occasionally mimic the nuclei dysplastic nuclei (Fig. 5.**4**).

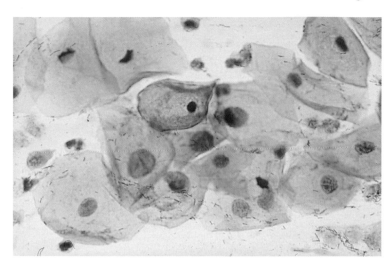

Fig. 5.**2a** **Normal intermediate cells.** Distinct nuclear swelling and smooth nuclear envelopes. (**DD:** mild dysplasia.) ×500.

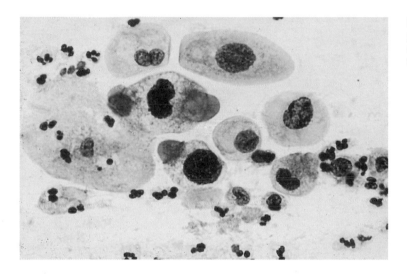

Fig. 5.**2b** **Moderate and severe dysplasia.** Hyperchromatic nuclei, irregular chromatin structure, and first signs of karyorrhexis. (**DD:** disturbed cell maturation.) ×630.

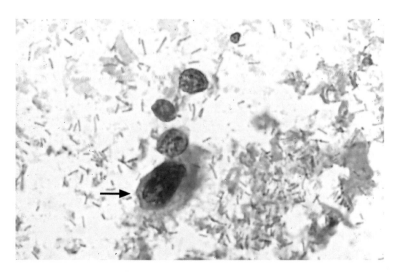

Fig. 5.**3** **Moderate dysplasia.** Lactobacillus-induced cytolysis of a cell with abnormal nucleus and partially disintegrating cytoplasm (⟶). Naked nuclei of normal intermediate cells are visible nearby. (**DD:** severe dysplasia.) ×630.

5

Fig. 5.**4 Superposition of cells.** A normal para-basal cell (⟶) lies on top of the marginal cyto-plasm of a superficial cell. The summary effect of the staining mimics an abnormal nucleus. (**DD:** naked nucleus of a dysplastic cell.) ×400.

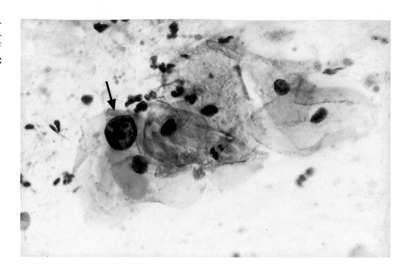

Fig. 5.**5 a Normal intermediate cells.** Distinct nu-clear swelling due to degeneration, but nuclear en-velopes are still smooth. Normal superficial cells with karyopyknosis are seen nearby. (**DD:** moderate dysplasia.) ×630.

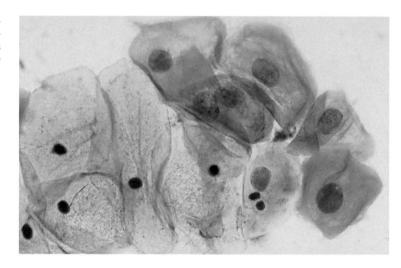

Fig. 5.**5 b Moderate metaplastic dysplasia.** Vac-uolation of the cytoplasm and abnormal nuclei with first signs of degeneration. (**DD:** glandular dys-plasia.) ×400.

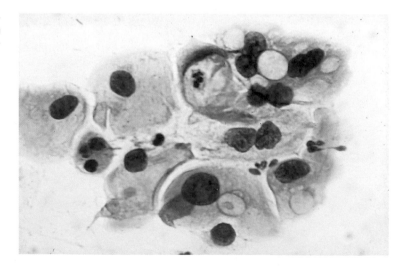

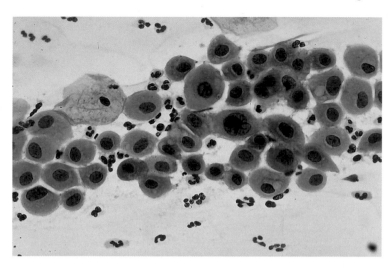

Fig. 5.**6 a** **Parabasal cells in a postpartum smear.** Strong basophilic staining and well-defined cell borders. The nuclei are active, enlarged, and hyperchromatic, and the cytoplasm is partly vacuolated. (**DD:** moderate dysplasia.) ×400.

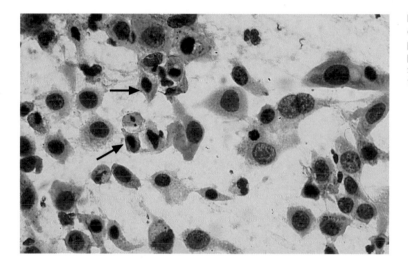

Fig. 5.**6 b** **Large-cell squamous carcinoma of the cervix.** The dissociated polymorphic cells show partly still intact, partly degenerated nuclei and partly keratinized cytoplasm (⟶). The cells vary in size, and the cytoplasmic borders are often ill-defined. (**DD:** severe dysplasia.) ×630.

Cell degeneration may also cause swelling of the nuclei. These nuclei lose their chromatin structure and show reduced staining, i. e., they become **hypochromatic**, a feature that differentiates them from those of intact dysplastic cells (Fig. 5.**5 a, b**).

Atrophic postpartal cells sometimes show intense cytoplasmic staining, enlarged and hyperchromatic nuclei, and a shift in the nucleocytoplasmic ratio. The resulting cell image resembles that of severe dysplasia (Fig. 5.**6 a, b**).

Parabasal cells that show hyaline degeneration and thus resemble small horn cells are often observed in connection with **atrophic vaginitis**, although their round shape suggests that they are benign (Fig. 5.**7a**). Polymorphism of horn cells is more pronounced with carcinoma (Fig. 5.**7b**). Atrophic parabasal cells are

sometimes long and spindle-shaped and resemble cells of highly differentiated keratinizing squamous carcinoma (Fig. 5.**8 a, b**).

In old age, the smears of atrophic cells occasionally show much enlarged, and sometimes keratinized, single cells; this may be accompanied by nuclear enlargement and simultaneous degeneration. Obviously, these changes represent sporadic disturbances in cell maturation, since they are no longer detected after estrogen replacement therapy. Nevertheless, these cells are frequently confused with cells derived from dysplasia or carcinoma (Figs. 5.**9 a, b**–5.**12 a, b**; see also Fig. 3.**18**, p. 62).

Atrophic cells present in smears taken after vaginal estrogen application may, in rare cases, exhibit accelerated maturation leading to miniature superficial cells (see Fig. 3.**19**, p. 63),

5

Fig. 5.**7 a** **Severe inflammation.** Numerous parabasal cells with enlarged, hyperchromatic nuclei and coarse chromatin structure. The cytoplasm has well-defined borders and partly shows hyaline degeneration (→). The background of the preparation indicates inflammation. (**DD:** squamous carcinoma.) ×250.

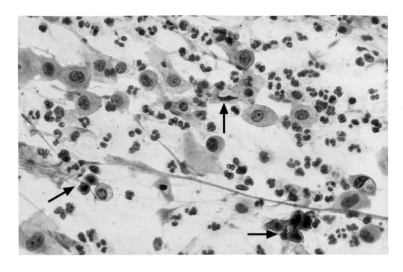

Fig. 5.**7 b** **Invasive squamous carcinoma of the cervix.** The dissociated polymorphic cells have predominantly condensed, hyperchromatic nuclei. The cytoplasm shows signs of keratinization. Normal superficial cells are visible at the top. (**DD:** inflammation.) ×400.

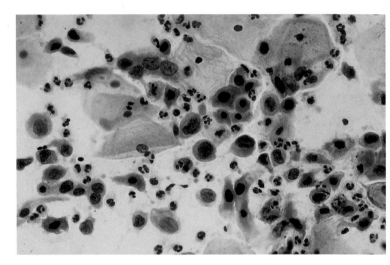

Fig. 5.**8 a** **Atrophic spindle-shaped parabasal cells** in a loose, polarized cluster, exhibiting isomorphic nuclei with regular chromatin structure. (**DD:** keratinizing squamous carcinoma.) ×400.

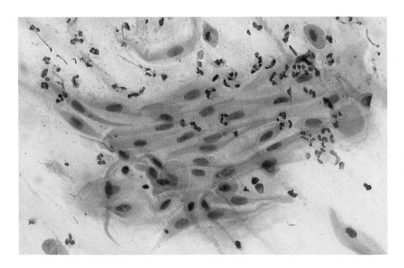

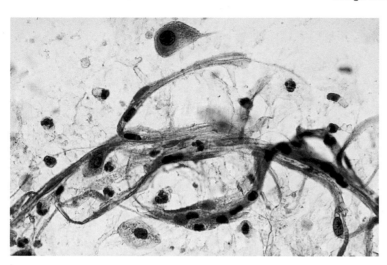

Fig. 5.**8 b** **Keratinizing squamous carcinoma of the cervix.** Abnormal spindle cells in a loose irregular cluster. The cells exhibit hyperchromatic, condensed, and pyknotic nuclei. (**DD:** fibrocytes.) ×400.

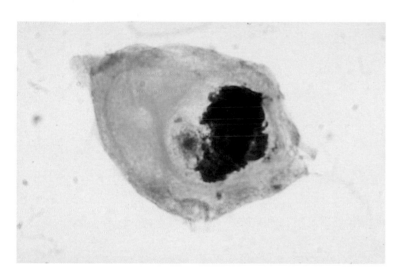

Fig. 5.**9 a** **Disturbed maturation of a parabasal cell** associated with atrophy. The much enlarged, hyperchromatic nucleus shows karyorrhexis. (**DD:** severe dysplasia.) ×1250.

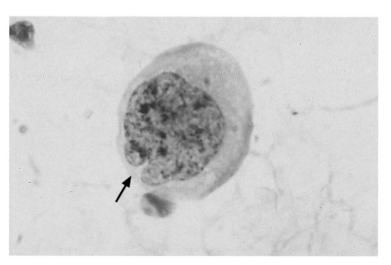

Fig. 5.**9 b** **An abnormal squamous cell** associated with cervical carcinoma. The much enlarged, irregularly structured nucleus shows a cytoplasmic invagination (⟶). (**DD:** disturbed cell maturation.) ×1250.

5

Fig. 5.**10 a** **Disturbed maturation of parabasal cells** associated with senile atrophy. The nuclei are enlarged and condensed. (**DD:** severe dysplasia.) ×630.

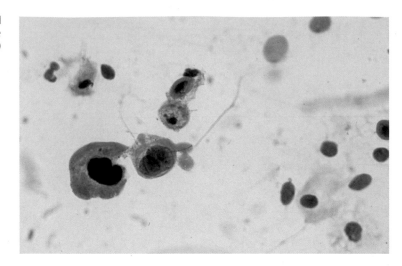

Fig. 5.**10 b** **Moderate and severe dysplasia.** Binuclear cell with hyperchromatic nuclei and signs of degeneration in the form of karyorrhexis. (**DD:** disturbed cell maturation.) ×630.

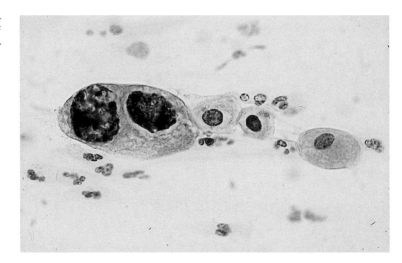

Fig. 5.**11 a** **Degenerated nuclei** associated with atrophy. The nucleus in the center shows pronounced karyorrhexis and resembles a mitotic figure (⟶). (**DD:** severe dysplasia.) ×1250.

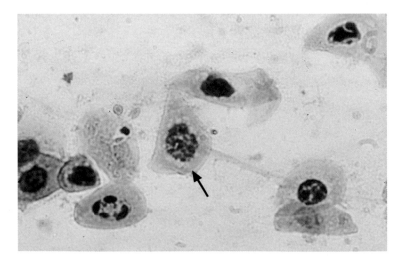

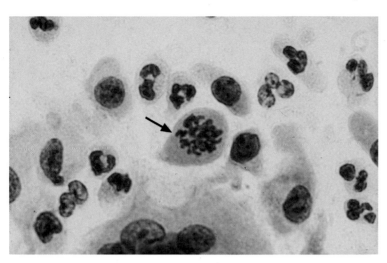

Fig. 5.**11 b Abnormal glandular cells** associated with uterine corpus carcinoma. The cell in the center is undergoing mitosis (⟶). (**DD:** degenerated parabasal cells.) ×1250.

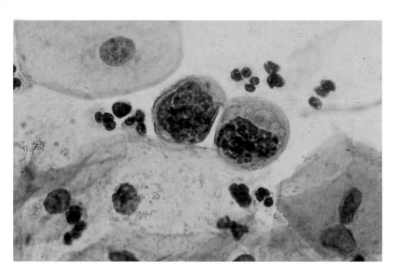

Fig. 5.**12 a Karyorrhexis with remnants of coarse chromatin** associated with severe dysplasia. Normal intermediate cells are visible nearby. (**DD:** degenerated basal cells.) ×1000.

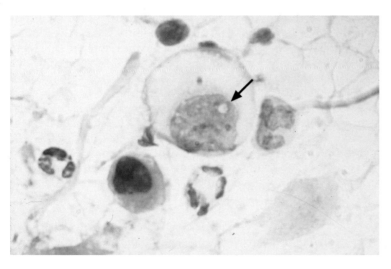

Fig. 5.**12 b Karyorrhexis** associated with squamous carcinoma of the cervix. The nucleus in the center shows signs of disintegration resembling a hole (⟶). (**DD:** degenerated parabasal cells.) ×1250.

5

Fig. 5.**13 a** **Blue blobs** associated with atrophy. The hyperchromatic mucus condensations on top of parabasal cells mimic abnormal naked nuclei (→). (**DD:** naked nuclei associated with squamous carcinoma or adenocarcinoma.) ×500.

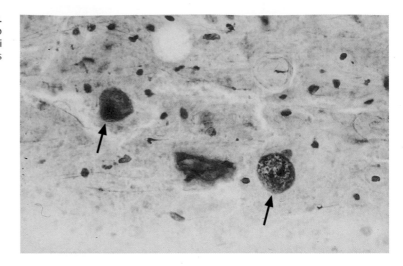

Fig. 5.**13 b** **Abnormal naked nuclei** associated with uterine corpus carcinoma. Hyperchromatic nuclei with coarse chromatin structure (→) (**DD:** blue blobs.) ×1000.

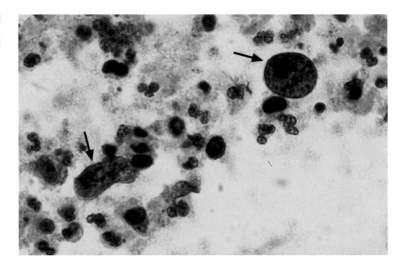

thus creating an image resembling that of dyskeratocytes or mild dysplasia.

Areas of condensed mucus ("blue blobs") observed on top of parabasal cells during atrophy are easily confused with naked nuclei of adenocarcinoma or squamous carcinoma (Figs. 5.**13 a, b**, 5.**14 a, b**).

Degenerating cells with vacuoles may strongly resemble koilocytes or cells with vacuoles due to chlamydial infection (Figs. 5.**15 a, b**, 5.**16 a, b**).

Technical Factors

Technical damage to the cell preparation may also give rise to misinterpretations. **Fixation errors** and **staining errors** may cause pseudoeosinophilia and nuclear swelling, thus mimicking HPV infection or even mild dysplasia. Dehydration artifacts sometimes cause structural changes to the cytoplasm, and these resemble indirect signs of HPV infection and, in particular, the "cracked cytoplasm" phenomenon (Fig. 5.**17 a, b**).

Improper **smear techniques** cause squeezing of the cells and lead to longitudinal extension of cells and nuclei, thus making the cells look like dysplastic spindle cells (Fig. 5.**18 a, b**).

Occasionally, there is also retrograde **transmission** of horn cells from the vulva into the vagina (Fig. 5.**18 c**).

Even years or decades after irradiation of the genital region, smears show actinic cell changes (gigantism or bizarre cytoplasmic processes), degenerative or inflammatory cells in the background, pseudoeosinophilia, amphophilia, and phagocytosis. Although the nu-

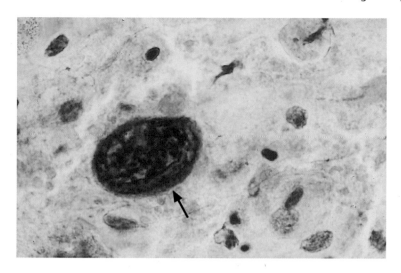

Fig. 5.**14 a** A **blue blob** associated with atrophy. The hyperchromatic, irregular mucus condensation (⟶) on top of a parabasal cell mimics an abnormal naked nucleus. (**DD:** naked nucleus associated with squamous carcinoma or adenocarcinoma.) ×1000.

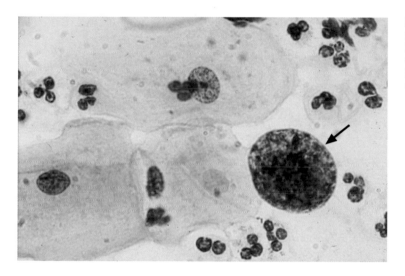

Fig. 5.**14 b An abnormal naked nucleus** associated with squamous carcinoma of the cervix. Hyperchromatic nucleus with irregular, coarse chromatin structure (⟶). Also shown are normal intermediate cells. (**DD:** reactive enlargement of endocervical nucleus.) ×1000.

clei are mostly degenerate in spite of their enlargement, problems with differentiating radiation effects from dysplastic or even carcinomatous processes may arise (Fig. 5.**19 a, b**).

Smears taken during or after **cytostatic chemotherapy** frequently show enlarged, hyperchromatic, and condensed nuclei, thus resembling those of dysplastic lesions (Figs. 5.**20 a, b**, 5.**21 a, b**).

Macrocytosis resulting from **folic acid deficiency** usually involves proportional enlargement of the nucleus (Fig. 5.**22a**), but it may be confused with cell changes induced by radiation (Fig. 5.**22b**) or HPV infection.

Inflammation

The most common error is the classification of **inflammatory reactions** of the squamous epithelium as dysplastic processes, as they may be associated with considerable nuclear changes, particularly hyperchromatism and coarse chromatin structure (Figs. 5.**23**, 5.**24 a, b**). In case of doubt, the following signs indicate benign rather than malignant changes:

▶ Perinuclear halos
▶ Regular chromatin structure
▶ Hyperchromatic and smooth nuclear envelopes
▶ Presence of chromocenters or nucleoli
▶ Diffuse changes of the overall cell image

However, **severe granular inflammation** may lead to such polymorphic cell images that

5

Fig. 5.**15 a** **Vacuolated cytoplasmic degeneration** of intermediate cells. The vacuoles are of various size, and the nuclei are degenerate. (**DD:** koilocytosis.) ×1000.

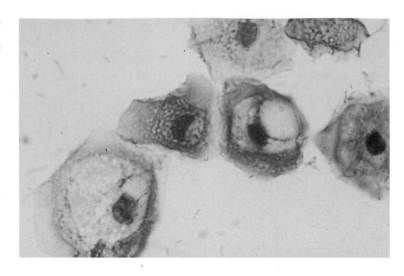

Fig. 5.**15 b** **Koilocytosis associated with mild dysplasia.** The nuclei show signs of degeneration. Normal superficial cells are visible on the right. (**DD:** cytoplasmic degeneration.) ×400.

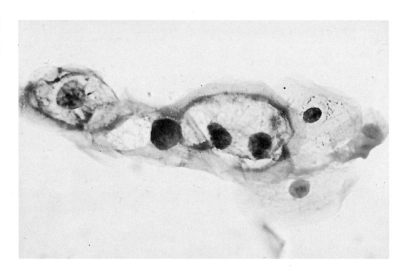

Fig. 5.**16 a** **Vacuolated cytoplasmic degeneration** of intermediate cells with condensed nuclei. (**DD:** koilocytosis.) ×630.

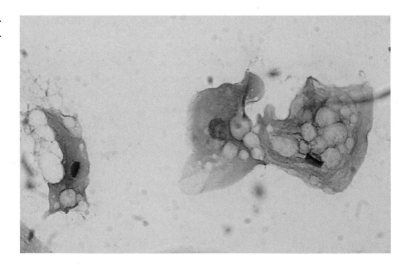

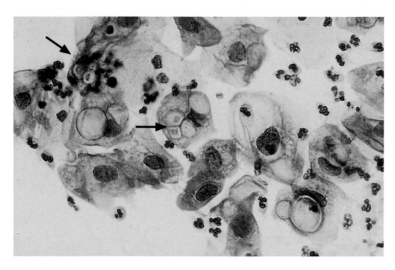

Fig. 5.**16 b** **Cytoplasmic vacuolation of parabasal cells** associated with chlamydial infection. Several intravacuolar inclusion bodies are visible (⟶). (**DD:** koilocytosis.) ×500.

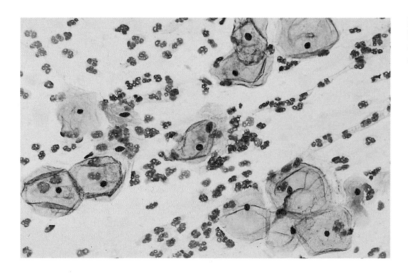

Fig. 5.**17 a** **Cytoplasmic degeneration.** Thickening of the cell borders, translucency of the central cytoplasm, and pseudoeosinophilia. (**DD:** cracked cytoplasm associated with HPV infection.) ×320.

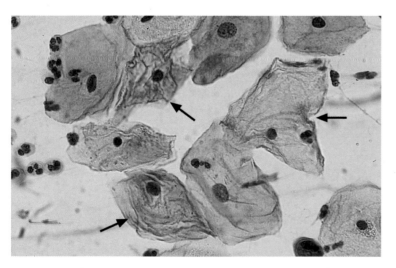

Fig. 5.**17 b** **Cracked cytoplasm of superficial cells.** Intracytoplasmic ridges associated with suspected HPV infection (⟶). Some normal superficial cells are also present. (**DD:** cytoplasmic degeneration.) ×630.

5

Fig. 5.**18 a** **Squeezed cells** caused by abrupt smear technique. The artificial deformation of cells leads to spindle-shaped cytoplasm and elongated nuclei. (**DD:** abnormal spindle cells.) ×400.

Fig. 5.**18 b** **Abnormal spindle cells** associated with mild vulvar dysplasia. The nuclei are slightly enlarged and degenerated. (**DD:** surface effect.) ×400.

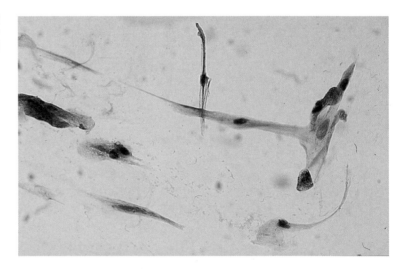

Fig. 5.**18 c** **Dyskeratotic cell cluster** (⟶) associated with retrograde transmission of vulvar carcinoma cells into the atrophic cervical smear. Intermediate cells and parabasal cells are visible in the background. ×400.

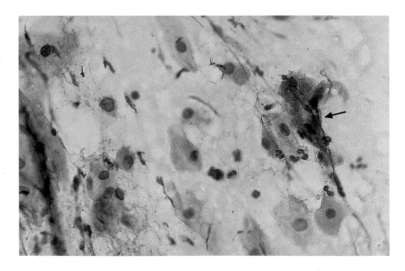

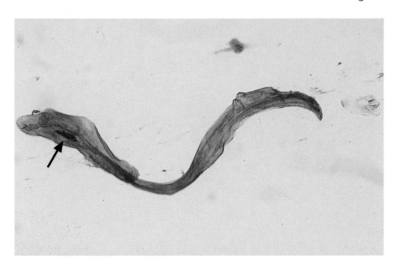

Fig. 5.**19a** **Macrocytotic spindle cell** after radio-therapy, showing mild amphophilia and karyo-pyknosis (→). (**DD:** squeezed cells.) ×320.

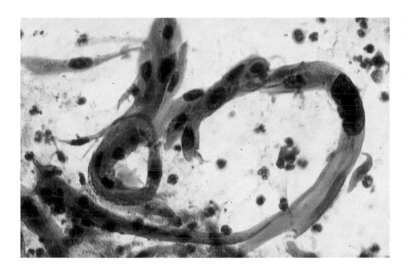

Fig. 5.**19b** **Macrocytotic abnormal spindle cells** associated with keratinizing squamous carcinoma of the cervix. The nuclei are much enlarged, hyper-chromatic, and polymorphic. (**DD:** cell image after radiotherapy.) ×250.

differentiation from squamous carcinoma is no longer possible. On the other hand, it may also happen that a malignant process becomes masked by concurrent inflammation (Fig. 5.**25a, b**).

An inflammation-induced **tendency to ker-atinize** is often difficult to distinguish from HPV-induced dyskeratosis, keratinizing squa-mous carcinoma, or adenoacanthoma (Figs. 5.**26a, b**–5.**28a, b**).

Trichomonads may look exactly like degenerating denucleated parabasal cells (Fig. 5.**29a, b**).

Vacuoles caused by *Chlamydia* occasionally show similarities to dysplastic cell-in-cell con-figurations (Fig. 5.**30a, b**).

Herpes simplex infection causes numerous cellular phenomena that may lead to misinter-pretation. The size and hyperchromatism of nuclei may reach proportions similar to those in carcinomas or other malignancies (Figs. 5.**31a, b**, 5.**32a, b**), and some virus-induced giant cells resemble postradiation cells (Fig. 5.**33a, b**).

In the context of inflammatory processes, **macrophages** that phagocytose leukocytes may look very much like cells of highly dif-ferentiated adenocarcinoma (Figs. 5.**34a, b**, 5.**35a, b**).

In addition, it can sometimes be problematic to differentiate **histiocytic giant cells** from malignant processes (Figs. 5.**36a, b**, 5.**37a, b**).

Certain nonepithelial cells that appear during chronic inflammation, such as **fibro-cytes, fibroblasts**, or **smooth muscle cells**, re-semble abnormal squamous epithelial cells because of their spindle-shaped structure (Figs. 5.**38a, b**–5.**42a, b**).

5

Fig. 5.**20 a Cell image after chemotherapy.** Intermediate cells show enlarged nuclei with partial, central condensation (——>). (**DD:** moderate dysplasia.) ×250.

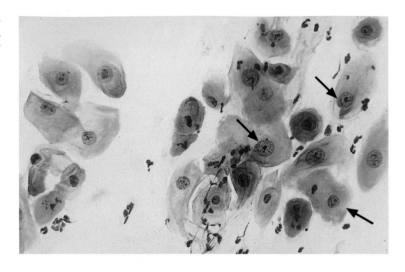

Fig. 5.**20 b Severe dysplasia.** The nuclei are enlarged, hyperchromatic, and their chromatin structure is irregular. Individual cells are polynuclear and show signs of keratinization (——>). (**DD:** postradiation dysplasia.) ×630.

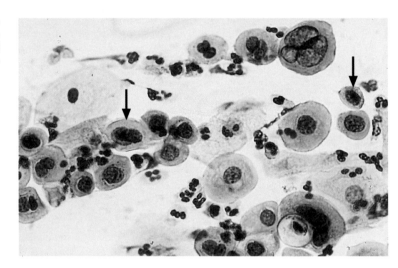

Fig. 5.**21 a Cell image indicating folic acid deficiency.** The enlarged cells show cytoplasmic vacuolation and proportional nuclear enlargement, and some are binuclear. (**DD:** cell image after radiotherapy.) ×400.

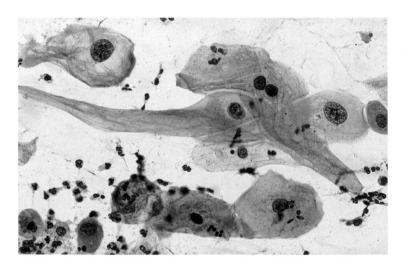

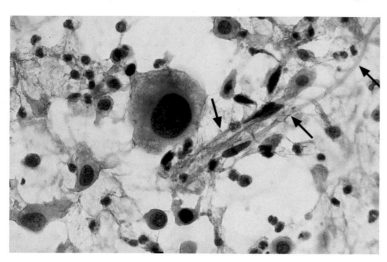

Fig. 5.**21 b** **Keratinizing squamous carcinoma of the cervix.** The cells and their nuclei vary considerably in size. The abnormal nuclei are partly still intact, partly condensed. Spindle cells are also visible (⟶). (**DD:** cell image after radiotherapy.) ×400.

Fig. 5.**22 a** **Cell image indicating folic acid deficiency.** The cells are enlarged and show proportional nuclear enlargement. (**DD:** cell image after radiotherapy.) ×630.

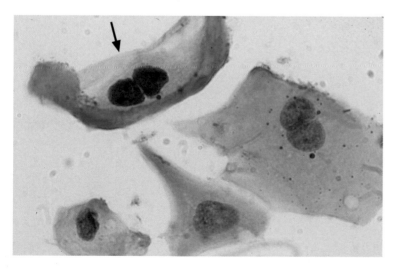

Fig. 5.**22 b** **Postradiation dysplasia.** Radiation effects include cell enlargement, vacuolation of the cytoplasm, and binuclear cells. Individual cells show cytoplasmic clearing resembling that of koilocytes (⟶). The abnormal changes include enlarged, hyperchromatic, and polymorphic nuclei. (**DD:** benign radiation effects.) ×1000.

5

Fig. 5.**23 False-positive findings.** They are caused by changes in epithelial proliferation, as illustrated here for **mild dysplasia**. Parabasal cells exfoliate as a result of insufficient proliferation in case of atrophy, or as a result of excessive proliferation in case of inflammation. If there is a simultaneous infection with high-risk types of HPV, maturation is no longer possible. As a result, only immature dysplastic cells exfoliate, although this is only a CIN I lesion (mild dysplasia).

Cytology	Parabasal cell	Immature dysplastic cell	Super-ficial cell	Mature dysplastic cell	Para-basal cell	Immature dysplas-tic cell
Superficial cell layer			◇	◆	●	★
Intermediate cell layer			●	★	●	★
Intermediate cell layer			●	★	●	●
Parabasal cell layer	●	★	●	★	●	●
Basal cell layer	●	★	●	★	●	★
Histology	Atrophy	Atrophy and mild dysplasia	Normal	Mild dysplasia	Inflam-mation	Inflam-mation and mild dysplasia

Fig. 5.**24a Acute inflammation.** Intermediate cells with enlarged, hyperchromatic nuclei and coarse chromatin structure. The background of the preparation indicates inflammation. (**DD:** moderate dysplasia.) ×500.

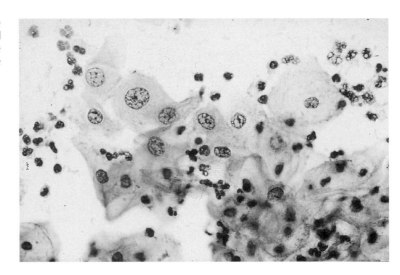

Fig. 5.**24b Mild dysplasia.** Enlarged and hyperchromatic nuclei with irregular chromatin structure and first signs of nuclear condensation. (**DD:** inflammatory reaction.) ×630.

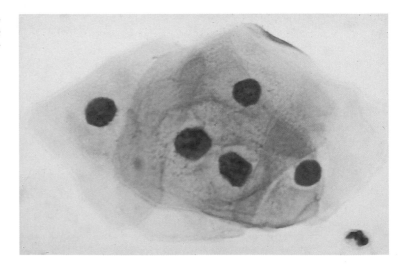

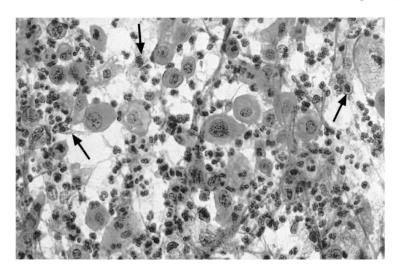

Fig. 5.**25 a**　**Severe inflammation.** Increased numbers of parabasal cells with much-enlarged, hyperchomatic nuclei, increased nucleocytoplasmic ratio, and coarse chromatin structure. The polymorphism includes anisocytosis and anisokaryosis, and some cells show signs of keratinization (⟶). The background of the preparation indicates inflammation. (**DD:** squamous carcinoma.) ×250.

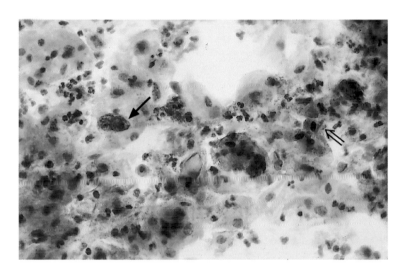

Fig. 5.**25 b**　**Invasive squamous carcinoma of the cervix.** Polymorphic dysplastic cells. An enlarged, hyperchromatic nucleus is seen on the left (⟶) and an abnormal keratinizing cell on the right (⟹). The background of the preparation indicates severe inflammation. (**DD:** inflammation.) ×250.

Epithelial regeneration may be associated with considerable nuclear enlargement and macronucleolus formation. This creates such a polymorphic character that confusion with poorly differentiated malignancies is possible (Figs. 5.**43 a, b**–5.**46 a, b**).

The surface effect caused by regeneration usually produces spindle-shaped horn cells (parakeratosis), the nuclei of which are not enlarged, unlike those seen in dysplasia and carcinoma (Figs. 5.**47 a, b**, 5.**48 a, b**). The cells are sometimes arranged like layers of an onion, thus forming keratin pearls. This makes it difficult to distinguish them from cells changed by HPV or from keratinized squamous carcinoma cells (Fig. 5.**49 a, b**).

5

Fig. 5.26 a Inflammatory reaction. Keratinized superficial cells with slightly enlarged nuclei. The background of the preparation indicates inflammation. (**DD:** keratinizing dysplasia.) ×500.

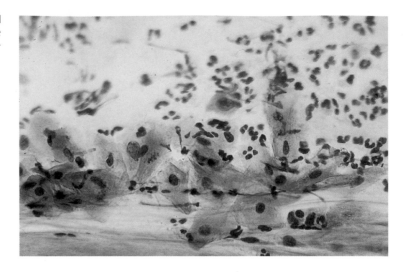

Fig. 5.26 b Dyskeratocytes associated with HPV infection. Small keratinized cells with enlarged, hyperchromatic, and condensed nuclei (⟶). Normal superficial cells are visible nearby. (**DD:** surface effect.) ×630.

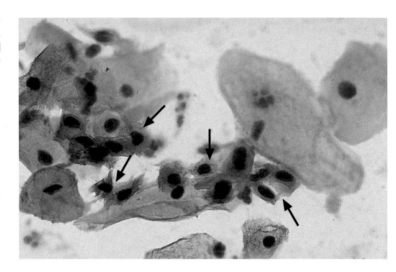

Fig. 5.27 a Acute inflammation. Pseudoeosinophilic cells with signs of keratinization, especially in the center (⟶). The cell borders are partly blurred, and the nuclei are enlarged and hyperchromatic, while some are degenerated. (**DD:** mild keratinizing dysplasia.) ×400.

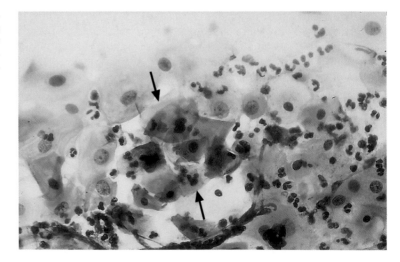

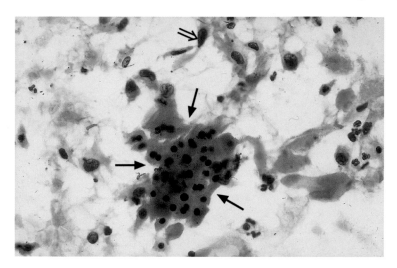

Fig. 5.**27 b Squamous carcinoma cells associated with invasive cervical carcinoma.** The cells in the center show signs of keratinization (→). By contrast, the surrounding cells show basophilic staining and are partly spindle-shaped (⟹). All nuclei are hyperchromatic and condensed. (**DD:** keratinizing dysplasia.) ×400.

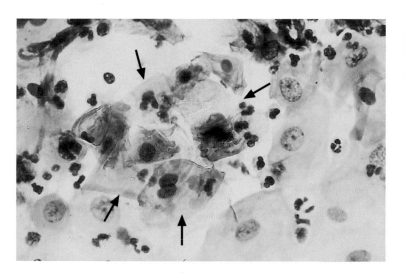

Fig. 5.**28 a Acute inflammation.** The pseudo-eosinophilic cells in the center of the image have blurred cell borders (→). The nuclei of the surrounding intermediate cells are enlarged and partly exhibit nucleoli. (**DD:** moderate dysplasia.) ×630.

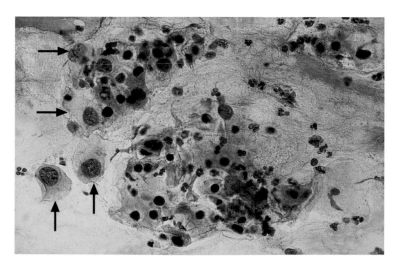

Fig. 5.**28 b Adenocarcinoma of the endocervix** with a squamous epithelial component. In addition to abnormal glandular cells (→), there are keratinizing squamous cells with hyperchromatic, condensed nuclei. (**DD:** severe keratinizing dysplasia.) ×400.

5

Fig. 5.**29a** **Trichomonadal vaginitis.** Moderately well-preserved pathogens on the left (→) and normal superficial cells on the right. (**DD:** degenerated parabasal cells.) ×250.

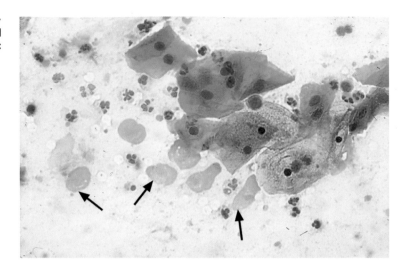

Fig. 5.**29b** **Degenerated parabasal cells** associated with atrophy. The nuclei show severe signs of degeneration and even signs of complete lysis. The background of the preparation indicates inflammation. A few preserved parabasal cells are visible on the lower right (→). (**DD:** trichomonadal vaginitis.) ×400.

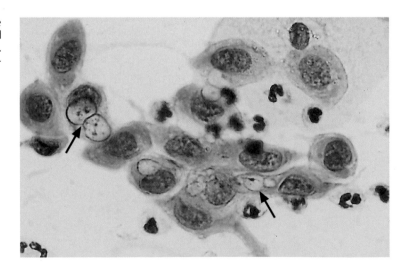

Fig. 5.**30a** **Chlamydial infection.** Immature metaplastic cells containing poorly defined vacuoles of various size and inclusion bodies (→). The nuclei are active and enlarged. (**DD:** severe dysplasia.) ×1000.

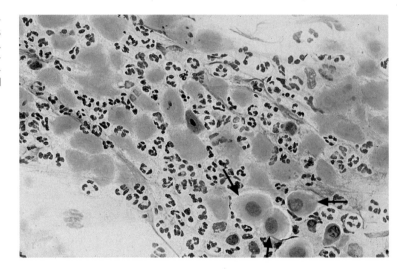

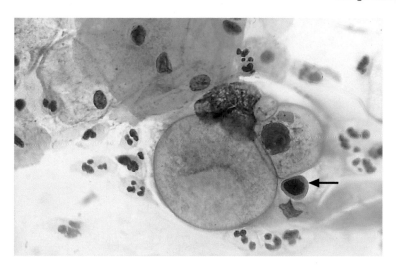

Fig. 5.**30 b** **Severe dysplasia of the cervix.** Vacuolation and cell-in-cell configuration. The nuclei are enlarged and hyperchromatic or condensed (—►). Normal superficial cells are visible at the top. (**DD:** macrophages.) ×400.

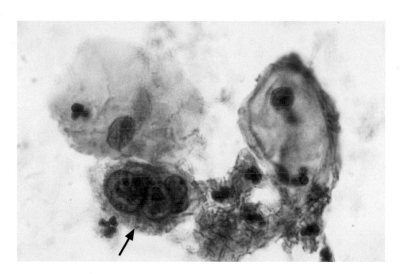

Fig. 5.**31 a** **Herpes simplex infection.** Polynuclear parabasal cell with much-enlarged nuclei and hyperchromatic nuclear envelope (—►). Normal intermediate cells are seen nearby. (**DD:** severe dysplasia.) ×1000.

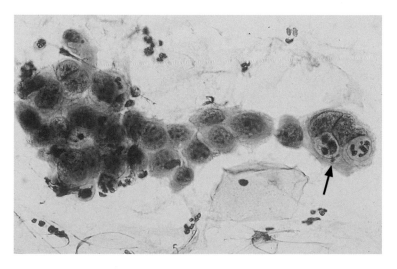

Fig. 5.**31 b** **Invasive squamous carcinoma of the cervix.** An irregular cluster of abnormal cells with enlarged and hyperchromatic nuclei, irregular chromatin structure, and occasional presence of nucleoli. Phagocytosing cells are visible on the right (—►). A normal superficial cell is seen at the bottom. (**DD:** regenerative epithelium, ×500).

5

Fig. 5.**32 a** **Herpes simplex infection.** A polynuclear giant cell showing nuclear molding but no recognizable chromatin structure. (**DD:** cluster of abnormal glandular cells.) ×1000.

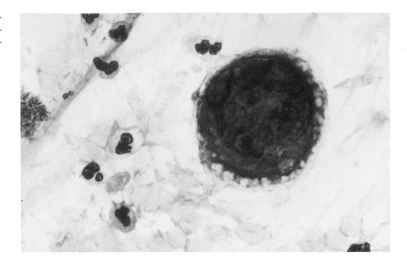

Fig. 5.**32 b** **Malignant melanoma of the cervix.** Intermediate and parabasal cells with moderately enlarged and hyperchromatic nuclei. A parabasal cell is loaded with coarse melanin pigment granules (⟶). (**DD:** moderate dysplasia.) ×630.

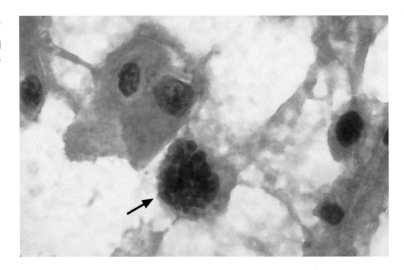

Fig. 5.**33 a** **Herpes simplex infection.** Polynuclear giant cell with many irregular hyperchromatic nuclei that exhibit signs of degeneration. Normal superficial cells are visible nearby. (**DD:** cell image after radiotherapy.) ×200.

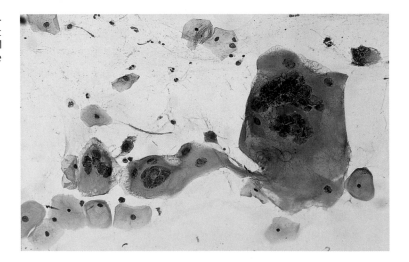

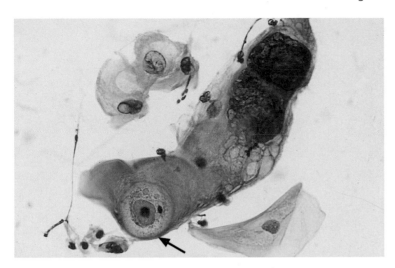

Fig. 5.**33 b Postradiation dysplasia.** A polynuclear giant cell with extremely enlarged, hyperchromatic nuclei and irregular chromatin structure. A cell-in-cell configuration (bird's eye cell) is visible in the lower part (→). (**DD:** cervical carcinoma.) ×500.

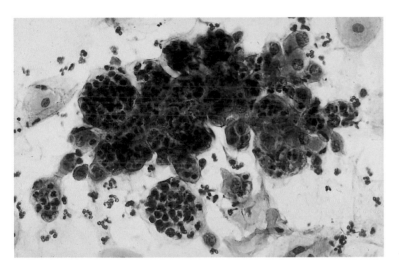

Fig. 5.**34 a A cluster of macrophages associated with chronic inflammation.** The macrophages have incorporated numerous leukocytes. Normal intermediate cells are visible at the periphery of the cluster. (**DD:** adenocarcinoma.) ×250.

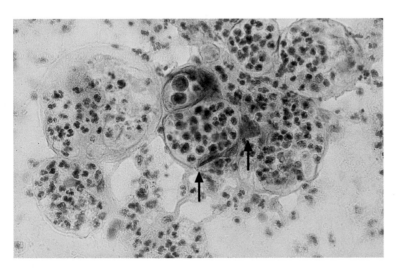

Fig. 5.**34 b Abnormal phagocytosing glandular cells** associated with uterine corpus carcinoma. The cells have incorporated numerous leukocytes. The nuclei are hypochromatic and pushed to the cell margin (→). (**DD:** macrophages.) ×400.

5

Fig. 5.**35 a** **A macrophage associated with chronic inflammation.** It has incorporated masses of leukocytes, and its nuclei are pushed to the cell margin (→). (**DD:** adenocarcinoma of the endocervix.) ×1000.

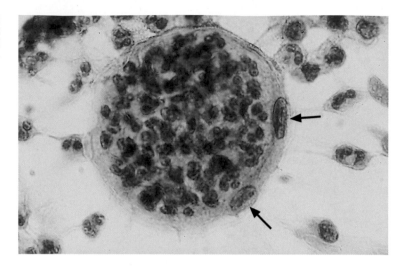

Fig. 5.**35 b** **Adenocarcinoma of the endocervix.** A compact cluster of abnormal glandular cells with irregular, three-dimensional arrangement of nuclei that are slightly hyperchromatic and exhibit micronucleoli. (**DD:** regenerative epithelium.) ×1000.

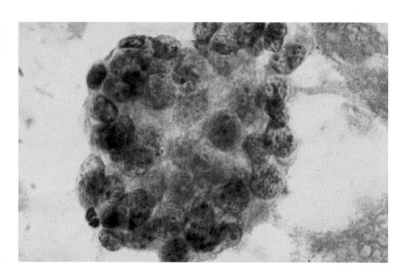

Fig. 5.**36 a** **A histiocytic giant cell** with numerous isomorphic, hyperchromatic, and condensed nuclei in a two-dimensional arrangement. (**DD:** adenocarcinoma.) ×630.

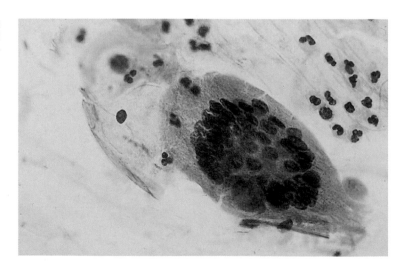

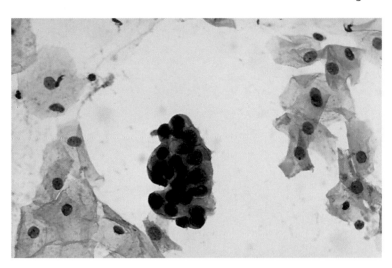

Fig. 5.**36 b** **Ovarian carcinoma.** An irregular three-dimensional cluster of abnormal glandular cells with enlarged, hyperchromatic, and condensed nuclei. Normal intermediate cells are visible nearby. (**DD:** cluster of degenerated endocervical cells.) ×400.

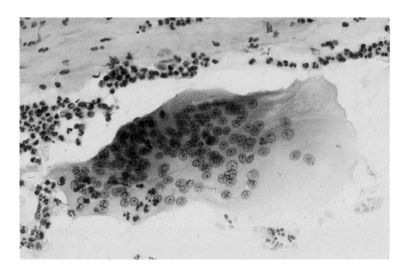

Fig. 5.**37 a** **A histiocytic giant cell.** The cytoplasm is amphophilic, and there are masses of regularly structured, nucleolus-containing nuclei in a two-dimensional arrangement. (**DD:** adenocarcinoma.) ×250.

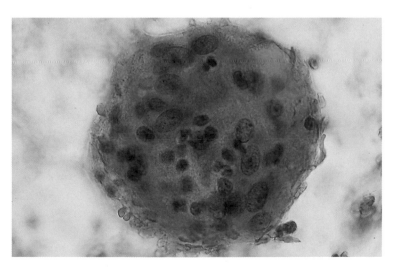

Fig. 5.**37 b** **Abnormal polynuclear giant cell** associated with adenocarcinoma of the endocervix. The numerous, irregularly arranged nuclei are partly intact, partly degenerated. (**DD:** histiocytic giant cell.) ×630.

5

Fig. 5.**38 a** **Fibrocytes associated with chronic inflammation.** The cell body is elongated and spindle-shaped, and the nuclei are intact and hyperchromatic (⟶). The background of the preparation shows lactobacillus-induced cytolysis. (**DD:** squamous carcinoma.) ×300.

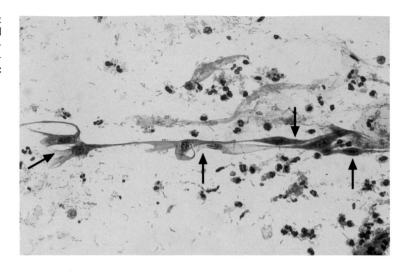

Fig. 5.**38 b** **Leiomyosarcoma of the uterine corpus.** The degenerated abnormal spindle cells are irregularly arranged and show hyperchromatic, condensed nuclei in a three-dimensional arrangement. (**DD:** fibrocytes.) ×400.

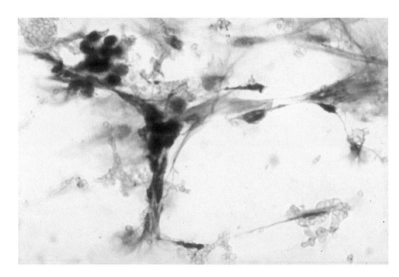

Fig. 5.**39 a** **Fibrocytes** with spindle-shaped cell body and regularly structured nucleus (⟶). The background of the preparation shows lactobacillus-induced cytolysis. (**DD:** keratinizing squamous carcinoma.) ×400.

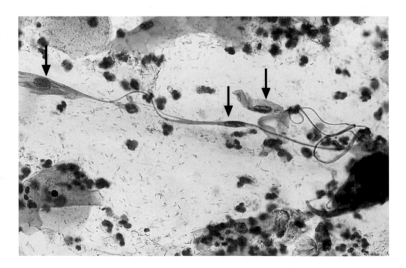

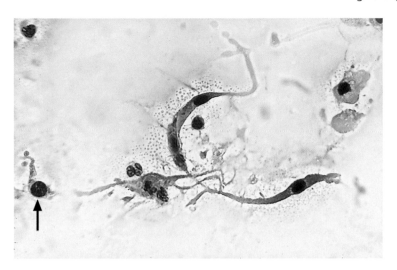

Fig. 5.**39 b** **Abnormal spindle cells** associated with keratinizing squamous carcinoma of the cervix. The cytoplasm is amphophilic, and the nuclei are hyperchromatic and condensed. An intact nucleus is visible on the lower left (→). There is blood in the background of the preparation. (**DD:** fibrocytes.) ×400.

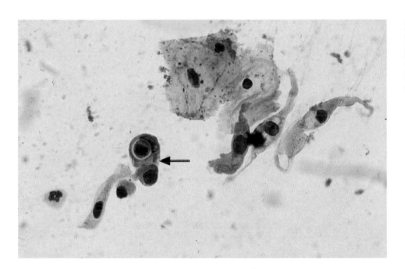

Fig. 5.**40 a** **Fibroblasts** associated with chronic inflammation. The cells are relatively short and plump, and the nuclei are partly intact, partly condensed. A cell-in-cell configuration is present on the left (→). Normal intermediate cells are visible at the top. (**DD:** squamous carcinoma.) ×630.

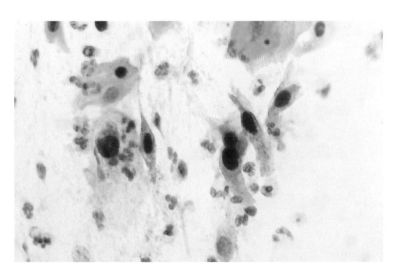

Fig. 5.**40 b** **Abnormal spindle cells** associated with severe cervical dysplasia. The nuclei are enlarged, hyperchromatic, and mostly condensed. The background of the preparation indicates inflammation. (**DD:** fibroblasts.) ×400.

5

Fig. 5.**41 a** **A smooth muscle cell** associated with submucous myoma. The cytoplasm is elongated and tadpole-shaped, and the small nucleus is regularly structured. Normal superficial cells are visible nearby. (**DD:** squamous carcinoma.) ×500.

Fig. 5.**41 b** **An abnormal spindle cell** associated with invasive squamous carcinoma of the cervix. The nucleus is much enlarged and has a nucleolus. More polymorphic dysplastic cells are seen at the bottom. The background of the preparation indicates inflammation. (**DD:** cell image after radiotherapy.) ×500.

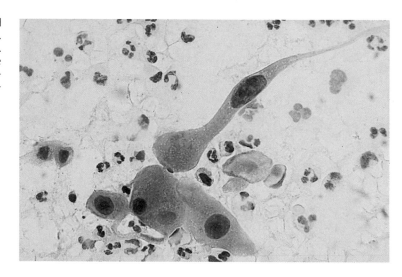

Fig. 5.**42 a** **Fibrocytes** associated with chronic inflammation. The cytoplasm is elongated and spindle-shaped. The cells vary in size, and the nuclei are partly condensed. The background of the preparation indicates inflammation. (**DD:** squamous carcinoma.) ×400.

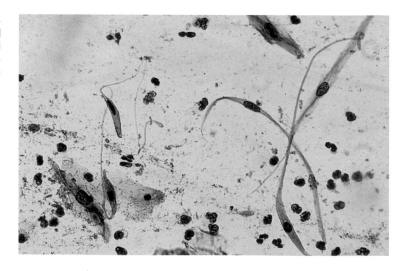

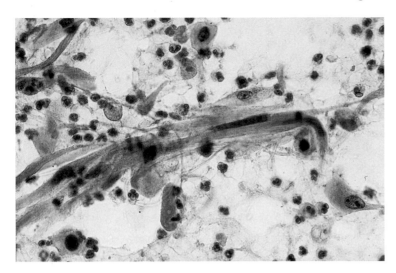

Fig. 5.**42 b** **Abnormal spindle cells** associated with keratinizing squamous carcinoma of the cervix. Their irregularly arranged nuclei are hyperchromatic. More polymorphic dyskaryotic cells are seen in the vicinity. The background of the preparation indicates inflammation. (**DD:** cell image after radiotherapy.) ×630.

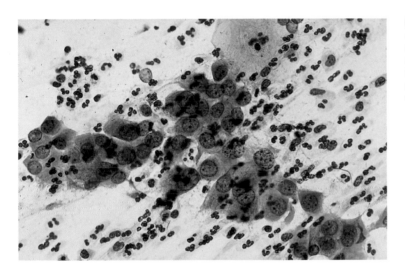

Fig. 5.**43 a** **Glandular regenerative epithelium.** A loose cell sheet with indistinct cell borders, foamy cytoplasm, and intact, hyperchromatic, nucleolus-containing nuclei. The background of the preparation indicates inflammation. A normal superficial cell is visible at the top. (**DD:** squamous carcinoma.) ×400.

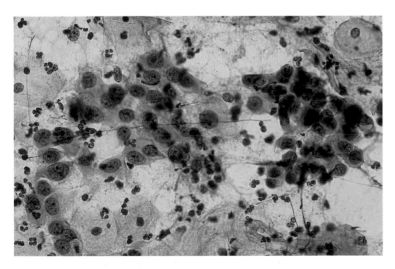

Fig. 5.**43 b** **Invasive squamous carcinoma of the cervix.** A loose cluster of abnormal cells with indistinct cell borders. The intact, hyperchromatic, and relatively isomorphic nuclei show an irregular chromatin structure and distinct nucleoli. Normal superficial cells are visible in the vicinity. (**DD:** regenerative epithelium.) ×400.

5

Fig. 5.**44 a Squamous regenerative epithelium.** Two-dimensional arrangement of cells with relatively distinct cell borders, much enlarged nuclei, and macronucleolus formation. Phagocytotic activity is visible at the bottom (→). (**DD:** squamous carcinoma or adenocarcinoma.) ×1000.

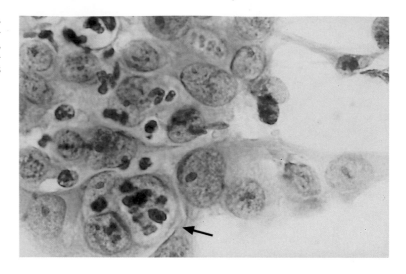

Fig. 5.**44 b Highly differentiated adenocarcinoma of the endocervix.** A syncytial cell sheet with a relatively distinct outline. The numerous irregularly arranged nuclei are active and show a relatively regular chromatin structure and macronucleolus formation. (**DD:** histiocytic giant cell.) ×630.

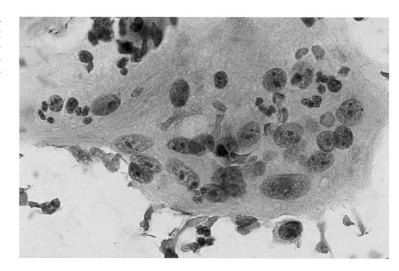

Fig. 5.**45 a Squamous regenerative epithelium.** Two-dimensional arrangement of cells with distinct borders, much enlarged nuclei, and macronucleolus formation. Normal superficial cells are visible nearby. (**DD:** squamous carcinoma or adenocarcinoma.) ×630.

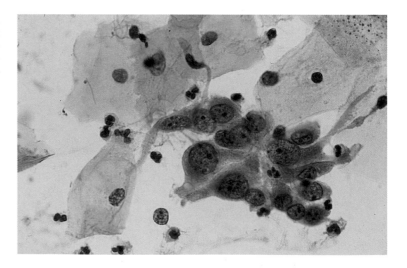

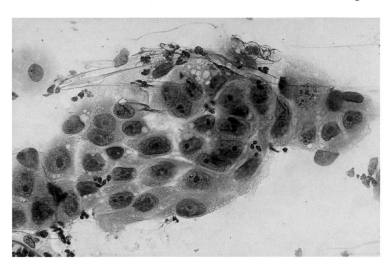

Fig. 5.**45 b** **Leiomyosarcoma of the uterine corpus.** Loose two-dimensional arrangement of cells with indistinct borders, foamy cytoplasm, and much enlarged, macronucleolus-containing nuclei. (**DD:** regenerative epithelium.) ×400.

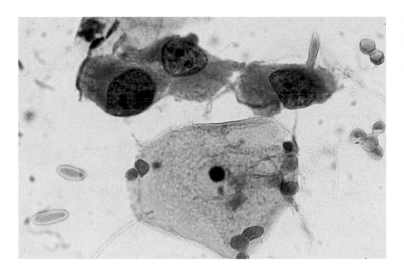

Fig. 5.**46 a** **Squamous regenerative epithelium.** Small group of cells with distinct borders and enlarged nuclei in which several eosinophilic macronucleoli are recognizable. A normal superficial cell is seen in the lower part of the picture. (**DD:** squamous carcinoma.) ×1000.

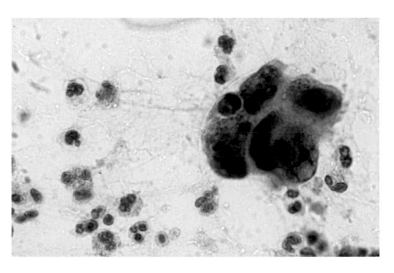

Fig. 5.**46 b** **Poorly differentiated squamous carcinoma of the cervix.** A small three-dimensional cluster of abnormal cells with distinct cell borders and much enlarged, abnormal nuclei that show signs of degeneration. (**DD:** regenerative epithelium.) ×1000.

5

Fig. 5.**47 a** **Surface effect.** Parakeratotic, orangophilic horn cells with condensed nuclei. Normal superficial cells are in the vicinity. (**DD:** keratinizing dysplastic cells.) ×400.

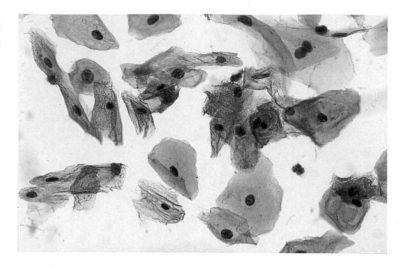

Fig. 5.**47 b** **Keratinizing dysplastic cells associated with severe cervical dysplasia.** The irregularly arranged cells have enlarged, hyperchromatic, and condensed nuclei. A few nonkeratinized immature dysplastic cells are also present (⟶). A normal superficial cell is seen on the upper left. (**DD:** surface effect.) ×500.

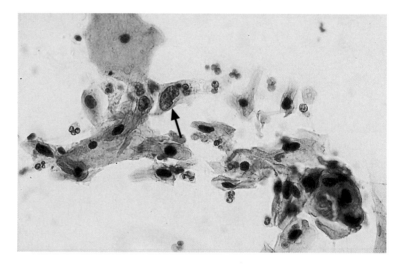

Fig. 5.**48 a** **Surface effect.** A compact cluster of spindle-shaped parakeratotic cells. Normal superficial cells are visible nearby. (**DD:** keratinizing dysplastic cells.) ×630.

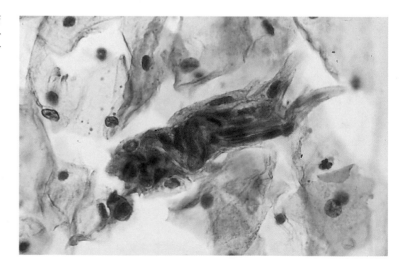

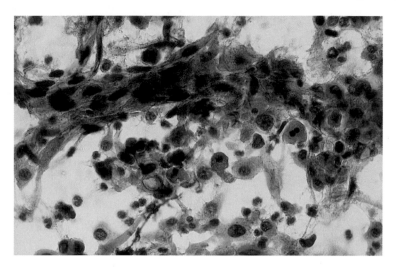

Fig. 5.**48 b** **Keratinizing squamous carcinoma of the cervix.** A group of abnormal keratinized spindle cells in irregular arrangement, showing enlarged, condensed nuclei. Numerous other polymorphic dysplastic cells are visible in the vicinity. (**DD:** severe inflammation.) ×400.

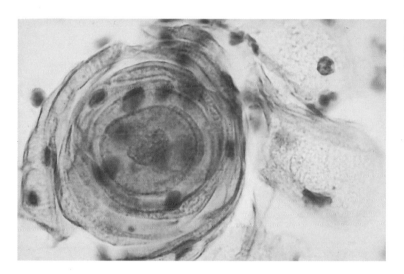

Fig. 5.**49 a** **Benign keratin pearl** caused by the regenerative surface effect. The parakeratotic cells are arranged like layers of an onion. (**DD:** keratinizing dysplastic cells.) ×630.

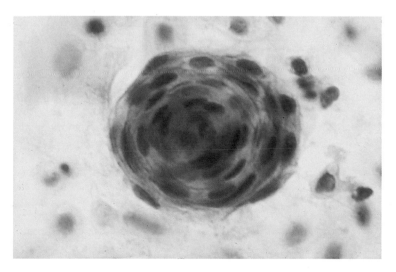

Fig. 5.**49 b** **Malignant keratin pearl** associated with keratinizing squamous carcinoma of the cervix. The abnormal spindle cells are arranged like layers of an onion, and they have enlarged, condensed nuclei. (**DD:** benign keratin pearl.) ×630.

5

Fig. 5.**50 a** **Vulvar dystrophy.** Orthokeratotic and parakeratotic cells with slightly enlarged nuclei and cytoplasm. (**DD:** keratinizing squamous carcinoma of the vulva.) ×400.

Fig. 5.**50 b** **Keratinizing squamous carcinoma of the vulva.** Severe dyskeratosis. The cells show pronounced anisocytosis and cytoplasmic polymorphism, while the condensed nuclei are only slightly enlarged. Solitary anaplastic tumor cells are also visible (——►). (**DD:** inflammatory reaction.) ×630.

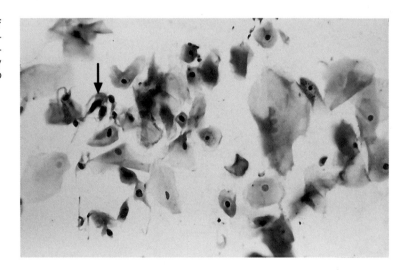

Vulva

Physiological Factors

Vulvar smears taken during the ovulation phase or during postmenopausal estrogen replacement therapy may show such an increase in the **parakeratotic index** that an inflammatory process is suspected.

Dystrophy

Dystrophic vulvar lesions exfoliate orthokeratotic and parakeratotic cells. When the horn cells are larger than normal, problems with differentiation from highly differentiated car-

cinoma may arise (Fig. 5.**50 a, b**). A wide range in cell size, however, indicates a carcinoma.

Inflammation

In vulvitis, **parakeratotic cells with enlarged nuclei** occasionally differ little from moderate dyskeratocytes derived from severe precancerous lesions. The parakeratotic cells are usually uniform in size, and detection of inflammatory cells in the background may be useful for establishing a diagnosis (Figs. 5.**51 a–c**, 5.**52 a, b**).

Vulvar condylomas may give rise to **dyskeratotic cells** that resemble those of precancerous lesions (see Fig. 3.**199**, p. 164).

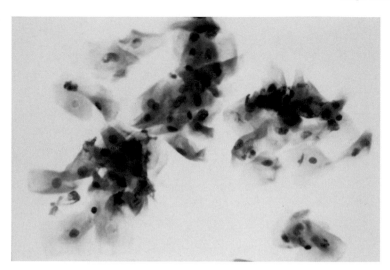

Fig. 5.**51 a** **Candida-induced vulvitis.** Diffusely arranged, isomorphic parakeratotic cells with mild nuclear swelling. (**DD:** mild dyskeratosis.) ×400.

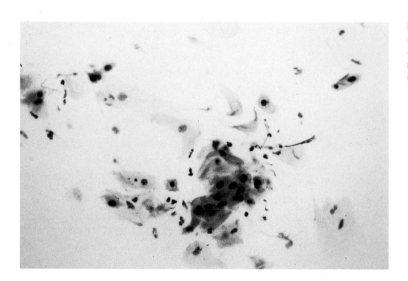

Fig. 5.**51 b** **Carcinoma in situ of the vulva.** Moderate dyskeratosis. The relatively small cells have clearly enlarged, hyperchromatic, and condensed nuclei. Normal intermediate cells are visible nearby. (**DD:** inflammatory reaction.) ×380.

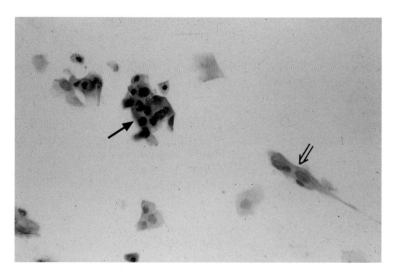

Fig. 5.**51 c** **Keratinizing squamous carcinoma of the vulva.** Moderate dyskeratosis. A cluster of relatively small cells with hyperchromatic, condensed nuclei (⟶). A solitary cell with severe dyskeratosis and increased cytoplasmic volume is also seen (⟹). (**DD:** carcinoma in situ of the vulva.) ×650.

5

Fig. 5.**52 a** **Candida-induced vulvitis.** The iso-morphic parakeratotic cells in the center clearly have swollen nuclei (—►). Admixed vaginal superficial and intermediate cells are visible nearby. The background of the preparation indicates inflammation. (**DD:** moderate dyskeratosis.) ×400.

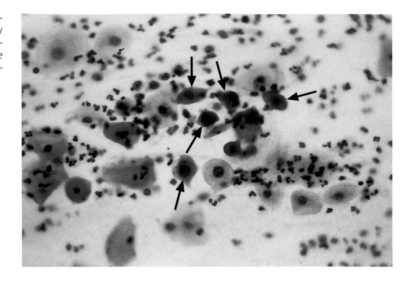

Fig. 5.**52 b** **Carcinoma in situ of the vulva.** Moderate dyskeratosis. The cells have relatively small bodies and enlarged, hyperchromatic, and condensed nuclei. (**DD:** inflammatory reaction.) ×630.

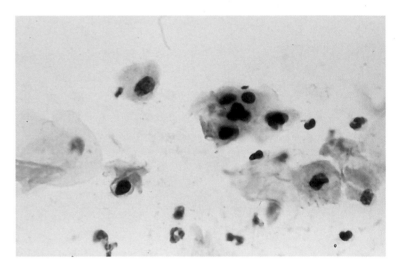

Glandular Epithelial Lesions

Differential cytological problems are slightly less common in the context of glandular epithelium than they are for squamous epithelium, but they usually have more severe consequences because a waiting attitude involving regular control smears is risky. Cells derived from glandular precancerous lesions are hard to distinguish from the cells of invasive adenocarcinoma.

Endocervical problem cells occurring during sexual maturation need to be assessed more stringently than endometrial problem cells, since adenocarcinoma of the endocervix is much less age-dependent than carcinoma of the uterine corpus, which occurs almost exclusively after the menopause.

Endocervix

Physiological Factors

Normal, well-preserved **endocervical cells** are sometimes confused with cells from microinvasive squamous carcinoma or adenocarcinoma when they show secretory activity and occur in a special arrangement. However, there are usually clear differences in the nuclear criteria, and abnormal cells generally show a higher degree of dissociation than normal cells (Figs. 5.**53 a, b**–5.**57 a, b**).

More often, degenerative and enlarged **endocervical naked nuclei** lead to confusion with precancerous or malignant processes, especially when they occur isolated in the smear. Only minor problems arise, however,

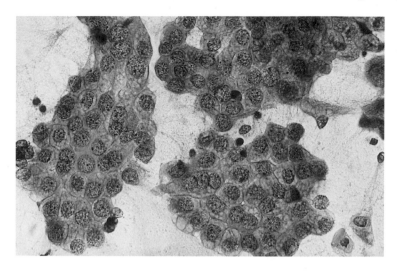

Fig. 5.**53 a Secretory endocervical cells.** A regularly structured, two-dimensional sheet of cells viewed from above. The cells exhibit slightly vacuolated cytoplasm and active, vesicular nuclei. (**DD:** highly differentiated adenocarcinoma.) ×630.

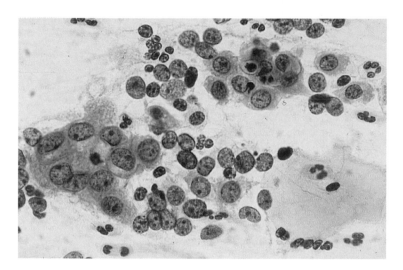

Fig. 5.**53 b Microinvasive squamous carcinoma of the cervix.** Abnormal, nucleoli-containing nuclei, seen partly in loose cell clusters and partly as naked nuclei. A normal superficial cell is visible on the lower right. (**DD:** regenerative epithelium.) ×630.

when it is possible to assign them to still intact clusters of endocervical cells (Fig. 5.**58 a, b**).

Technical Factors

Squeezing of endocervical epithelial cell clusters is common and leads to problems in distinguishing the resulting effects from malignant changes. Mechanical damage to the sensitive columnar epithelial cells occurs easily, particularly when endocervical biopsies are taken with an instrument such as the Cytobrush.

Cytoplasmic vacuolation caused by **radiation** may resemble the cell-in-cell configurations observed with dysplasia, especially when nuclear anomalies are present at the same time (Fig. 5.**59 a, b**).

5

Fig. 5.**54 a** **Secretory endocervical cells.** Regularly structured, two-dimensional sheet of cells with slightly enlarged, active nuclei and prominent nucleoli. Two normal superficial cells are visible at the bottom. (**DD:** highly differentiated adenocarcinoma.) ×630.

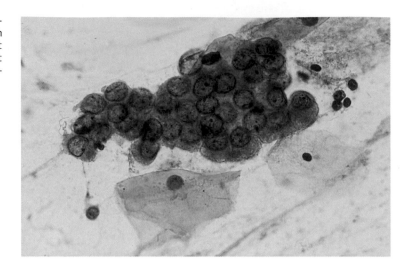

Fig. 5.**54 b** **Poorly differentiated uterine corpus carcinoma.** The abnormal naked nuclei are hyperchromatic; they show an irregular chromatin structure and prominent nucleoli. The background of the preparation shows blood cells. (**DD:** cervicitis.) ×400.

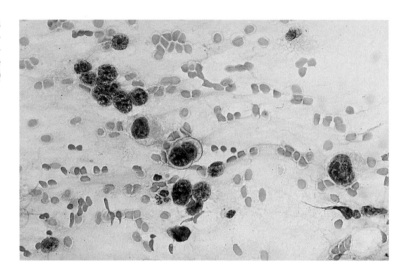

Fig. 5.**55 a** **Secretory endocervical cells.** A single layer of cells in palisade arrangement, showing slightly enlarged, basal, vesicular nuclei. (**DD:** highly differentiated adenocarcinoma.) ×630.

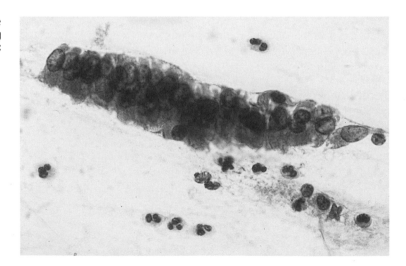

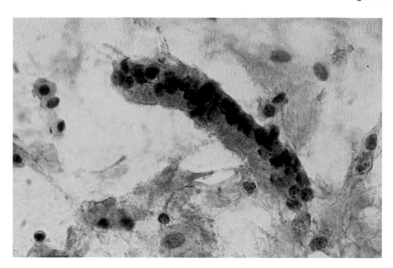

Fig. 5.**55 b** **Highly differentiated uterine corpus carcinoma.** Palisade arrangement of abnormal glandular cells showing pseudostratification, anisonucleosis, and hyperchromatic nuclei. (**DD:** cervicitis.) ×630.

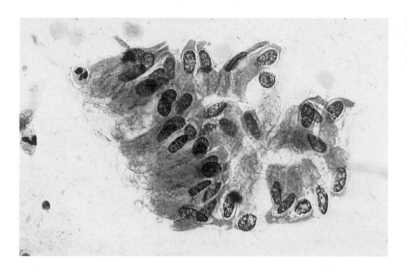

Fig. 5.**56 a** **Endocervical peg cells.** An irregular, loose cluster of cells with indistinct cytoplasmic borders and basal nuclei. (**DD:** highly differentiated adenocarcinoma.) ×630.

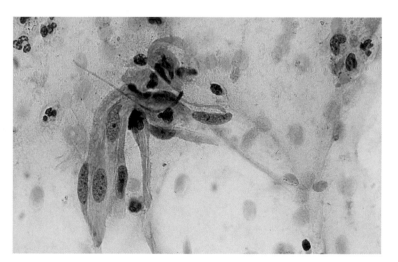

Fig. 5.**56 b** **Highly differentiated uterine corpus carcinoma.** Irregularly arranged abnormal columnar cells with indistinct cytoplasmic borders and enlarged, relatively regularly structured nuclei with chromocenters. (**DD:** cervicitis.) ×630.

5

Fig. 5.**57 a** **Secretory endocervical cells.** A regular, acinus-shaped, single-layered sheet of cells with indistinct cytoplasmic borders and basal, condensed nuclei. (**DD:** highly differentiated adenocarcinoma.) ×630.

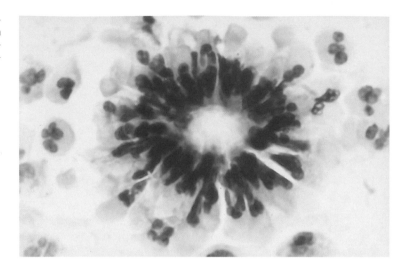

Fig. 5.**57 b** **Highly differentiated uterine corpus carcinoma.** An irregular, acinus-shaped, pseudostratified cluster of abnormal columnar epithelial cells. The cytoplasm has indistinct borders, and the nuclei are enlarged and hyperchromatic with relatively regular chromatin structure and nucleoli. (**DD:** cervicitis.) ×630.

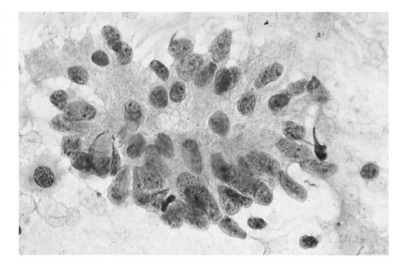

Fig. 5.**58 a** **Endocervical naked nuclei.** A compact cluster of endocervical cells with several naked nuclei dissociated from this cluster. The naked nuclei show severe swelling but regular chromatin structure (——▶). (**DD:** adenocarcinoma.) ×200.

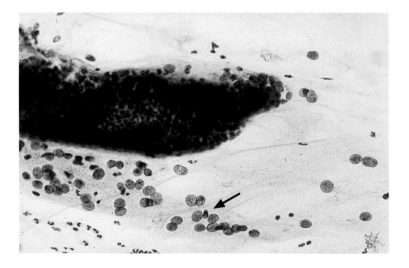

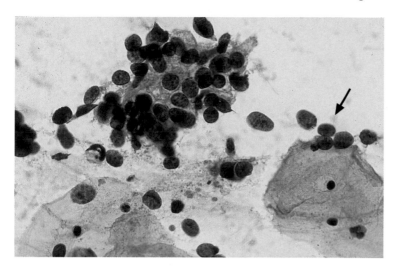

Fig. 5.**58b** **Adenocarcinoma of the endocervix.** An irregular, three-dimensional cluster of abnormal glandular cells with several naked nuclei dissociated from the cluster. The nuclei are enlarged and hyperchromatic, but regularly structured (⟶). (**DD:** cervical polyp.) ×630.

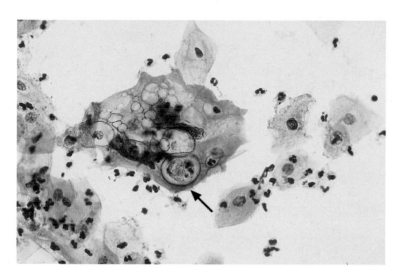

Fig. 5.**59a** **Cell image after radiotherapy.** Syncytial endocervical cell cluster, exhibiting cytoplasmic vacuolation, a cell-in-cell configuration (⟶), and irregular arrangement of nuclei. Normal superficial and intermediate cells are visible nearby. (**DD:** abnormal giant cell.) ×400.

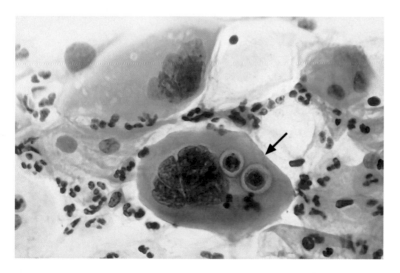

Fig. 5.**59b** **Severe cervical dysplasia.** Abnormal multinuclear giant cells with enlarged, irregularly structured nuclei and cell-in-cell configurations (⟶). (**DD:** Cell image after radiotherapy.) ×400.

5

Fig. 5.**60 a** **Severe cervicitis.** An irregular, three-dimensional cluster of endocervical glandular cells. The nuclei are much enlarged, polymorphic, hyperchromatic, and contain nucleoli (→). The background of the preparation indicates inflammation. (**DD:** adenocarcinoma of the endocervix.) ×400.

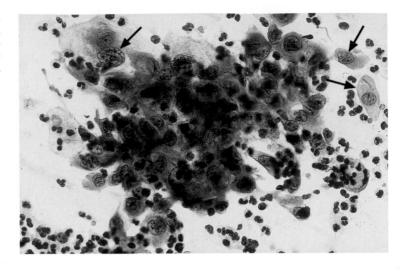

Fig. 5.**60 b** **Intravaginal perforation of a rectal carcinoma.** Irregular, diffuse arrangement of enlarged and abnormal nuclei that are partly preserved and nucleoli-containing (→), partly degenerated. A few cells contain vacuoles (⟹). Normal intermediate cells are visible on the lower left. (**DD:** inflammation.) ×400.

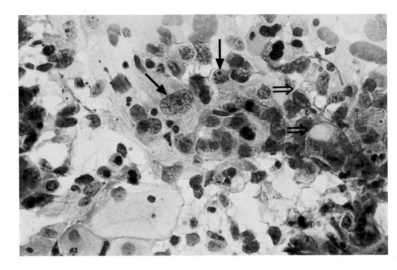

Fig. 5.**61 a** **Tubal metaplasia.** Irregular, three-dimensional cluster of endocervical glandular cells. Enlarged and hyperchromatic nuclei, macronucleolus formation, and phagocytosis on the lower right (→). The background of the preparation contains blood. (**DD:** adenocarcinoma of the endocervix.) ×400.

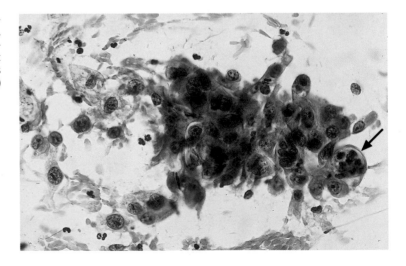

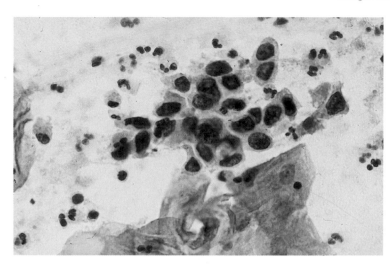

Fig. 5.**61 b** **Adenomatous endometrial hyperplasia.** Loose, irregular cluster of glandular cells with enlarged, hyperchromatic, and polymorphic nuclei. Normal superficial cells are visible at the bottom. (**DD:** endometritis.) ×630.

Inflammation

Acute **inflammation of the endocervical epithelium** is occasionally accompanied by pronounced cellular polymorphism and nuclear enlargement. This makes differentiation from malignant lesions impossible (Fig. 5.**60 a, b**).

The rare **tubal metaplasia** is also easily confused with dysplastic changes (Fig. 5.**61 a, b**).

During endocervical inflammation, there may be increased secretory activity associated with intense **cytoplasmic vacuolation**, which may occasionally reach proportions similar to that of malignant processes (Figs. 5.**62 a, b**, 5.**63 a, b**).

In the day-to-day practice of a cytology laboratory, **metaplastic transformation** of the endocervical glandular epithelium probably represents the most common problem for differential diagnosis. Immature metaplastic cells, in particular, are often confused with CIN III lesions because their nuclei tend to undergo secondary changes, such as condensation, shrinking, and hyperchromatism of the nuclear envelope. These changes cause artificial polymorphism of the nuclei (Figs. 5.**64**–5.**66**). The detection of nucleoli may occasionally be helpful, as they are not associated with metaplasia (Fig. 5.**67 a, b**). The cytoplasmic clearing of intermediate metaplastic cells sometimes leads to confusion with koilocytes (Fig. 5.**68 a, b**). Furthermore, metaplastic cells may form vacuoles and degenerate, thus creating problems with differentiating them from adenocarcinoma (Fig. 5.**69 a, b**).

Regenerative epithelium of the endocervix occasionally has a strong polymorphic character because nuclear polarization is often absent. This makes it difficult to distinguish regenerative epithelium from glandular or squamous epithelial malignancies (Fig. 5.**70 a, b**).

Glandular **endocervical malignancies** frequently cannot be distinguished from poorly differentiated squamous carcinoma. Naked nuclei derived from adenocarcinoma are difficult to tell apart from those derived from squamous carcinoma, although they have often a more regular chromatin structure (Figs. 5.**71 a, b**–5.**73 a, b**).

5

Fig. 5.**62 a** **Acute inflammation.** The regular, pseudoeosinophilic sheet of endocervical glandular cells shows vacuolar degeneration. The degenerating nuclei are often pushed to the cell margin (signet-ring cells) (→). The background of the preparation indicates inflammation. (**DD:** uterine corpus carcinoma.) ×630.

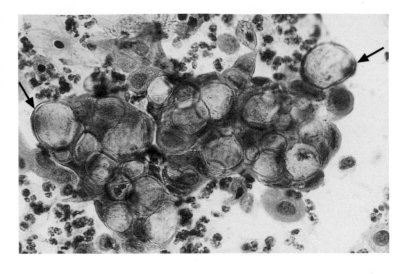

Fig. 5.**62 b** **Moderately differentiated uterine corpus carcinoma.** An irregular, three-dimensional cluster of glandular cells showing vacuolar degeneration. The degenerated nuclei are often pushed to the cell margin (signet-ring cells) (→). The background of the preparation indicates inflammation. (**DD:** endometritis.) ×500.

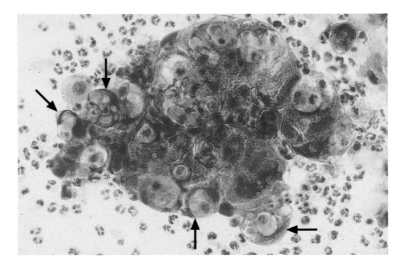

Fig. 5.**63 a** **Glandular regenerative epithelium.** The cells exhibit vacuolated cytoplasm, enlarged nuclei, and distinct nucleoli. The background of the preparation indicates inflammation. (**DD:** adenocarcinoma.) ×790.

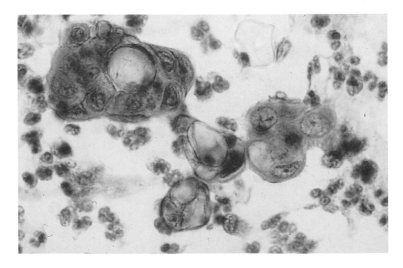

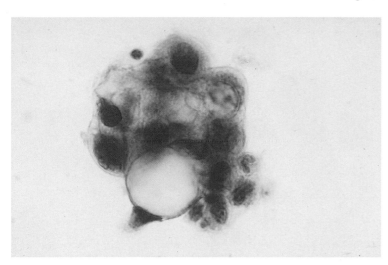

Fig. 5.**63 b** **Abnormal glandular cells.** An irregular, three-dimensional cell cluster found in the puncture fluid of ovarian carcinoma, showing severe cytoplasmic vacuolation and enlarged, polymorphic nuclei. (**DD:** cluster of granulosa cells.) ×790.

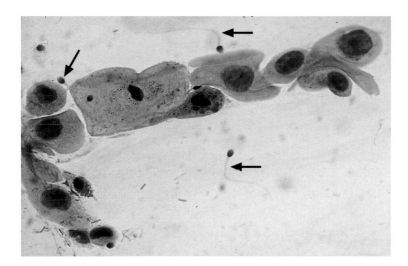

Fig. 5.**64 a** **Intermediate metaplastic cells.** A loose, elongated cell sheet showing slightly enlarged, degenerated nuclei. Spermatozoa are seen nearby (—►). (**DD:** intermediate dysplastic cells.) ×630.

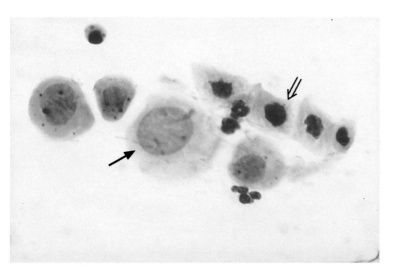

Fig. 5.**64 b** **Immature dysplastic cells** associated with severe cervical dysplasia. Lysis of degenerated nuclei (—►) and condensed, pyknotic nuclei (⟹). (**DD:** immature metaplastic cells.) ×400.

5

Fig. 5.**65 a** **Intermediate metaplastic cells.** Degenerative changes, such as nuclear shrinkage with the formation of perinuclear halos (→), lead to polymorphism of the nuclei. Two normal superficial cells are also visible (⇒). (**DD:** moderate dysplastic cells.) ×400.

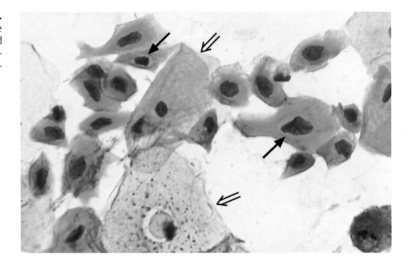

Fig. 5.**65 b** **Squamous carcinoma in situ cells** associated with small cell squamous cervical carcinoma in situ. The nuclei are enlarged, hyperchromatic, polymorphic, and condensed. A normal intermediate cell is visible at the bottom. (**DD:** immature metaplastic cells.) ×250.

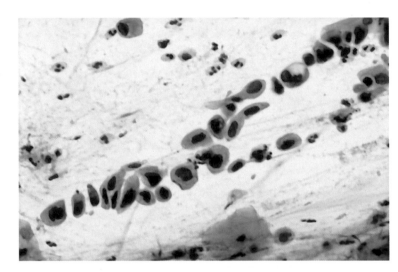

Fig. 5.**66 a** **Immature and intermediate metaplastic cells.** Degenerative changes, such as nuclear shrinkage with formation of perinuclear halos (→), lead to polymorphism of the nuclei. (**DD:** severe dysplastic cells.) ×250.

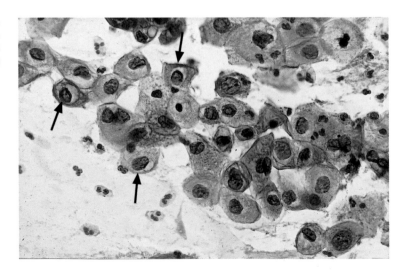

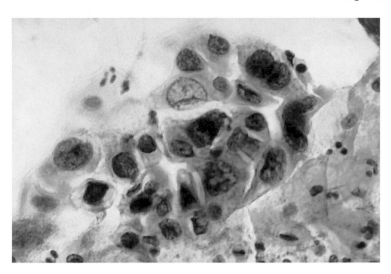

Fig. 5.**66 b** **Severe dysplastic cells** associated with severe cervical dysplasia. Lysis of the cytoplasm, pronounced nuclear polymorphism, irregular chromatin structure, and partly degenerated nuclei. (**DD:** immature metaplastic cells.) ×400.

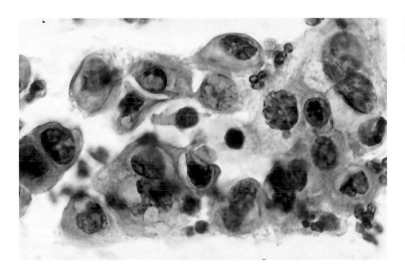

Fig. 5.**67 a** **Immature metaplastic cells.** Degenerative changes, such as nuclear shrinkage with formation of perinuclear halos, lead to polymorphism of the nuclei. (**DD:** severe dysplastic cells.) ×630.

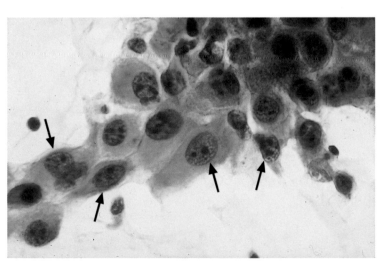

Fig. 5.**67 b** **Invasive squamous carcinoma of the cervix.** A loose, two-dimensional sheet of cells with distinct cytoplasmic borders, enlarged nuclei, irregular chromatin structure, and macronucleoli (⟶). (**DD** regenerative epithelium.) ×400.

5

Fig. 5.**68 a** **Intermediate metaplastic cells.** Thickening of the cell margins, central clearing of the cytoplasm, and slightly swollen, condensed nuclei. (**DD:** koilocytotic dysplasia.) ×630.

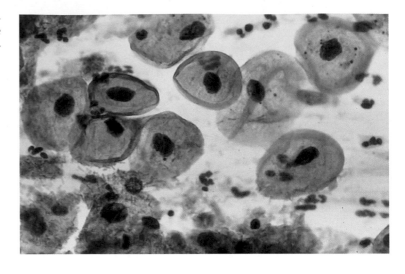

Fig. 5.**68 b** **Moderate koilocytotic dysplastic cells.** The nuclei are enlarged, hyperchromatic, and condensed. Normal intermediate cells are visible nearby. (**DD:** intermediate metaplastic cells.) ×630.

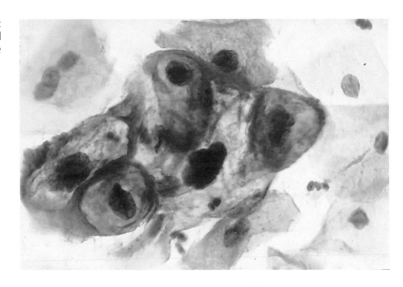

Fig. 5.**69 a** **Intermediate metaplastic cells.** Vacuolated degeneration and slightly swollen nuclei due to acute inflammation. The background of the preparation indicates inflammation. Normal superficial cells are visible in the center. (**DD:** adenocarcinoma.) ×630.

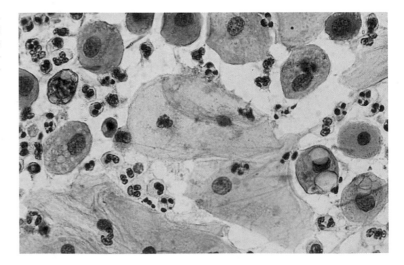

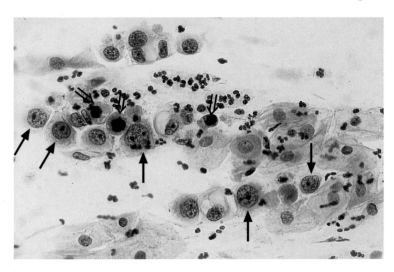

Fig. 5.**69b** **Ovarian carcinoma.** Dissociated abnormal glandular cells in the cervical smear. Their cytoplasm has indistinct borders and is partly vacuolated. The nuclei are partly active with irregular, hyperchromatic structure and nucleoli (→) and partly degenerated, pyknotic, and condensed (⇒). Normal intermediate cells are visible nearby. (**DD:** endometritis.) ×400.

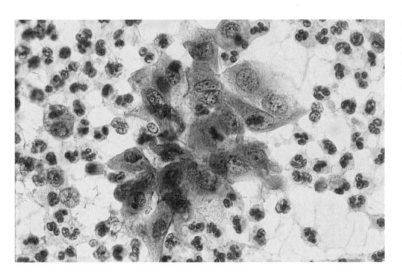

Fig. 5.**70a** **Glandular regenerative epithelium.** Irregular arrangement of cells with indistinct cytoplasmic borders, much enlarged nuclei, and nucleolus formation. The background of the preparation indicates inflammation. (**DD:** adenocarcinoma of the endocervix.) ×630.

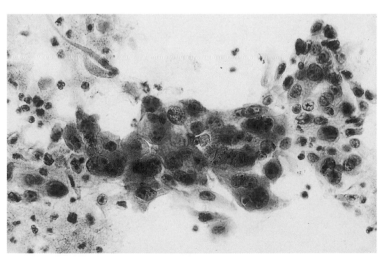

Fig. 5.**70b** **Invasive squamous carcinoma of the cervix.** An irregular, loose cluster of cells with partly indistinct cytoplasmic borders, enlarged and hyperchromatic nuclei, and macronucleolus formation. (**DD:** regenerative epithelium.) ×320.

5

Fig. 5.**71 a** **Adenocarcinoma in situ of the endocervix.** A loose cluster of abnormal glandular cells showing sparse cytoplasm and enlarged, hyperchromatic nuclei with irregular chromatin structure. Normal superficial cells are visible at the bottom. (**DD:** squamous carcinoma in situ.) ×630.

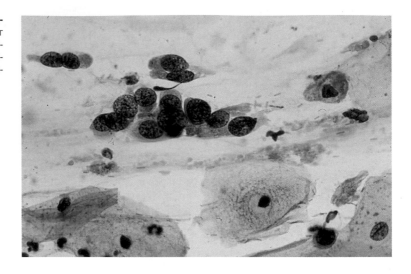

Fig. 5.**71 b** **Invasive cervical carcinoma.** The squamous carcinoma cells have little cytoplasm and indistinct borders. There are large abnormal naked nuclei (⟹), small cells with condensed nuclei (→), and individual keratinizing cells (➡). The background of the preparation indicates inflammation. (**DD:** adenocarcinoma of the endocervix.) ×500.

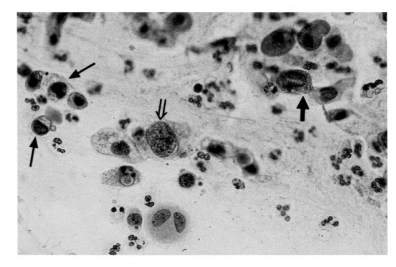

Fig. 5.**72 a** **Adenocarcinoma in situ of the endocervix.** Enlarged, dissociated naked nuclei (→). The nuclear structure is relatively regular, and some nuclei contain nucleoli. Several normal intermediate cells are visible nearby. (**DD:** degenerated nuclei of endocervical cells.) ×400.

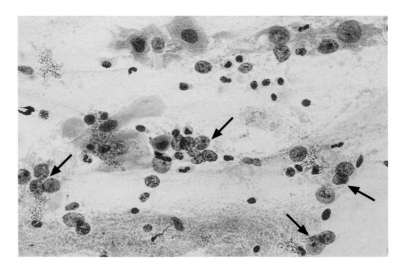

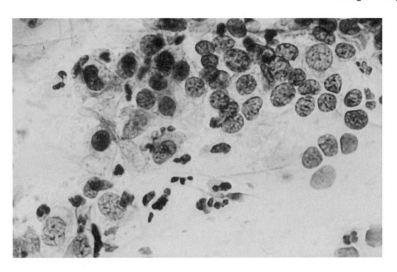

Fig. 5.**72 b** **Severe cervical dysplasia.** Enlarged, dissociated naked nuclei with hyperchromatic and irregular chromatin structure. (**DD:** adenocarcinoma in situ of the endocervix.) ×630.

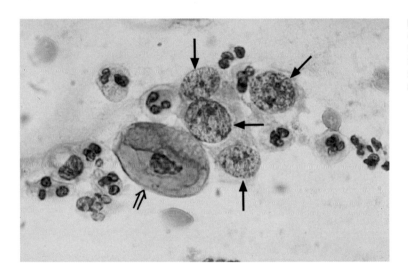

Fig. 5.**73 a** **Poorly differentiated uterine corpus carcinoma.** Enlarged, intact naked nuclei with relatively regular chromatin structure and nucleolus formation (──►). A normal parabasal cell (══►) is seen nearby. (**DD:** microinvasive cervical carcinoma.) ×1000.

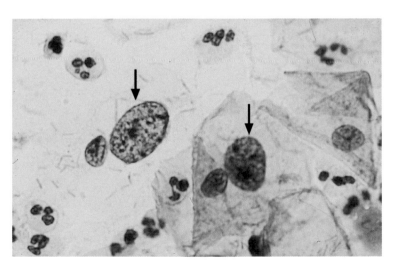

Fig. 5.**73 b** **Severe cervical dysplasia.** Enlarged, active naked nuclei with irregular chromatin structure and chromocenter formation (──►). Normal intermediate cells are visible nearby. (**DD:** adenocarcinoma.) ×1000.

5

Fig. 5.**74 a** **Endometrial cells of the late proliferative phase.** A loose cell cluster with slightly hyperchromatic nuclei and nucleolus. Normal superficial cells are seen on the lower left. (**DD:** uterine corpus carcinoma.) ×630.

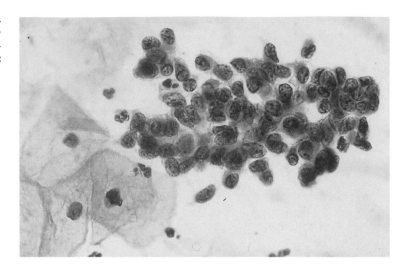

Fig. 5.**74 b** **Highly differentiated uterine corpus carcinoma.** A regularly structured, two-dimensional sheet of endometrial cells with slightly hyperchromatic nuclei and prominent nucleoli. (**DD:** normal endometrial cells.) ×400.

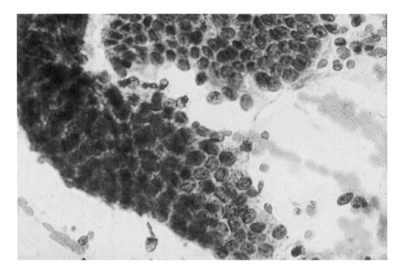

Fig. 5.**75 a** **Degenerated endometrial cells.** An irregular and compact cluster of glandular cells. The cell borders are distinct, the cytoplasm is partly vacuolated, and cell-in-cell configurations are detectable. The background of the preparation indicates inflammation. A normal superficial cell is seen on the upper left. (**DD:** adenocarcinoma.) ×630.

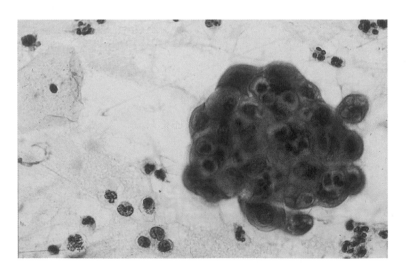

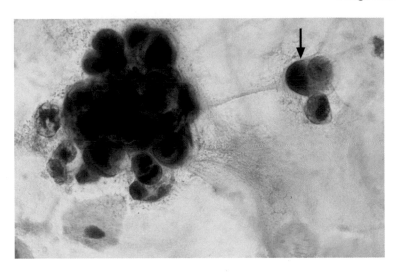

Fig. 5.**75 b** **Uterine corpus carcinoma.** An irregular, three-dimensional, compact cluster of degenerated glandular cells. The nuclei are enlarged, polymorphic, and condensed. A cell-in-cell configuration with still intact nuclei is visible on the upper right (→), and a normal intermediate cell is seen on the lower left. (**DD:** endometritis.) ×630.

Endometrium

Physiological Factors

During the late proliferative phase, the nuclei of endometrial cells may be slightly hyperchromatic and show nucleoli. They therefore look very much like those of highly differentiated carcinoma of the uterine corpus (Fig. 5.**74 a, b**).

Endometrial cells, like endocervical cells, are often only present in a degenerated state. If they are also vacuolated and the cell clusters are spherical, differentiation from uterine corpus carcinoma is almost impossible (Fig. 5.**75 a, b**).

Remnants of the decidua or the throphoblast, which may appear in the smear taken after an abortion, usually show extreme nuclear enlargement, polymorphism, and prominent nucleoli. Their differentiation from malignant tumors is therefore only possible on the basis of the patient's history (Fig. 5.**76 a, b**).

Technical Factors

Smears taken during the menstrual phase are often obscured by blood, and they have an inflammatory appearance due to the admixture of round cells, such as lymphocytes, histiocytes, and granulocytes. A cluster of endometrial cells exfoliated late during menstruation (exodus) may be confused with spherical cell clusters from ovarian carcinoma (Fig. 5.**77 a, b**).

Inflammation

Smears obtained from extracted IUDs often exhibit signs of an endometrial reaction to the IUD. The endometrial cell sheets show enlarged and hyperchromatic nuclei with a coarse chromatin structure. Occasionally nucleoli are also detected, which make differentiation from malignant lesions difficult (Fig. 5.**78 a, b**). The patient's age and history are important diagnostic clues, since uterine corpus carcinoma is rare during sexual maturity.

5

Fig. 5.**76 a** **Cytothrophoblast remnants associated with abortion.** A loose cluster of large, elongated cells with foamy cytoplasm, indistinct cell borders, and much enlarged, polymorphic, and hyperchromatic nuclei with nucleoli. The background of the preparation indicates inflammation. (**DD:** leiomyosarcoma.) ×400.

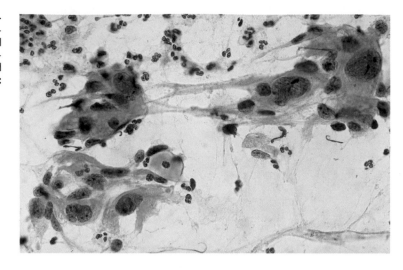

Fig. 5.**76 b** **Tubal carcinoma.** A loose, irregular cluster of abnormal glandular cells with foamy cytoplasm and enlarged, polymorphic, and hyperchromatic nuclei with nucleoli. (**DD:** glandular regenerative epithelium.) ×1000.

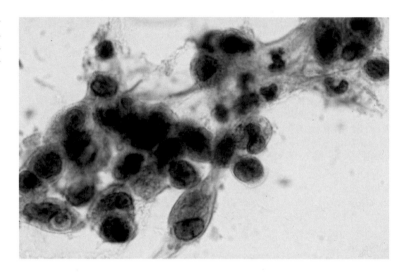

Fig. 5.**77 a** **Exodus.** A degenerated, compact, hyperchromatic cluster of endometrial cells with a peripheral layer of stromal cells. Normal superficial cells are visible in the background. (**DD:** ovarian carcinoma.) ×400.

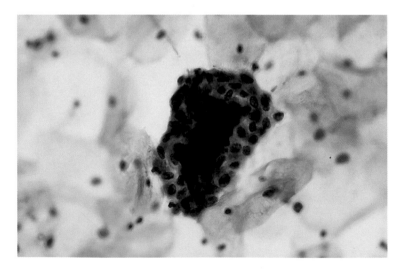

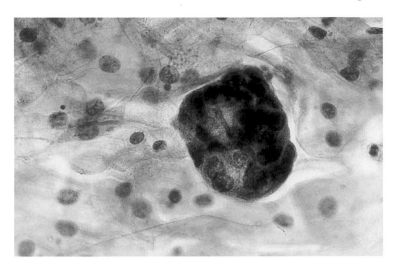

Fig. 5.**77 b** **Ovarian carcinoma.** A compact, irregular, three-dimensional cell cluster. The nuclei are enlarged and hyperchromatic with irregular chromatin structure, and they are partly intact, partly condensed. The spherical cell cluster shows a line of retraction. Normal superficial cells are visible in the background. (**DD:** exodus.) ×630.

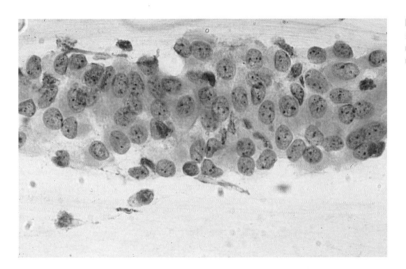

Fig. 5.**78 a** **Endometritis.** A loose, two-dimensional sheet of endometrial cells with enlarged, regularly structured nuclei and delicate nucleoli. (**DD:** uterine corpus carcinoma.) ×630.

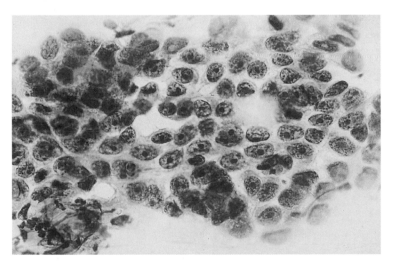

Fig. 5.**78 b** **Leiomyosarcoma of the uterine corpus.** A loose cluster of abnormal glandular cells with sparse cytoplasm, enlarged and hyperchromatic nuclei, irregular chromatin structure, and macronucleoli. (**DD:** regenerative epithelium.) ×500.

5

Small-cell Lesions _____

Diagnostic problems with small-cells play a major role during both sexual maturity and menopause. Small-cell squamous carcinomas and their precursors amount to about 20% of all cervical carcinomas (see p. 171), and distinguishing small cells becomes particularly important after the menopause, since highly differentiated endometrial carcinoma is the most common growth type of endometrial carcinoma.

Small cells not only derive from squamous epithelium and glandular epithelium; they also originate from the reticular cell system and include, in particular, small histiocytes, lymphocytes, as well as reserve cells and stromal cells.

An overview of benign and malignant small-cell lesions in the genital region is provided in Table 5.1, and a comparison of the different sizes of cells and nuclei occurring in cervical smears is presented in Fig. 5.79.

Physiological Factors

Basal and parabasal cells present in atrophic smears may occasionally give rise to confusion with small-cell CIN III lesions (Figs. 5.**80a, b**, 5.**81a, b**).

Normal **endometrial and stromal cells** frequently cause problems with the differential diagnosis of uterine corpus carcinoma (Figs. 5.**82a, b**–5.**84a, b**).

Inflammation

The most frequent small-cell change observed in cytological smears is **basal cell hyperplasia**, which is regularly observed during the regenerative phase of inflammation. It is often associated with degenerative changes of the nuclei. This creates considerable problems with the differential diagnosis of small-cell CIN III lesions because, in both cases, the cells usually occur as single cells and are evenly distributed over the smear (Figs. 5.**85a, b**–5.**89a, b**).

Reserve cells and **immature metaplastic cells** are slightly larger than basal cells, appear frequently in clusters, and occasionally still show features of glandular epithelium with secretory activity, such as vacuoles or indistinct cell borders (Figs. 5.**90a, b**–5.**95a, b**). Since they frequently show nuclear degeneration as well, there are problems with the differential diagnosis of small-cell CIN III lesions similar to those arising with basal cells. However, these problems arise predominantly with cell clusters.

Small-cell exudates with **endometritic reactions** are frequently observed in the presence of an IUD and may give rise to confusion with adenocarcinoma (Figs. 5.**96a, b**, 5.**97a, b**). Glandular endometrial hyperplasia is also difficult to distinguish from invasive glandular carcinoma (Fig. 5.**98a, b**).

Histiocytes may resemble malignant cells if their nuclei undergo degeneration, particularly when this is accompanied by vacuolation of the cytoplasm (Figs. 5.**99a, b**–5.**101a, b**).

Follicular cervicitis is associated with exfoliation of reticular cells, and these often show greatly enlarged and polymorphic nuclei. However, diagnostic clues usually include their confined localization in the smear, the regular structure of the cell clusters (germinal centers), and the benign criteria of their nuclei. (Figs. 5.**102a, b**–5.**104a, b**).

Table 5.**1**   Classification of small-cell genital lesions

	Squamous epithelial cells	Glandular epithelial cells	Reticular cells
Benign	▶ Basal cells ▶ Immature metaplastic cells	▶ Endometrial cells ▶ Endocervical cells ▶ Reserve cells	▶ Small histiocytes ▶ Stromal cells ▶ Lymphocytes
Malignant	▶ Small-cell carcinoma in situ ▶ Small dysplastic metaplastic cells (poorly differentiated) ▶ Small squamous carcinoma cells (poorly differentiated)	▶ Abnormal endometrial cells (highly differentiated) ▶ Abnormal endocervical cells (highly differentiated)	▶ Sarcoma cells ▶ Lymphoma cells

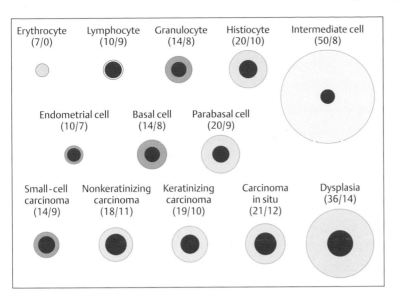

Fig. 5.**79** **Sizes of cells and nuclei.** A comparison of cell types observed in gynecological smears. (The diameters of the cells and their nuclei are given in μm.)

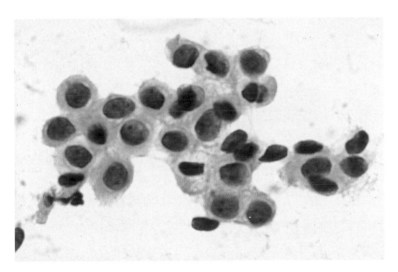

Fig. 5.**80 a** **Atrophic cell cohesion associated with poorly developed squamous epithelium.** Pseudosyncytium of cells, showing distinct cytoplasmic borders and slightly enlarged, isomorphic nuclei. (**DD:** severe dysplastic cells.) ×1000.

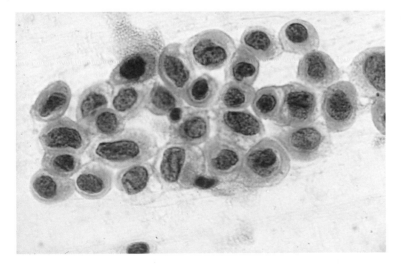

Fig. 5.**80 b** **Small-cell squamous carcinoma in situ of the cervix.** A loose sheet of squamous carcinoma in situ cells with distinct cytoplasmic borders and enlarged, irregularly structured nuclei. (**DD:** immature metaplastic cells.) ×630.

5

Fig. 5.**81 a** **Basal cells associated with atrophy.** Slightly enlarged nuclei and regular chromatin structure. (**DD:** squamous carcinoma in situ.) ×1600.

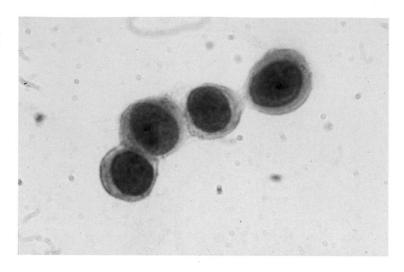

Fig. 5.**81 b** **Small-cell squamous carcinoma in situ of the cervix.** The squamous carcinoma in situ cells show cytoplasmic vacuolation, enlarged and hyperchromatic nuclei, and irregular chromatin structure. (**DD:** histiocytes.) ×630.

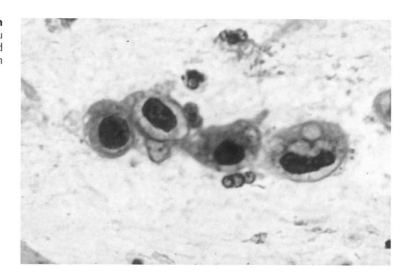

Fig. 5.**82 a** **Endometrial cells during menstruation.** Loose groups of cells with slightly hyperchromatic nuclei (⟶). Normal superficial cells are visible in the background. (**DD:** severe small-cell dysplasia.) ×400.

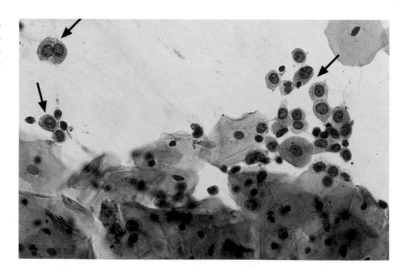

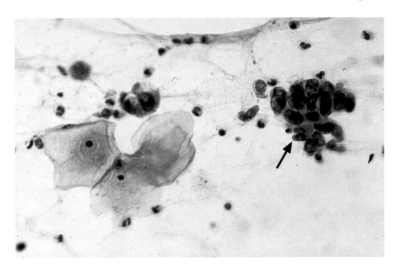

Fig. 5.82 b Highly differentiated uterine corpus carcinoma. An irregular, three-dimensional cluster of abnormal glandular cells (⟶). The nuclei are slightly enlarged and hyperchromatic. Normal superficial cells are seen on the left. (**DD:** normal endometrial cells.) ×400.

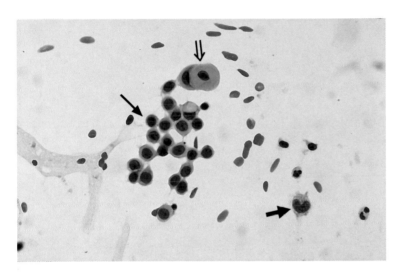

Fig. 5.83 a Endometrial atrophy. A loose, relatively regular sheet of glandular cells (⟶). The nuclei are slightly enlarged and hyperchromatic. A parabasal cell is seen at the top (⟹) and a histiocyte on the lower right (➡). (**DD:** highly differentiated uterine corpus carcinoma.) ×400.

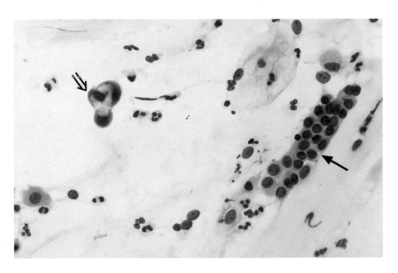

Fig. 5.83 b Highly differentiated uterine corpus carcinoma. A regularly structured, elongated sheet of abnormal endometrial cells with only slightly enlarged, hyperchromatic nuclei (⟶) and a small, vacuolated tumor cell cluster (⟹). (**DD:** normal endometrial cells.) ×400.

5

Fig. 5.**84a** **Endometrial stromal cells.** A loose cluster of cells with indistinct borders and foamy cytoplasm. The cells are partly spindle-shaped (deep layer) (⟶), partly round (superficial layer). The nuclei are regularly structured. (**DD:** stromal sarcoma of the endometrium.) ×400.

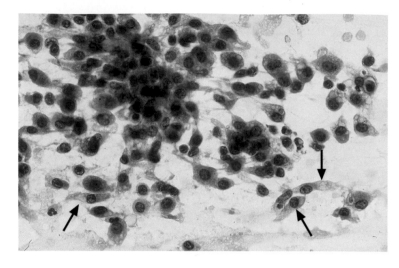

Fig. 5.**84b** **Uterine corpus carcinoma.** A loose, irregular cluster of abnormal glandular cells with hyperchromatic nuclei and formation of nucleoli. Several normal superficial cells are visible in the background. (**DD:** endometritis.) ×400.

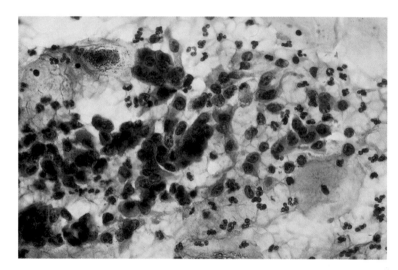

Fig. 5.**85a** **Basal cell hyperplasia.** Small round squamous cells, partly single, partly arranged in loose clusters (⟶). The nuclei are condensed. Numerous normal superficial cells are visible in the background. (**DD:** small-cell squamous carcinoma in situ.) ×250.

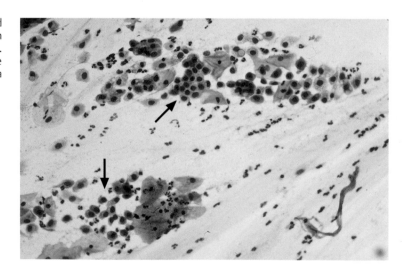

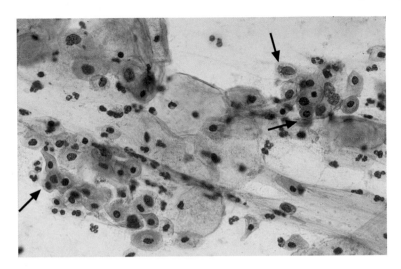

Fig. 5.**85 b** **Small-cell squamous carcinoma in situ of the cervix.** Diffusely arranged, small, squamous carcinoma in situ cells (→). The nuclei are only slightly enlarged, hyperchromatic, and partly condensed. Some cells show signs of keratinization. Numerous normal superficial cells are seen in the background. (**DD:** basal cell hyperplasia.) ×400.

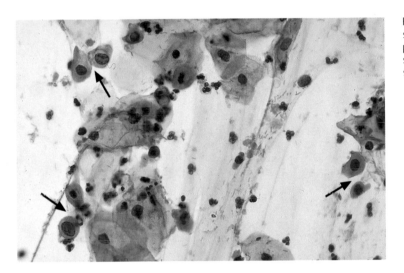

Fig. 5.**86 a** **Basal cell hyperplasia.** Dissociated, small round squamous cells (→). The nuclei are partly intact, partly condensed. Numerous normal superficial cells are seen in the background. (**DD:** severe small-cell dysplasia.) ×400.

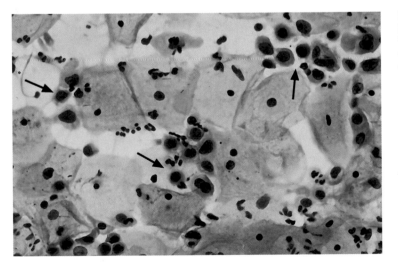

Fig. 5.**86 b** **Small-cell cervical carcinoma in situ.** Dissociated squamous carcinoma in situ cells (→) with enlarged and hyperchromatic nuclei. Numerous normal superficial cells are visible in the background. (**DD:** basal cell hyperplasia.) ×400.

Fig. 5.**87 a   Basal cell hyperplasia.** A loose sheet of small round squamous cells (→). The cytoplasm has distinct borders, and the active nuclei are regularly structured. Normal superficial cells are visible at the top. (**DD:** severe dysplasia.) ×400.

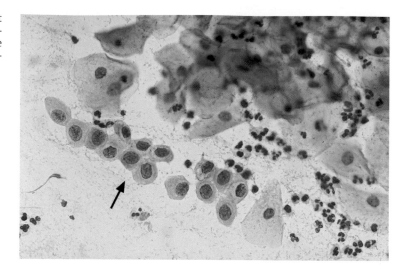

Fig. 5.**87 b   Severe small-cell dysplasia of the cervix.** A loose sheet of small severe dysplastic cells with enlarged, hyperchromatic nuclei and irregular chromatin structure (→). A cluster of abnormal naked nuclei is also present (⇒), and normal superficial cells are visible at the bottom. (**DD:** basal cell hyperplasia.) ×400.

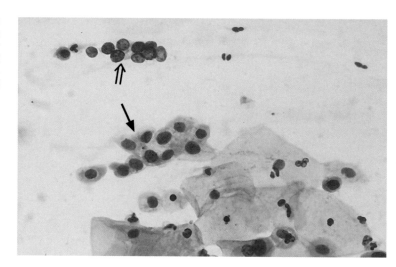

Fig. 5.**88 a   Basal cell hyperplasia.** Small, round squamous cells, partly single, partly arranged in loose clusters (→). The nuclei are slightly enlarged and partly condensed. Several normal superficial cells are visible in the center. (**DD:** severe small-cell dysplasia.) ×400.

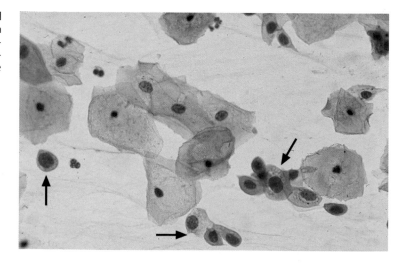

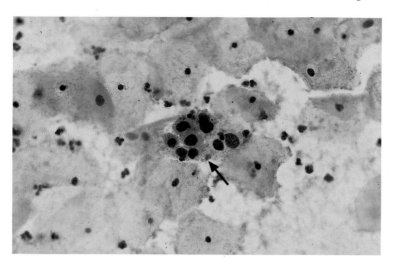

Fig. 5.**88 b Severe small-cell dysplasia of the cervix.** A loose cluster of small severe dysplastic cells with much enlarged, hyperchromatic, and condensed nuclei (→). Numerous normal superficial cells are visible in the background. (**DD:** degenerated endocervical cells.) ×400.

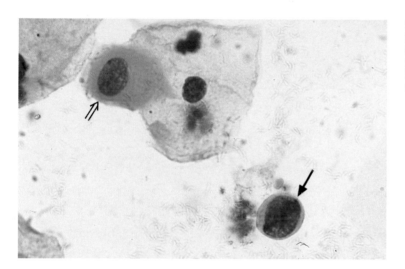

Fig. 5.**89 a Basal cell hyperplasia.** A small, round squamous cell with slightly enlarged, regularly structured nucleus (→). A normal parabasal cell (⟹) and intermediate cells are seen on the upper left. (**DD:** small-cell squamous carcinoma in situ.) ×1000.

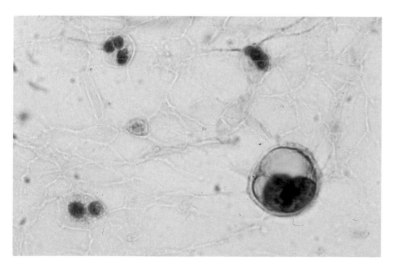

Fig. 5.**89 b Invasive squamous carcinoma of the cervix.** A small, binuclear dysplastic cell with cytoplasmic vacuolation and hyperchromatic, nucleolus-containing nuclei. (**DD:** cervicitis.) ×1000.

5

Fig. 5.**90 a** **Reserve cell hyperplasia.** A loose cluster of small, round cells which partly resemble metaplastic, partly glandular cells. The cytoplasm is partly vacuolated and has indistinct borders, the nuclei are slightly enlarged and regularly structured. Several normal intermediate cells are visible at the bottom. (**DD:** severe small-cell dysplasia.) ×400.

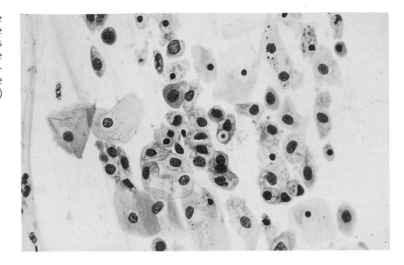

Fig. 5.**90 b** **Severe small-cell dysplasia of the cervix**. Squamous carcinoma in situ cells. The dissociated small cells have hyperchromatic and partly condensed nuclei that vary in size (anisokaryosis). Numerous normal superficial cells are visible in the background. (**DD:** basal cell hyperplasia.) ×400.

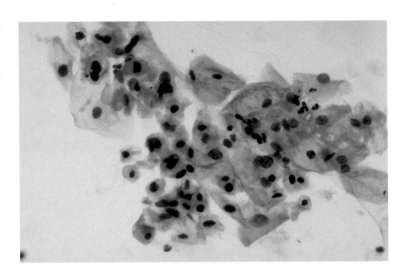

Fig. 5.**91 a** **Reserve cells.** A loose cluster of cells with indistinct cytoplasmic borders and partly degenerated nuclei. A normal intermediate cell is seen at the bottom. (**DD:** small-cell squamous carcinoma in situ.) ×400.

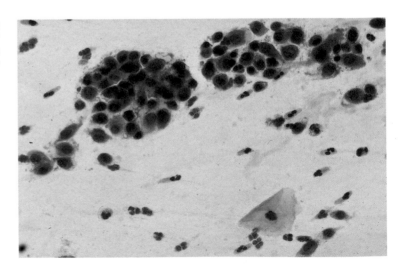

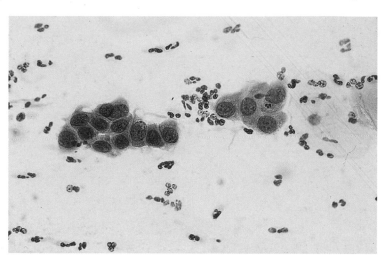

Fig. 5.**91 b** **Severe small-cell dysplasia of the cervix.** A loose sheet of small severe dysplastic cells with enlarged and hyperchromatic nuclei, irregular chromatin structure, and formation of chromocenters. (**DD:** immature metaplastic cells.) ×400.

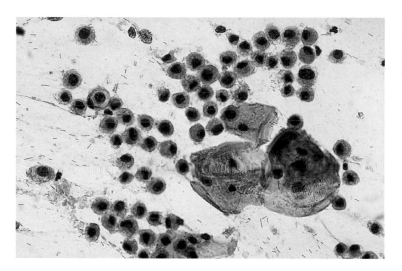

Fig. 5.**92 a** **Reserve cells.** A loose sheet of cells with foamy cytoplasm, moderately enlarged nuclei, and regular chromatin structure. Some nuclei are condensed and hyperchromatic. (**DD:** squamous carcinoma in situ.) ×400.

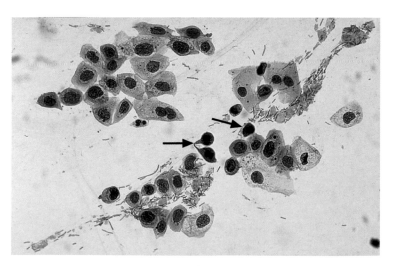

Fig. 5.**92 b** **Severe small-cell dysplasia of the cervix.** Loosely arranged, small severe dysplastic cells. The cytoplasm has distinct borders, and the nuclei are enlarged, hyperchromatic, and irregularly structured. Some cells show signs of keratinization (⟶). (**DD:** immature metaplastic cells.) ×400.

5

Fig. 5.**93 a** **Immature metaplastic cells.** A loose group of cells with slightly enlarged, hyperchromatic, and condensed nuclei. Several normal superficial cells are visible nearby. (**DD:** small-cell squamous carcinoma in situ.) ×400.

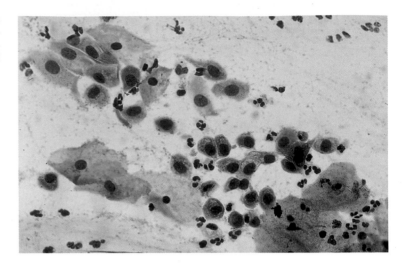

Fig. 5.**93 b** **Severe cervical dysplasia.** An irregular cluster of small dysplastic cells. The nuclei are enlarged, hyperchromatic, condensed, and polymorphic. Normal superficial cells are visible nearby. (**DD:** cervicitis.) ×400.

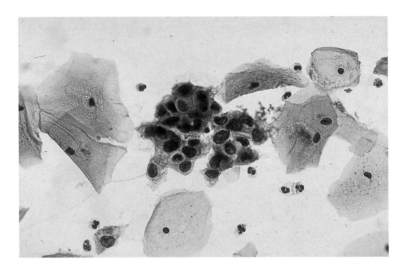

Fig. 5.**94 a** **Immature metaplastic cells.** A loose group of cells with vacuolated cytoplasm and hyperchromatic, condensed nuclei. Normal superficial cells are visible nearby. (**DD:** severe small-cell dysplasia.) ×400.

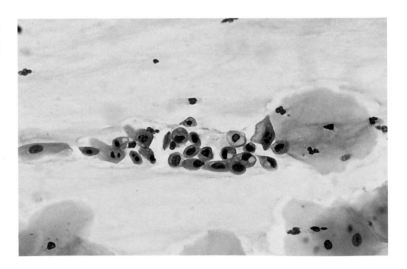

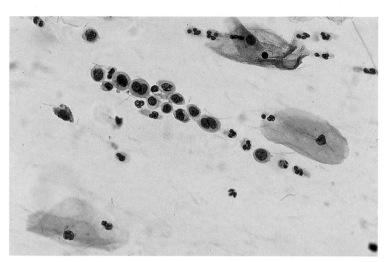

Fig. 5.**94 b** **Small-cell squamous carcinoma in situ of the cervix.** A loose, elongated cluster of small squamous carcinoma in situ cells. The nuclei are enlarged, hyperchromatic, and condensed. Single normal superficial cells are seen nearby. (**DD:** histiocytes.) ×400.

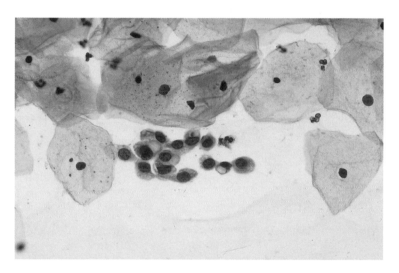

Fig. 5.**95 a** **Immature metaplastic cells.** A loose cluster of cells with partly vacuolated cytoplasm and slightly enlarged nuclei. Numerous normal superficial cells are seen at the top. (**DD:** severe dysplasia.) ×400.

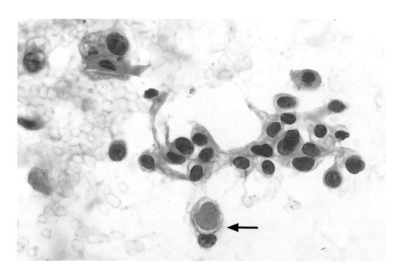

Fig. 5.**95 b** **Invasive small-cell squamous carcinoma of the cervix.** A loose cluster of small, spindle-shaped dysplastic cells. The cytoplasm has indistinct borders, and the nuclei are enlarged, hyperchromatic, and irregularly structured. A cell-in-cell configuration is visible at the bottom (→). (**DD:** cervicitis.) ×400.

5

Fig. 5.**96 a** **Endometritis.** A loose, irregular cluster of glandular cells with slightly enlarged, hyperchromatic nuclei. The background of the preparation indicates inflammation. (**DD:** highly differentiated uterine corpus carcinoma.) ×400.

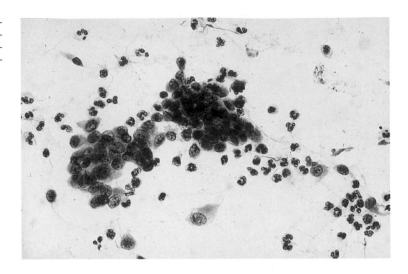

Fig. 5.**96 b** **Ovarian metastasis of mammary carcinoma.** A regularly structured cluster of glandular cells. The nuclei are fairly isomorphic but hyperchromatic and irregularly structured. Normal intermediate cells are visible nearby. (**DD:** normal endometrial cells.) ×400.

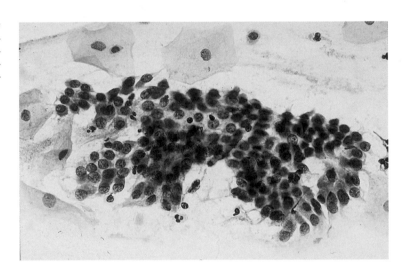

Fig. 5.**97 a** **Endometritis.** The small cluster of glandular cells in the center shows slightly hyperchromatic nuclei and a cell-in-cell configuration (⟶). Numerous normal superficial cells are seen in the vicinity. (**DD:** highly differentiated corpus carcinoma.) ×400.

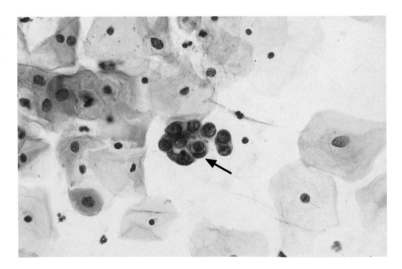

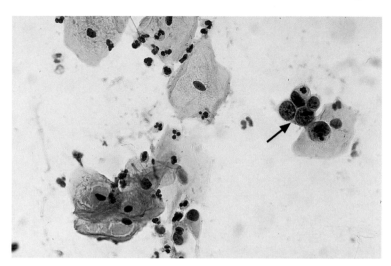

Fig. 5.**97 b** **Adenocarcinoma of the endocervix.** A small compact cluster of abnormal glandular cells (⟶). The nuclei are enlarged, hyperchromatic, and irregularly structured. Normal superficial cells are visible nearby. (**DD:** cervicitis.) ×400.

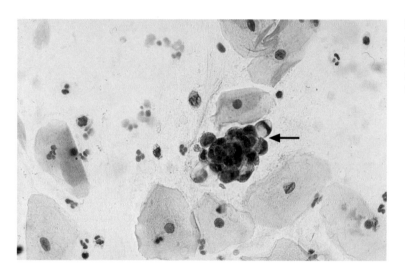

Fig. 5.**98 a** **Glandular endometrial hyperplasia.** An irregular, three-dimensional, compact cluster of glandular cells with vacuolated cytoplasm and degenerated nuclei (⟶). Normal superficial cells are visible nearby. (**DD:** highly differentiated uterine corpus carcinoma.) ×400.

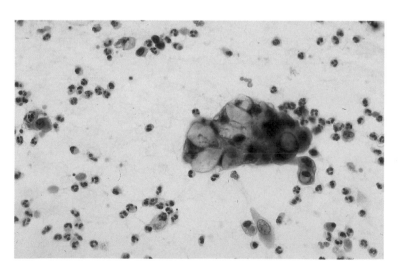

Fig. 5.**98 b** **Moderately differentiated uterine corpus carcinoma.** A compact cluster of abnormal glandular cells with vacuolated cytoplasm and degenerated nuclei. The background of the preparation indicates inflammation. (**DD:** cervicitis.) ×400.

5

Fig. 5.**99 a** **Histiocytes associated with chronic inflammation.** Loosely arranged histiocytes with foamy cytoplasm, indistinct cytoplasmic borders, and round isomorphic nuclei. (DD: severe small-cell dysplasia.) ×400.

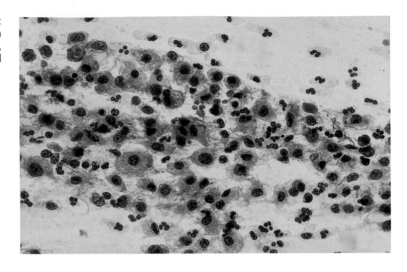

Fig. 5.**99 b** **Small-cell squamous carcinoma in situ of the cervix.** Small squamous carcinoma in situ cells. The loosely arranged cells show distinct cytoplasmic borders and enlarged, hyperchromatic, polymorphic nuclei. (**DD:** basal cell hyperplasia.) ×400.

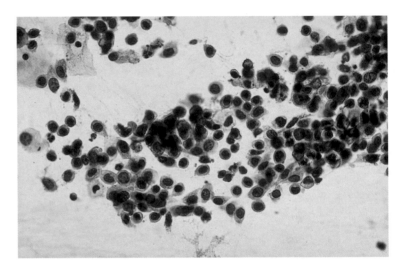

Fig. 5.**100 a** **Histiocytes.** An irregular cluster of cells with indistinct cytoplasmic borders and hyperchromatic, isomorphic nuclei. A small intermediate cell is seen at the top. (**DD:** uterine corpus carcinoma.) ×400.

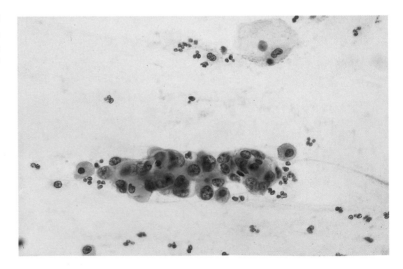

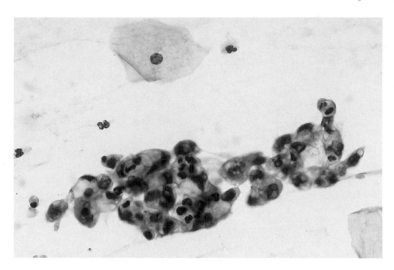

Fig. 5.**100 b Highly differentiated uterine corpus carcinoma.** An irregular cluster of vacuolated abnormal glandular cells. The nuclei are hyperchromatic and polymorphic. A few cell-in-cell configurations are visible. A normal intermediate cell is seen at the top. (**DD:** histiocytes.) ×500.

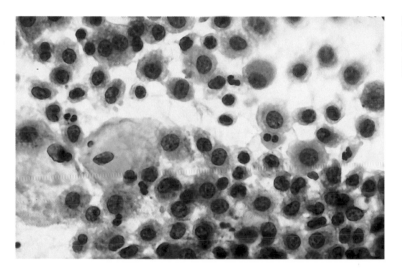

Fig. 5.**101 a Histiocytes.** Diffusely arranged cells with foamy cytoplasm, indistinct cytoplasmic borders, and round, isomorphic, hyperchromatic nuclei. Normal intermediate cells are visible on the left. (**DD:** uterine corpus carcinoma.) ×630.

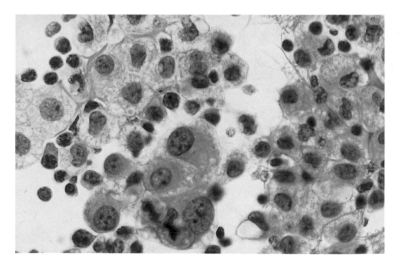

Fig. 5.**101 b Leiomyosarcoma of the uterine corpus.** Loosely arranged abnormal cells with vacuolated, foamy cytoplasm. The enlarged, hyperchromatic nuclei vary in size (anisokaryosis) and exhibit an irregular chromatin structure and formation of nucleoli. (**DD:** histiocytes.) ×630.

5

Fig. 5.**102 a   Follicular cervicitis.** Groups of dissociated reticular cells (——▶) with isomorphic, mostly condensed nuclei and hardly any cytoplasm visible. Normal intermediate cells are visible in the background. (**DD:** small-cell squamous carcinoma in situ.) ×400.

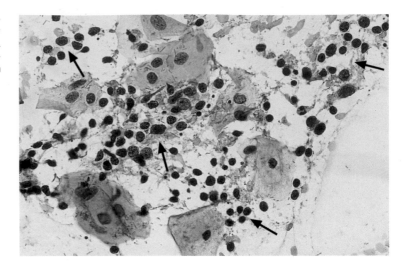

Fig. 5.**102 b   Highly differentiated uterine corpus carcinoma.** The small abnormal glandular cells (——▶) are partly dissociated, partly arranged in loose clusters. The nuclei are hyperchromatic and degenerated. The background of the preparation indicates inflammation and contains a few intermediate metaplastic cells. (**DD:** normal endometrial cells.) ×400.

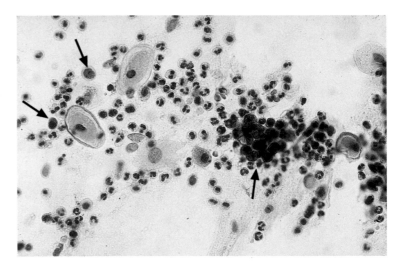

Fig. 5.**103 a   Follicular cervicitis.** Diffusely arranged reticular and lymphocytic cells with hardly any recognizable cytoplasm. The nuclei are hyperchromatic and regularly structured. A few normal superficial cells are also visible. (**DD:** small-cell squamous carcinoma in situ of the cervix.) ×400.

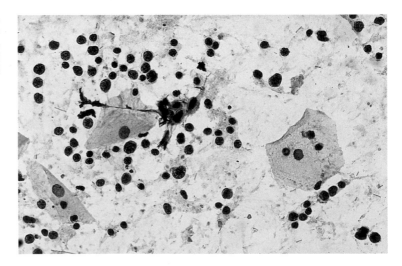

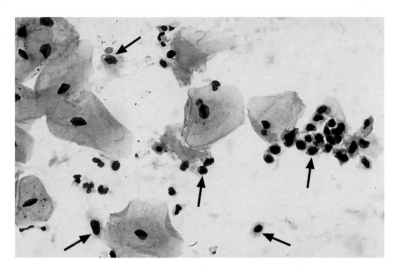

Fig. 5.**103 b Small-cell squamous carcinoma in situ of the cervix.** The small squamous carcinoma in situ cells are partly single, partly arranged in loose groups (⟶). The cytoplasm has distinct borders, and the nuclei are enlarged, hyperchromatic, and condensed. Normal superficial cells are visible on the left. (**DD:** basal cell hyperplasia.) ×400.

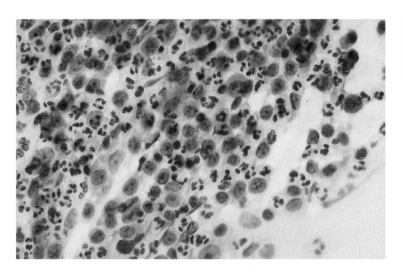

Fig. 5.**104 a Follicular cervicitis.** Diffusely arranged reticular cells with hardly recognizable cytoplasm. The hyperchromatic, relatively polymorphic nuclei show chromocenters. (**DD:** small-cell carcinoma in situ.) ×500.

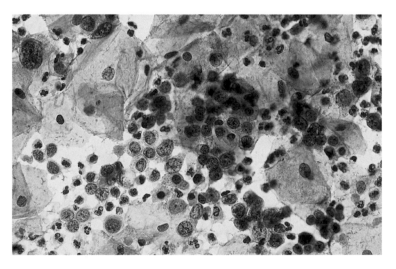

Fig. 5.**104 b Small-cell squamous carcinoma in situ of the cervix.** The dissociated small isomorphic naked nuclei are hyperchromatic and irregularly structured, and they show chromocenters. Normal intermediate cells are visible in the background. (**DD:** cervicitis.) ×400.

5

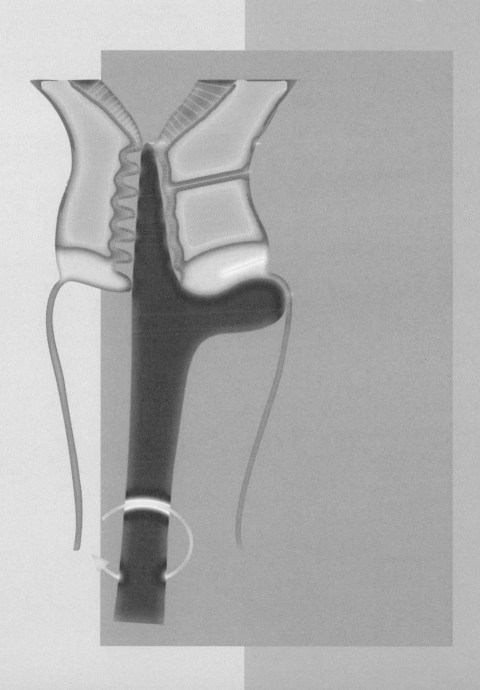

Specimen Collection and Processing

6

6 Specimen Collection and Processing

Portio, Vagina, and Endocervix

Cytology

Taking a smear for early detection of cervical cancer is probably the most important routine procedure carried out in a gynecological practice, because it can effectively prevent the development of cervical carcinoma. In Germany, this is considered a screening examination, meaning that it is carried out on clinically healthy women at least once a year and is paid for by the health insurance system.

The utmost care should be taken with such an examination, as any avoidable error during cell sampling may have legal consequences for the physician. Although the correct and efficient sampling of smear material should be an essential part of medical specialist training, it frequently happens that the quality of smears is unsatisfactory and the cytology laboratory has to advise office-based physicians on the correct sampling techniques.

The cells are sampled from the portio vaginalis or, in women with a history of hysterectomy, from the fornix of the blind vaginal pouch.

Prerequisites for Cell Sampling

Unfavorable conditions. The menstrual phase is not a suitable time for taking a smear, because blood, cell debris, leukocytes, and endometrial cells make it difficult to establish a diagnosis. Inflammation also has an unfavorable effect on the assessment and, if possible, should be treated before a smear is taken.

Although postmenopausal atrophy of the vagina is not an optimal condition, it is nevertheless often unavoidable because general hormonal therapy before cell sampling would be too uneconomic.

No medication should be applied intravaginally within 1–2 days before the smear sample is taken. Any interventions in the region of the portio vaginalis—such as cauterization, thermocoagulation, cryotherapy or laser therapy, polypectomy, curettage or conization—should occur more than 1 month before sampling. This interval also applies to sampling after delivery.

Favorable timing. Sufficient supply of estrogen to the squamous epithelium provides optimal preconditions for cell sampling, since it ensures differentiation of the cervical epithelium up to the superficial cell layer. During sexual maturity conditions are optimal in the middle of the cycle; after the menopause, they need to be induced by hormonal replacement therapy or topical estrogen application.

Site of cell sampling. The site of cell sampling, in addition to proper timing, is decisive for the efficiency of smear collection. Since about one third of all cervical carcinomas develop within the cervical canal, it is important to sample not only cells from the squamous epithelium of the portio vaginalis but also cells from the columnar epithelium of the endocervix.

Sampling Devices

Numerous different instruments are available for sampling cervical cells (Fig. 6.1). They are usually more effective when used assertively rather than in a way that is less damaging to the tissue.

Cotton swab. For a long time, until other devices were developed, cell sampling with a cotton swab was considered inexpensive and suitable. Newer techniques ensure that the specimen obtained is not only better preserved but also contains endocervical cells. In this respect a spatula is superior to a cotton swab (322). Because of its softness, the cotton swab only picks up cells already exfoliated from the surface of the tissue and subjected to cytolytic digestion by the vaginal secretion. The spatula, being much harder, removes cells from the superficial tissue layer, which are therefore preserved in good condition. Another disadvantage of the cotton swab is that it often does not

Fig. 6.**1** **Cervical sampling devices.** A selection of the numerous instruments used for sampling cervical cells. From top to bottom:
▶ plastic spatula (Szalay Cyto Spatula)
▶ endocervical brush (Cytobrush)
▶ cervical broom (Cervex Brush)
▶ platinum loop
▶ wooden spatula (Ayre spatula)
▶ cotton swab

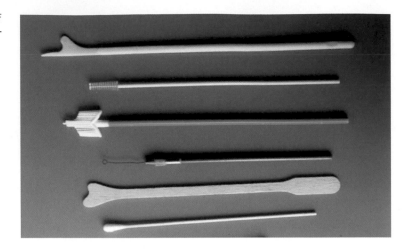

penetrate sufficiently into the endocervix. As a result, no columnar epithelial cells—or not enough of them—are harvested if the transformation zone between squamous epithelium and columnar epithelium lies inside the cervix.

Spatula. The spatula has developed from a simple **wooden spatula** (**Ayre spatula**) to the more efficient **plastic spatula** (e. g., the Szalay Cyto Spatula). The latter has a rough surface to ensure strong cell adhesion. Furthermore, the Szalay spatula has a tapered tip that permits safe penetration into the endocervix. There may be a slightly greater tendency for the epithelium to bleed when a spatula is used rather than a cotton swab; however, this is acceptable in view of the considerable advantages of the spatula method. The sampling does not cause pain, provided it is done by an experienced practitioner. There are no clear data available as to whether a plastic spatula may be safely reused; this depends essentially on its suitability for disinfection and sterilization.

Cervex Brush and Cytobrush. Other devices that also aim to collect sufficient amounts of endocervical cells include the cervical broom (Cervex Brush) and the endocervical brush (Cytobrush).
▶ The **Cervex Brush** consists of plastic lamellae of various lengths, with the shorter ones designed for gliding over the ectocervix and the longer ones for penetrating into the cervical canal. The disadvantage of this broom-shaped instrument is that large amounts of mucus may get stuck between the plastic lamellae, thus interfering with the mounting and evaluation of the smear.

▶ The **Cytobrush** consists of a conical sampler shaped like a bottle-brush. It is introduced into the cervix with a swirling movement, like a pipe cleaner. This device has the disadvantage of sampling only endocervical cells but no squamous epithelial cells. Furthermore, the fine nylon bristles often cause mechanical damage to the sensitive columnar epithelium, thus making evaluation of the smear difficult.

Other sampling devices, such as glass pipettes and platinum loops, no longer play a role these days. Some techniques, e. g., blind sampling with a tampon or the examination glove, have become obsolete because of their lack of accuracy.

Cell Sampling Method

Inserting the speculum. The routine gynecological examination consists of two parts: the visual assessment and the palpatory exploration of the pelvis. Pelvic examination usually introduces glove powder and lubricant into the vaginal cavity and wipes off any cells accumulating on the portio vaginalis. The visual part of the examination should therefore be done first. For this purpose, a specially formed metal speculum is inserted into the vagina to fully expose the portio vaginalis. Any interfering accumulation of mucus or discharge is carefully removed from the portio by means of a cotton swab.

Collecting the smear material. A suitable instrument, such as the Szalay Cyto Spatula, is introduced into the cervical canal until the shoulder of the spatula comes to rest on the

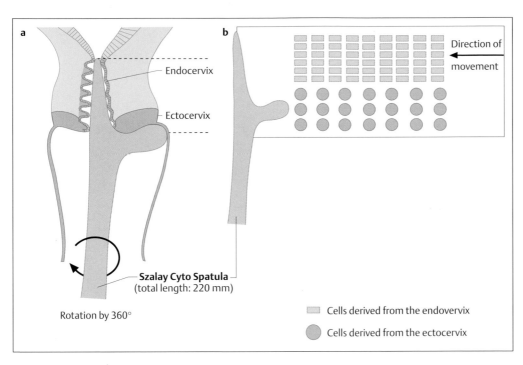

Fig. 6.2 **Szalay Cyto Spatula.**
a Application of the spatula in situ and demonstration of simultaneous sampling of cells from the endocervix and ectocervix.
b Arrangement of cells after smearing them onto the microscope slide.

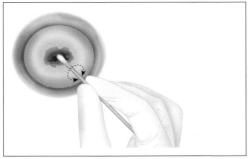

Fig. 6.3 **Using a cotton swab for sampling cervical cells.** The sampling from ectocervical and endocervical areas should be carried out with two separate cotton swabs. Although this type of cell sampling is no longer commonly used, the technique may be indicated in exceptional cases (e. g., severe atrophy of the genitals associated with reduced exfoliation of cells). The cotton swab should be rolled across the microscope slide to increase the yield of cells.

surface of the portio vaginalis. The spatula is rotated once or twice in one direction to collect cells (Fig. 6.**2 a**). It is then retracted, and any adherent mucus is carefully removed with absorbent paper. The collected cells are then transferred onto a microscope slide already prepared with the patient's name on its frosted part. The cells are evenly spread across the

slide by guiding the spatula tangentially. While doing so, it is important that the shoulder and tip of the spatula maintain the direction of movement, thus keeping ectocervical and endocervical cells separated (Fig. 6.**2 b**). To increase the yield of endocervical cells, the tip of the spatula may then be rolled over the corresponding area on the slide.

Difficult sampling conditions. In rare cases (e. g., severe atrophy, or scarring due to conization) the tip of the spatula does not completely penetrate into the cervix. If this is the case, the spatula should only be introduced as far as possible without using force or causing pain. The material collected in this way is transferred onto one half of the slide. The surface of the portio is then wiped with a cotton swab or regular spatula (Fig. 6.**3**), and the collected material is rolled or smeared onto the other half of the slide.

If a **cervical ectopy** is present and it is larger than the shoulder of the spatula, the sampling yields only cells derived from the endocervical epithelium. As a result, the smear cannot be evaluated because cells from the squamous epithelium are absent. If this is the case, additional cells from the squamous epithelium should be collected from the peripheral borders of the ectopy using a cotton swab or regular spatula. However, in the case of an ectopy, it is impossible for the cytologist to deter-

6

Fig. 6.**4 Cervical smear request/report form.** A typical blank form is divided into two fields:

a Patient particulars, clinical information, and site of cell collection

b Cytological findings, recommendations, and comments.

In the present example, the findings are reported as Pap classes according to the revised Munich nomenclature (Munich II) used in most European laboratories.

a

General information:

Name: Date of birth:

Address: Age:

Family physician: Condition:

Insurance:

Date of cell sampling: Date of last menstrual period:

_|_| _|_| _|_| (yy/mm/dd)   _|_| _|_| _|_| (yy/mm/dd)

Specimen no. of previous smear: _____ Result: _____

Clinical information:

Oral contraceptives ☐		HRT ☐		Vaginitis	☐
Vaginal bleeding ☐		IUD ☐		Pregnancy	☐
Status post radiation ☐		Cytostatics ☐		Status post hysterectomy	☐

Findings: _____

Site of cell sampling:

Portio/cervix ☐   Endometrium ☐   Mamma ☐   Vulva ☐   Vagina ☐

Other information: _____

Specific sampling technique: _____

Requesting physician:

b

Cytological Findings

Date:

Specimen No.:

Results:

Class I .... ☐	Class II ....☐	Class II W ....☐
Class III .... ☐	Class III E ....☐	Class III D ....☐
Class IV A .... ☐	Class IV B ....☐	Class V ....☐

No endocervical cells ☐   Satisfactory but limited ☐   Unsatisfactory ☐

Degree of epithelial differentiation _____ Atrophy ☐   Cytolysis ☐

Mixed flora ☐   Latobacilli ☐   Cocci ☐   Gardnerella ☐

Mycosis ☐   Trichomonas ☐   Herpes ☐   HPV ☐

Recommendations:

(a) Cytological follow-up

Immediately ☐   After vaginitis therapy ☐   After estrogen therapy ☐

After 1 month ☐   After 3 months ☐   After 6 months ☐

(b) Histological examination:

Conization ☐   Curettage ☐   Biopsy ☐

Other information: _____

Comments:

(Cell image, specific diagnosis, etc.)

[Cytology Laboratory: name, address, and contact details]

mine the origin of columnar cells because the glandular epithelium is localized on both ectocervix and endocervix.

Fixation

The Papanicolaou stain commonly used in cytodiagnostics requires immediate fixation of the smears (within a few seconds after sampling), since even short-term exposure to air causes artifacts that interfere with establishing a diagnosis.

Spray fixation. The fastest and simplest method is to use a spray fixative.

Fixation by immersion. In this method, the slides are immersed in absolute alcohol or a mixture of alcohol and ether in a staining dish. The fixative quickly becomes diluted with water and needs to be replaced at regular intervals to avoid fixation errors. Certain adulterants of alcohol, such as toluene or pyridine, may also lead to fixation errors. To avoid any carry-over of cells, slide surfaces that carry cells must not touch each other when placed into the dish. Fixation should take at least 30 minutes. The slides are then taken out,

dried, packaged, and shipped to the cytology laboratory.

Shipment

The slides are transported in unbreakable containers, which are usually made of cardboard, plastic, wood, or expanded polystyrene. Each shipment must be accompanied by a cytological examination form, either the one for **cancer prevention** provided by the health authorities in Germany, or the one for **curative treatment** specially designed by the cytology laboratory (Fig. 6.**4**).

In addition to the patient's name, date of birth, and address, the following information is absolutely essential: date of sampling, date of the last menstrual period, results of previous Pap tests, hormonal or cytostatic therapy, inflammation, current use of an IUD, previous gynecological surgery or radiotherapy, pregnancy, vaginal bleeding. Furthermore, the site of sampling (cervix, vagina, endometrium, mamma, or vulva) and the specific sampling technique (spatula, cotton swab, brush, or puncture) must be indicated on the form.

Table 6.**1** Papanicolaou stain (modified for automated processing)

Step	Set time (minutes)
1 Hydration	
▶ Ethanol 50 %	1
▶ Distilled water	1
2 Hematoxylin (staining)	2
3 Partial destaining	
▶ Running tap water	1
▶ Hydrochloric acid 0.25 %	1
▶ Running tap water	1
4 Dehydration	
▶ Ethanol 70 %	1
▶ Ethanol 80 %	1
▶ Ethanol 96 %	1
5 Orange G (staining)	2
6 Rinsing	
▶ Ethanol 96 %	1
▶ Ethanol 96 %	1
7 Polychrome (staining)	2
8 Rinsing	
▶ Ethanol 96 %	1
▶ Ethanol 96 %	1
Complete dehydration	
▶ Xylene	2
▶ Xylene	2

Receipt of Specimens

Once the shipment has been received by the cytology laboratory, the smear samples are first assigned to their accompanying examination forms. Every specimen is given an **identification number**, which must appear both on the slide and on the examination form. Nowadays, paging machines or flat-bed printers are often used to facilitate processing.

The patients' names and identification numbers are recorded in a laboratory log book or entered into an electronic data processing system. Recently, electronic scanners have been used which automatically read the address labels on the examination forms and assign numbers to them.

Staining

Nowadays, specimens for gynecological cytodiagnosis are stained almost exclusively according to a procedure empirically developed by **Papanicolaou** (Table 6.**1**). This procedure is relatively expensive, and both the aqueous and

alcoholic solutions employed may damage the cells if not applied properly.

Hydration and nuclear staining. Because the smears are fixed in alcohol, it is necessary to rehydrate them in a descending series of diluted alcohol so that they can be stained in an aqueous **hematoxylin** solution. This dye stains mainly the heterochromatin of the nuclei. Subsequent washing in dilute hydrochloric acid is used to bleach the undesired staining of the cytoplasm, which releases the dye more readily than the nuclei do. The hydrochloric acid is then eliminated by rinsing the samples in tap water.

Dehydration and staining of the cytoplasm. The samples are subsequently dehydrated using an ascending alcohol series and further processed with alcoholic dyes that specifically stain the cytoplasm. **Orange-G** produces a red stain and **polychrome** a blue stain of the cytoplasm. All residual water must be removed by incubation first in absolute alcohol and then in xylene before the samples are mounted. Complete dehydration is essential because it prevents optical artifacts (e. g., the cornflake artifact) that would interfere with the interpretation of the smear.

Manual vs. automatic staining. The **manual staining procedure**, which is appropriate for small laboratories, requires a series of staining dishes arranged in tandem. The microscope slides are placed into special metal holders that fit into the staining dishes. **Automatic stainers**, which save a lot of time and ensure a uniform rendition of colors, are used in almost all large laboratories.

Staining quality. All staining and washing solutions must be regularly renewed, usually once a week depending on the number of specimens processed. To maintain a uniform staining quality, continuous monitoring and quality control carried out by a cytologist is essential. There are numerous sources of error, such as improper cleaning of dishes, inadequate filtration of solutions, etc., and these may affect the quality of the stain considerably.

Alternative stains. Several attempts have been made to replace the traditional Papanicolaou stain by less costly methods (Pap stain alternatives). So far, however, this has only be possible at the expense of reduced quality. Hence, alternative procedures are not recommended.

6

Mounting

To protect and preserve the specimens, it is essential that they are embedded in a translucent medium and mounted with a **coverslip**. Unmounted specimens cannot be evaluated, since the presence of a glass coverslip is taken into account when calculating the optical path of the microscope.

An epoxy **mounting medium** is used to attach the coverslip to the specimen. It is essential that no air bubbles are formed during this process, as any inclusion of air will result in optical artifacts that may interfere with the evaluation of the smear. Mounting is done either **manually** or by means of an **automatic mounting machine** that uses glass or plastic coverslips.

Workplace Conditions

Most of the solutions employed in the Papanicolaou stain cause irritation of skin and mucosae. Direct contact and the inhalation of fumes should therefore be avoided. Proper **ventilation** of the laboratory and a **fume hood** at the workplace are essential.

Gloves should be worn for all manual work, in order to prevent chemical or allergic skin reactions.

Archiving and Waste Disposal

The completed examination forms and the processed smears must be filed chronologically, in binders and collection boxes respectively, which are then stored in filing cabinets. In Germany, the legal **period of safe keeping** is 10 years. If no electronic data bank is available, the laboratory's log books should allow access to old specimens as promptly as possible. Documents that are no longer subject to legal safe keeping must be destroyed according to current data protection laws.

Used staining solutions are stored in containers and collected at regular intervals by waste disposal companies. Alcoholic solutions may be recycled by means of distillation.

Reporting Results and Communicating with the Requesting Physician

In Germany, the cytological laboratory must provide the sender of the specimen with all material required for shipping and packing; this includes prepaid postage and microscope slides, but does not include fixative and sampling devices. The completed examination forms are then returned to the sender by regular post or courier. More modern techniques permit electronic **data transmission** by email, if the physician's practice is connected to the appropriate network.

Nevertheless, it is advisable to establish a personal contact with the referring physician to ensure a high level of quality with respect to sampling and evaluation. Positive findings, in particular, should be communicated by telephone if they require histological confirmation or show special features.

It is essential to have a manual or automated system in place for sending regular reminders to the requesting physician for **follow-up cases** or **cases requiring confirmation** if no specimens have been received within certain deadlines.

Histology

Biopsy

Portio vaginalis. When a suspicious lesion is discovered by colposcopy, and especially when this coincides with positive cytological findings, small pieces of tissue may be removed by means of biopsy forceps (Fig. 6.**5**). The tissue samples are about the size of a grain of rice, and they are cone-shaped because of the bite function of the forceps. Although this intervention is painless and does not require general or local anesthesia, such bite biopsies from the portio vaginalis should only be done in exceptional cases. Their diagnostic reliability is clearly inferior to that of biopsies obtained by conization, and there is no guarantee that the localization of a histologically positive area corresponds to that of the colposcopically suspicious area. Furthermore, a biopsy does not constitute definitive treatment, whereas conization does (see also p. 212 f.). A tissue specimen obtained by bite biopsy is not much larger than a grain of rice. Because its surface cannot be marked, the direction of sectioning cannot be indicated during the embedding process. Hence, tangential or horizontal sections may result, thus interfering with the histological

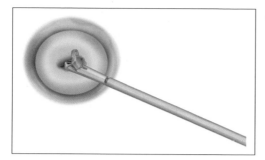

Fig. 6.**5** **Bite biopsy** in the area of the portio vaginalis.

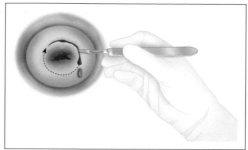

Fig. 6.**6** **Cervical conization (cold knife cone biopsy).** Excision of ectocervix and endocervix with a scalpel. The anatomical conditions are almost completely restored after a few weeks because the tissue in this region has a remarkable ability to regenerate.

evaluation. The resulting pathological findings are often doubtful.

Vagina. A biopsy forceps is nevertheless indispensable in gynecological practice. Its area of application concerns not so much the portio vaginalis as the vaginal mucosa. Suspicious lesions often appear as erythroplakia or leukoplakia, particularly in the fornix of the blind vaginal pouch created by hysterectomy, with the cytological smear usually providing the first indication of dysplasia. When using a biopsy forceps, the vaginal mucosa needs to be held with a bullet forceps or surgical tweezers to prevent the instrument from sliding away. Hemostasis is usually not required, though it may be carried out with a special pen or solution, if needed. The tissue samples are then fixed in formalin and sent to a pathology laboratory.

Cervical Conization

Principle. Conization involves the excision of a cone-shaped tissue sample from the center of the portio vaginalis where the opening of the cervical canal is located (Fig. 6.**6**). Excision is carried out by means of a scalpel, laser, or electrical loop.

Type of incision. The angle of the incision depends on the localization and size of the lesion to be excised. If the lesion is intracervical, the instrument should cut at a steep angle and deep into the portio vaginalis, thus keeping the defect on the surface relatively small. If the lesion is ectocervical, the angle of the incision depends on the spread of the lesion. It is there-

fore recommended that the iodine test is carried out before conization to ensure the complete surgical removal of the resulting iodine-negative area. The shallower the angle of the incision, the larger the defect on the surface and the resulting wound.

Curettage. After removal of the cone biopsy, the cervical canal is dilated by means of Hegar dilators, and the curette is inserted into the uterine cavity for the obligatory curettage. This additional intervention is easily done and extends the diagnostic possibilities, since it also allows for histological examination of the glandular epithelium of both endocervix and endometrium.

Hemostasis. Bleeding from the surgical wound of the portio vaginalis is stopped by electrical coagulation; sutures may be required in rare cases. The entire intervention is done as an outpatient procedure, even though it requires general or local anesthesia in most cases. As a rule, regeneration of the wound is satisfactory and does not interfere with future pregnancies.

Biopsy samples. The excised cone of tissue is marked according to its topography to help the pathologist with orientation. Both tissue cone and curettage sample are fixed in formalin and sent for histological evaluation to a pathology laboratory.

6

Fig. 6.**7** **Endometrial sampling devices.** A selection of the numerous instruments used for sampling intrauterine cells. From top to bottom:
▶ Endoshave
▶ Uterobrush
▶ Endopap
▶ Prevical
▶ Spirette
▶ Endorette
▶ Mi-Mark endometrial cell sampler (helix and brush)
▶ Pistolet

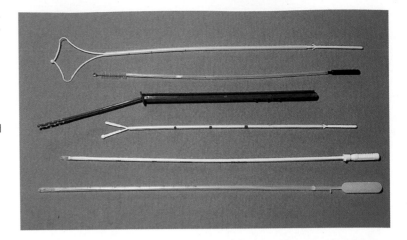

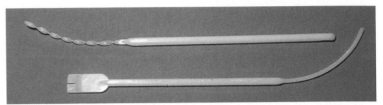

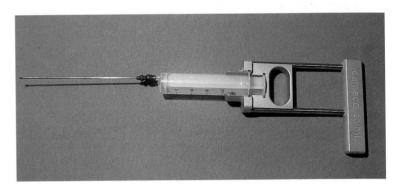

Endometrium

Cytology

Cervical smear. Unlike the Pap test for early detection of cervical cancer, intrauterine cell sampling is not considered a screening test. It is only done when mandated by clinical symptoms. At the most, only half of uterine corpus carcinomas are detected or suspected by routine cervical smears. As this type of cancer almost exclusively occurs after the menopause, any endometrial cells present in the routine smear during this period of life should receive special attention. Detection of endometrial cells after the menopause in the absence of hormone replacement is as important as vaginal bleeding, and requires further examination. Cyclic appearance of such cells

during hormone replacement therapy, however, is not a cause for concern.

Ultrasound measurement of endometrial thickness. This method has increasingly gained acceptance as a diagnostic aid and has essentially replaced the less reliable techniques of intrauterine cell sampling. If the endometrium is about 1 mm thick, it is very unlikely that a uterine corpus carcinoma is present. If the cervical smear is negative at the same time, a single postmenopausal bleed usually does not prompt a request for curettage; rather, a waiting attitude with observation of the clinical course is advised.

Intrauterine cell sampling. Numerous methods for intrauterine cell sampling have been developed in the past, but none of them has

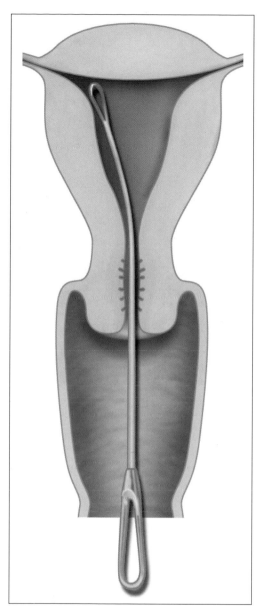

Fig. 6.8 Fractionated curettage (exploratory curettage). Scraping of the uterine cavity for sampling endometrial tissue by means of a curette.

pose being not so much the early detection of cancer as the detection of inflammatory reaction of the endometrium.

Sampling techniques. Only techniques that are both easy and inexpensive should be considered for the targeted sampling of intrauterine cells. Thin plastic tubes are used for this purpose, through which a spiral, loop, brush, or sponge is inserted into the uterine cavity and withdrawn after several rotations. Such sampling devices include, among others, the Uterobrush, Endopap, Endorette, and Mi-Mark helix (Fig. 6.7). The cells collected are smeared onto a slide and then fixed. Older procedures, in which the cells were directly sampled by means of needles, or aspirated from the uterine cavity by means of special irrigation techniques (Pistolet or Vabra aspiration, jet washing), are expensive and no longer in use.

Importance. It should be taken into account that the application of these techniques is painful and expensive, and their reliability is limited. For technical reasons, no cells or only insufficient amounts are sampled in 20% of all cases. Although the cytodiagnostic hit rate is pretty high (about 90%), nothing can be said about the sensitivity of these methods because they are not used as screening tests.

Histology

Fractionated curettage. To collect cells from the uterus, the cervical canal is first dilated by means of Hegar dilators so that the curette may be inserted into the uterine cavity. However, this procedure is painful and requires general or at least local anesthesia.

First, the endocervix and then the uterine cavity are systematically scraped with the curette. This approach is called **fractionated curettage** (Fig. 6.8). The fractioning of samples is required for separate histological examination of the endocervix and endometrium (see also p. 235). Hemostasis is not required.

The endocervical and endometrial tissue samples are separately fixed in formalin and sent for analysis to a pathology laboratory.

been widely used. If sampling becomes necessary, curettage is usually preferred; it is only a minor intervention but yields more reliable results. If the patient has limited tolerance for surgery or general anesthesia, and the ultrasound findings are suspicious, intrauterine cell sampling may nevertheless be appropriate.

Intrauterine devices. The cells adhering to an extracted IUD are often smeared onto a slide for opportunistic reasons, and the specimen is sent for diagnostic evaluation, with the pur-

6

Ovary

Cytology

Ovarian puncture. Cytological samples are collected by ovarian puncture guided either by ultrasound or by laparoscopy (Fig. 6.**9**). Ovarian cysts that measure more than 3–4 cm in diameter, persist for more than 3 months, and are thick-walled or multilocular, need to be examined (see also p. 240). The submitted material should be processed within a few hours to prevent autolysis of the cells. The puncture fluid is centrifuged, and the sediment is transferred onto a slide and fixed.

Cervical smear. The morphological structure of suspicious glandular cells in the cervical smear may also give rise to suspicions of ovarian carcinoma, thus prompting further measures. Since the cytological distinction from uterine corpus carcinoma is usually difficult, both curettage and laparoscopy should be done if ovarian carcinoma is suspected.

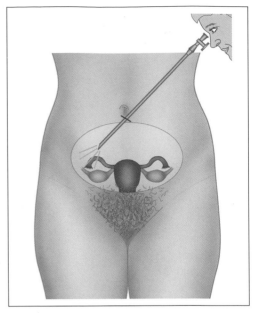

Fig. 6.**9** **Laparoscopic inspection** of the female internal genitals.

Histology

Histological examination in case of suspicious ovarian findings is possible through laparoscopic **biopsy** or by laparotomy with partial or complete **resection** of the ovary.

Vulva

Cytology

Cotton swab. Unlike the situation with the cervix, a cotton swab is well suited for sampling cells from the vulva, since the collected horn cells are resistant and do not undergo lysis. Furthermore, cell sampling is more efficient this way than with the spatula.

The cotton swab used for sampling should be **mechanically** manufactured, not handmade, in order to avoid contamination with horn cells from the hands. To improve cell adhesion, the cotton swab is **moistened** with water or saline (Fig. 6.**10 a**). If it is too wet, not enough cells are sampled because they are washed away. The moist cotton swab is vigorously rubbed several times over the suspicious skin area to increase the yield of cells and then rolled across a slide (Fig. 6.**10 b**).

Fixation is carried out by spraying or immersion. The accompanying examination form must list the macroscopic findings, the patient's history, and the exact site of sampling.

Other instruments for cell sampling, such as spatula, scalpel, or adhesive tape, have not gained acceptance because of the low cell yield and unfavorable cell orientation.

Histology

Biopsy

Punch biopsy. For the keratinized vulvar skin, a punch biopsy is more effective and gentler than a **bite biopsy** (Fig. 6.**11**, 6.**13 a**). Because of its thickness, the epidermis requires deeper penetration than the mucosal epithelium. When using a biopsy forceps, a large surface defect is created because of the bite-like type of incision; the result is a cone-shaped specimen, with relatively little material being collected from the deeper layers of the epidermis.

By contrast, the **punch cylinder** creates a skin defect with a constant diameter of 4 mm and collects sufficient tissue from the deeper layers because of the vertical nature of the incision (Fig. 6.**12**, 6.**13 b**).

After the application of local anesthetic, the punch cylinder is advanced into the skin with a rotating motion and then retracted. The piece

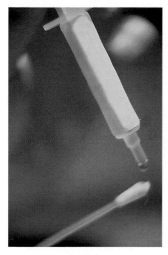

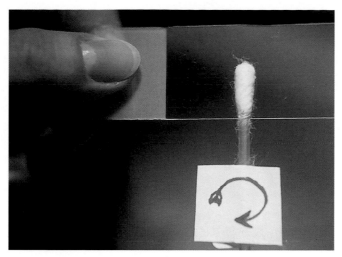

Fig. 6.**10   Collection of cells from vulvar skin.**

a Moistening a cotton swab for the sampling of cells.

b Rolling the cell-containing cotton swab across the microscope slide to increase the yield of cells.

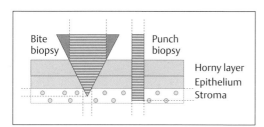

Fig. 6.**11   Vulvar bite biopsy vs. punch biopsy.** Comparison of the differences in surface defect and stromal portion.

of tissue, which is about the size of a grain of rice, is cut off from the stroma using scissors. It is placed onto a small piece of absorbent card, with its stromal side down for topographic orientation. The skin defect is closed with a single suture and disinfected. As a rule, there is no need for a dressing.

The biopsy with the attached card is subsequently fixed in formalin and sent to a dermatology laboratory.

Tumor Excision

Circumscribed lesions should be removed after the area has been stained with **toluidine blue**. The staining creates sharp borders between the diseased skin areas and the normal skin (Fig. 6.**14**), thus facilitating wide excision. Depending on the size of the lesions, either general or local anesthetic is required.

The **surgery** is usually done in an outpatient clinic. In most cases, the lesion is excised with a scalpel. The piece of skin is placed onto a piece of cardboard with the stromal side down, fixed in formalin, and sent to the pathology laboratory. The wound is taken care of, disinfected, and covered with sterile dressing.

Fig. 6.**12   Disposable punch cylinder** for collecting a tissue specimen from vulvar skin.

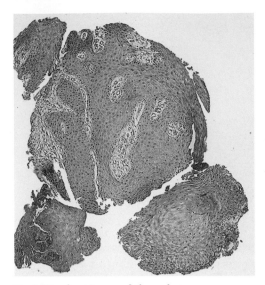

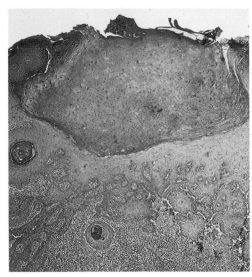

Fig. 6.13 a, b   Biopsy of the vulva

a   Bite biopsy. Because the type of incision is cone-shaped and the vertical orientation of the excised material is absent, only horizontal tissue sections are possible. This severely limits the diagnostic evaluation. HE stain, ×40.

b   Punch biopsy. Because the incision is cylindrical and the vertical orientation of the excised material is preserved, vertical tissue sections are possible. This creates optimal conditions for diagnostic evaluation. HE stain, ×40.

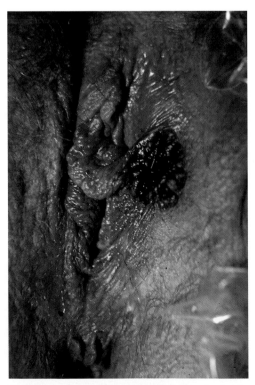

Fig. 6.14   Preoperative toluidine blue test. The left labium contains a large circular area (about 2 cm in diameter) which stains intensely blue.

Analysis
and Efficiency

7 Analysis and Efficiency

Cytological Findings

Screening Process

Even in today's era of technology and automation, visual analysis of the cytological specimens is first and foremost a job to be done within a specific time frame by a human being sitting at the light microscope.

For differentiating morphological features of individual cells (in particular, the size of cells and nuclei), it is important to chose an **optical scale** that is small enough to ensure sufficient visual acuity for this purpose. For effective screening within a short period of time, however, the scale should not fall below a certain minimum. To meet these two demands, screening the slides at 100-fold magnification (using a 10× eyepiece and a 10× objective) has become the standard for routine diagnostic procedures (368). At this magnification, the visual field contains about 100 cells on average, and these can usually be evaluated simultaneously at a single glance.

The **screening activity** consists of visually analyzing the entire cell material present on the microscope slide, and this involves looking at a long series of individual visual fields. The slide is manually advanced by the mechanical stage (cross-table) of the microscope, moving it along the horizontal and vertical axes by the diameter of one visual field at a time. Once the edge of the slide is reached, the slide is shifted horizontally by one visual field and then moved vertically in the other direction. By using this criss-cross type of examination, it is possible to read an average of 500 visual fields per slide, like a series of columns.

When the visual analysis is carried out by a highly trained and experienced cytologist (cytotechnologist), the average **analysis time** is 3 minutes per slide, provided the specimen is in perfect technical condition and does not contain pathological changes. If cellular abnormalities are present, it becomes necessary to re-examine individual visual fields in more detail, thus spending several additional seconds looking at specific fields, and to use higher magnifications of up to 400× for analyzing morphological details (in particular, nuclear envelopes and chromatin structures), which further lengthens the screening time considerably.

The personal **screening pattern** of the cytologist may be electronically monitored and thus objectively evaluated by means of sensors installed in the microscope. This pattern may be used for quality control, if necessary (334).

Diagnostic Evaluation

After visual analysis of a smear, the cytotechnologist suggests a diagnosis which must be confirmed or corrected by the director in charge of the laboratory.

The analysis of a cytological specimen is followed by the diagnostic interpretation, namely, whether the findings are normal or pathological. In the case of pathological findings, one has to decide whether they are benign or malignant.

Benign lesions usually have an effect on the **overall cell image**; they are recognized by the microbial flora and diffuse cellular changes visible throughout the entire smear. By contrast, malignancies manifest themselves predominantly as abnormalities of individual cells (or groups of cells) in the **single cell image**. Benign and malignant lesions may occur simultaneously, and this may complicate the diagnostic evaluation. Sometimes it is impossible to decide on the type of pathological change, or whether the findings are pathological at all.

A high degree of **visual acuity** and experience in morphological analysis are required for making a correct assessment. Furthermore, there is a certain subjective **latitude** which may also fluctuate over time. Especially when cellular changes are doubtful and suspicious, the cytotechnologist and the laboratory director should meet and discuss the diagnostic evaluation directly at the discussion microscope. In the end, however, it is the director of the laboratory who makes the **final decision**.

Documentation of Results

There have been numerous suggestions as to how the results of the cervical smear analysis should be documented, ranging from brief numerical classifications to extensive verbal descriptions. On the whole, the Bethesda System (TBS) and the German classification system (Munich nomenclature) have gained acceptance, either in their original versions or as some modifications thereof (e. g., TBS-3, Munich II).

Prerequisites. Much effort has been directed towards documenting the results of the diagnostic evaluation in a reporting system that is reproducible and easy for clinicians to understand. Almost all these systems have in common that they first report whether sufficient cell material was collected from the sampling site, whether the condition of the smear specimen was acceptable for evaluation, and whether endocervical cells were present. Comments on the degree of squamous epithelial proliferation and on the microbial flora are obligatory as well, since they make it possible to draw conclusions about atrophic or inflammatory factors.

The most important component of a report is the final information as to whether or not the smear is normal or pathological. In the case of **pathological findings**, the cytologist comments on whether or not certain benign or malignant lesions are suspected. Any inconclusive or doubtful results require an explanation and a repeat smear examination.

Numerical classification. A numerical classification supplemented by verbal comments is preferred in Europe, if only for reasons of quality control (365). Here, the different degrees of the severity of pathological changes are subdivided into classes, using the roman numerals I–V. This scheme was first introduced by Papanicolaou in 1963 (276). Assignment of a smear to class I means that the findings are normal, to class II that inflammatory changes are present, and to class III that the findings are doubtful. Assignment to class IV indicates suspicion of a carcinoma, and assignment to class V indicates the presence of an invasive carcinoma.

Munich Nomenclature. After precancerous stages had been described in detail in the 1950s and 1960s, first in terms of histology and then in terms of cytology, it became necessary to integrate these stages into the existing reporting system. The first Munich nomenclature (Munich I) was thus created in 1975, and the second (Munich II) followed in 1989. An important part of the new system was the creation of further subclasses, such as III D, IV a, and IV b, thus allowing for a more accurate distinction between cervical lesions (363) (Table 7.**1**).

Now it is possible to discriminate between a low-grade precancerous lesion (class III D), which includes both mild and moderate dysplasia, and a high-grade precancerous lesion (class IV a), or between a potentially invasive carcinoma (class IV b) and a clearly invasive one (class V). Inflammatory or degenerative changes, which are not clearly distinguishable from mild dysplasia, were assigned to class II with the comment "repeat Pap test after treatment," and suspicious glandular cells were assigned to class III with the comment "atypical glandular cells."

The Bethesda System (TBS). The Bethesda classification was introduced in 1989 and updated in 2001. It consists exclusively of **verbal documentation** and is mainly used in the United States (Table 7.**2**) (241). The main difference from the Munich classification is the assignment of moderate dysplasia to high-grade precancerous lesions. The subdivision of doubtful findings into those with squamous cells (atypical squamous cells of undetermined significance, ASC-US) and glandular cells (atypical endocervical cells and atypical endometrial cells) represents a certain discrepancy, but it does not constitute a fundamental difference. It is worth mentioning, however, that clinical recommendations deduced from the diagnostic evaluation have been integrated to a larger extent into the Munich nomenclature than into the Bethesda System.

Modified nomenclatures. So far, there is no internationally binding terminology, since different countries have very different ideas about the suitability and practicability of various reporting systems. The decision to use a particular system should depend, above all, on the benefits for the patient or her physician. For example, it is of little practical importance to document the presence of certain cell populations, such as blood cells, histiocytes, metaplastic or regenerative epithelial cells; the clinician has little use for this information, and it requires no specific action. Rather, it is far

Table 7.**1** **Munich Nomenclature.** Summary of the Munich II reporting system commonly used in German-speaking countries for the cytological analysis of Pap smears (second version of 1989, including the 1997 Freiburg Supplement which recommends colposcopy)

Adequacy of specimen	Stage of epithelial development	Microorganisms
▶ Satisfactory for evaluation ▶ Satisfactory but limited ▶ Unsatisfactory If a smear is of limited or unsatisfactory quality, specify the reason: ▶ Not enough cells ▶ Poor fixation ▶ Severe degenerative changes ▶ Obscuring inflammation ▶ Obscuring blood ▶ Obscuring superimposed cells ▶ No endocervical cells	According to Schmitt (stages 1–4)	▶ Lactobacilli (with or without cytolysis) ▶ Mixed flora ▶ Cocci ▶ Gardnerella ▶ Mycosis ▶ Trichomonads ▶ Others

Class	Criteria of classification	Recommendation
I	Normal cell image, according to age, including mild degenerative changes and bacterial cytolysis	
II	▶ Distinct inflammatory changes of cells derived from the cervical squamous and columnar epithelia; cells from regenerative epithelium, immature metaplastic cells, severe degenerative changes, keratotic cells ▶ Normal endometrial cells, also after menopause ▶ Special cell image, such as follicular cervicitis, changes due to IUD, signs of HPV infection without distinct nuclear changes, signs of infection with herpesvirus or cytomegalovirus	Possibly cytological follow-up, with interval depending on clinical findings—perhaps after treatment of inflammation or hormonal application
III	Inconclusive findings: ▶ Severe inflammatory, degenerative, or iatrogenic changes; a clear distinction between benign and malignant changes is not possible ▶ Suspicious cells derived from a glandular epithelium; their origin from carcinoma cannot be excluded completely; if possible, indicate whether they may be of endometrial, endocervical or extrauterine origin	Short-term cytological follow-up or immediate histological examination, depending on clinical and colposcopic findings
III D	Cells derived from mild to moderate dysplasia (signs of HPV infection should be mentioned in particular)	Colposcopic and cytological follow-up after 3 months
IV a	Cells derived from severe dysplasia or carcinoma in situ (signs of HPV infection should be mentioned in particular)	Colposcopic and cytological follow-up, histological examination
IV b	Cells derived from severe dysplasia or carcinoma in situ; their origin from invasive carcinoma cannot be excluded	Colposcopic and cytological follow-up, histological examination
V	Cells derived from malignant tumors ▶ Cells of squamous carcinoma (keratinizing or nonkeratinizing) ▶ Cells of adenocarcinoma; if possible, indicate whether they may be of endometrial, endocervical, or extrauterine origin ▶ Cells of other malignant tumors	Colposcopic and cytological follow-up, histological examination

more helpful to describe a specific condition, such as inflammation, atrophy, or menstruation, especially when these conditions affect the quality of the smear image. By suitable measures it is possible to prevent such interference from occurring in the repeat smears. Modified classifications are used in many countries, and in our laboratory we use a modified Munich II nomenclature (Table 7.**3**).

7

Table 7.**2** **Bethesda Nomenclature.** Summary of the Bethesda reporting system commonly used in the United States for the cytological analysis of Pap smears (1989 version, slightly modified)

Adequacy of specimen	**Within normal limit (negative)**

▶ Satisfactory for evaluation
▶ Satisfactory but limited
▶ Unsatisfactory
 	
If a smear is of limited or unsatisfactory quality, specify the reason:
▶ Not enough cells
▶ Poor fixation
▶ Cytolysis or degeneration
▶ Obscuring inflammation
▶ Obscuring blood
▶ Obscuring artifacts
▶ No endocervical cells
▶ Cells not representative for site of sampling
	Benign cellular changes
	
	▶ Inflammation (including regeneration)
	▶ Atrophy
	▶ Radiation
	▶ IUD user
	▶ Endometrial cells (in women over 40)
	▶ Glandular cells after hysterectomy
	
	Epithelial cell abnormalities
	
	Squamous cell
	▶ Atypical squamous cells
	  – of undetermined significance (ASC-US)
	  – cannot exclude HSIL (ASC-H)
	▶ Low-grade squamous intraepithelial lesion (LSIL), including HPV, mild dysplasia/CIN I
	▶ High-grade squamous intraepithelial lesion (HSIL), including moderate dysplasia/CIN II, severe dysplasia/CIN III, or carcinoma in situ/CIS
	▶ Invasive squamous carcinoma
	
	Glandular cell
	▶ Atypical endocervical cells
	▶ Atypical endometrial cells
	▶ Others
	▶ Endocervical adenocarcinoma in situ
	▶ Invasive adenocarcinoma of the endocervix or endometrium, or of extrauterine origin (e. g., ovary)

Specimen type and method of analysis ▶ Conventional Pap smear ▶ Liquid-based Pap smear ▶ Conventional analysis ▶ Automated analysis	
Microorganisms ▶ Bacteria (*Gardnerella*, *Actinomyces*, others) ▶ Fungi (*Candida*, others) ▶ Trichomonads ▶ Viruses (herpes simplex virus, others)	
	Other malignancies (specify, e. g., lymphoma)

Appropriate Responses

The cytological findings should be reported to the requesting physician in writing, using clear and unambiguous wording to avoid any uncertainties arising. If the smear is doubtful or positive, the laboratory should request adequate follow-up measures.

Improving the quality of the smear. If the specimen is **unsatisfactory** for evaluation, the smear should be repeated as soon as possible (within a month).

If the specimen is **satisfactory but limited** by interfering factors, a slightly longer follow-up interval of 3–6 months may be acceptable. It is essential that the physician is informed of the reasons why the evaluation of the smear is limited, so that any shortcomings may be corrected the next time a smear is collected. The most common reasons include incorrect sampling techniques, unsuitable sampling devices, sampling at the wrong time or at the wrong site, and also incorrect fixation techniques.

If such problems occur repeatedly, the cytologist should contact the physician directly.

Avoiding effects caused by inflammation and atrophy. Even when using correct sampling and fixation techniques, patient-specific factors interfering with the smear image may not be avoided completely. These are usually inflammatory and degenerative processes, which cause enlargement of the nuclei and thus lead to problems of differentiation from malignant lesions.

If these uncertainties apply only to **mature squamous cells** (class II W), a control smear soon after adequate topical treatment is sufficient. However, if they apply to **immature squamous cells** or **glandular cells** (class III, III E), controls after shorter intervals (within 1 month), or

Table 7.**3** **Modified Munich Nomenclature.** Summary of the modified Munich II reporting system used in our laboratory for the cytological analysis of Pap smears

Adequacy of specimen	Stage of epithelial development	Microorganisms
▶ Satisfactory for evaluation ▶ Satisfactory but limited ▶ Unsatisfactory ▶ Atrophy ▶ Cytolysis ▶ No endocervical cells	Modified Schmitt classification (stages 0–4)	▶ Lactobacilli ▶ Mixed flora ▶ Cocci ▶ *Gardnerella* ▶ Mycosis ▶ Trichomonads ▶ Herpes simplex virus ▶ Signs of HPV infection

Class	Criteria of classification	Recommendation
I	Normal image with enough representative cells; no damage due to technical, degenerative, or inflammatory factors	
II	Just enough cells, or only partly representative cells, and/or damage due to technical, degenerative, or inflammatory factors	
II W	Interpretation of cells is difficult because of their specific morphology or changes caused by technical, degenerative, or inflammatory factors; impossible to distinguish from mild or moderate dysplastic cells	Repeat smear after 1–6 months, possibly after treatment of vaginitis or after estrogen therapy
III	Interpretation of cells is difficult because of their specific morphology or changes caused by technical, degenerative, or inflammatory factors; impossible to distinguish from severe dysplastic cells or malignant tumor cells	Repeat smear after 1 month, possibly after treatment of vaginitis or after estrogen therapy; or histological examination (curettage, conization, exploratory excision), possibly colposcopy and HPV analysis
III E	Detection of endometrial cells or other atypical glandular cells after menopause without hormone replacement therapy, or in case of abnormal bleeding during hormone replacement therapy	Ultrasonography or histological examination (curettage, possibly conization) and/or laparoscopy
III D	Detection of mild or moderate dysplastic cells, possibly with signs of HPV infection	Repeat smears at intervals of 3–6 months, with colposcopy and possibly also HPV analysis; after a course of more than a year, possibly histological examination (conization, curettage, exploratory excision)
IV A	Detection of severe dysplastic cells or squamous carcinoma in situ cells, possibly with signs of HPV infection	Histological examination (conization, curettage, exploratory excision); under special circumstances (pregnancy, post partum, severe inflammation) possibly repeat smears at control intervals of one month, with colposcopy and possibly HPV analysis
IV B	Detection of questionable squamous carcinoma cells	Histological examination (conization, curettage, exploratory excision) after colposcopy and possibly HPV analysis
V	Detection of squamous carcinoma cells, abnormal glandular cells, or other malignant tumor cells	Histological examination (conization, curettage, exploratory excision) after colposcopy and possibly HPV analysis; possibly laparoscopy

even an immediate histological examination, are required. If repeat smears continue to be doubtful, HPV analysis may be helpful in making a decision regarding further measures.

Topical and/or systemic treatment with antibiotics, antimycotics, antiviral agents, or hormones is usually necessary before taking a control smear.

Follow-up intervals after positive findings. Since most mild and moderate dysplastic lesions of the cervix show spontaneous regression, **monitoring** of these cases at short intervals (3 months) is justified. Nevertheless, these lesions have a strong tendency to reoccur; on the other hand, some dysplastic lesions are not detected by means of a single smear due to the

7

limited sensitivity of the method. Therefore, whenever there is a positive finding of class III D, it is expedient to ask for at least two negative repeat smears before returning to the usual screening interval.

If the findings persist over several years, it should be decided on a case-by-case basis if and when a **histological examination** is necessary. This decision depends on several factors, such as age, mental state, the desire to have children, and outcome of the HPV analysis. For an infection with high-risk HPV, histological examination is preferred, but for a low-risk infection follow-up smears at relatively long intervals (6 months) are recommended (see p. 213).

Histological examination after positive findings. Any suspicious results call for the following follow-up measures:

▶ Suspected **severe precancerous lesions** (class IV A): As a rule, immediate histological examination is required, as long as there are no special circumstances, such as an existing pregnancy or another exceptional situation. In special cases, monthly checks within the foreseeable future are acceptable (see also pp. 212 f., 340 f.).
▶ Suspected **invasive squamous carcinoma** (classes IV B or V): Immediate histological examination is required, and a tissue sample is removed either by conization or by punch biopsy, depending on the clinical situation. If there is already an **advanced cervical carcinoma**, a punch biopsy is usually preferred because it creates better conditions for subsequent surgery. Further measures, such as surgery or radiotherapy, depend on the clinical classification of the stage of the carcinoma (see also pp. 212 f., 340 f.).
▶ Suspected **endocervical adenocarcinoma:** Conization is recommended here because the tumor's localization corresponds to that of squamous carcinoma.
▶ Suspected **corpus uteri carcinoma:** Curettage is indicated here. If atypical glandular cells are detected (class III E), vaginal ultrasonography may be carried out as an alternative to curettage (see also pp. 235, 344).
▶ Suspected **ovarian carcinoma**: Laparoscopy is required in addition to curettage.
▶ Suspected **vulvar carcinoma:** A punch biopsy or exploratory excision is required (see also pp. 250 ff., 344 ff.).

Quality Assurance

The quality of cytological findings depends on the diagnostic and organizational skills of the laboratory personnel. The use of modern technology supports the performance of the cytology laboratory. The quality of the services provided should be regularly checked by appropriate control measures.

Individual Qualifications

Director of the laboratory. It is essential that the medical or scientific education of the **pathologist** includes cytological training that meets the requirements for providing cytological services. In Germany, these requirements are stipulated by the German medical association (Bundesärztekammer) and Germany's federal association of panel doctors (Kassenärztliche Bundesvereiningung) (167). Obligatory practical examinations at the microscope have been required since 1992. Beyond that, the personal qualifications have to be maintained by regular participation in continuing education and training programs (at least once a year). In the United States, the qualifications of cytologists are stipulated by the Clinical Laboratory Improvement Amendments (CLIA) of 1988 (5).

Specially trained laboratory personnel. Within a minimum of 1–2 years of training, a **cytotechnologist** should acquire a qualification standard that is equivalent to that of a **cytotechnologist** trained at a recognized cytology school. This is important, because the laboratory director is permitted to transfer certain steps of the cytology service to non-medical assistants, but only when these individuals have the required qualifications (334, 364). Regional and national professional associations offer special courses and certification programs such as those offered by the German Society of Cytology, the International Academy for Cytology, or the European Federation of Cytological Societies. It is essential also for cytology assistants to maintain qualification status by regular participation in continuing education and training programs. In the United States, the requirements are stipulated by CLIA 1988 (5).

Technical Qualification

Optimal performance of a cytology laboratory can only be achieved when the technical facilities meet a **high quality standard**. The laboratory area should be adequate for the workload and have ergonomic and quiet workstations. Also essential are high-resolution microscopes equipped with a discussion bridge and a photographic camera, automated staining and mounting techniques, and computerized data processing and archiving.

Organizational Qualification

Workplace of the cytology assistant. The normal activities of the cytology assistant include processing of the incoming and outgoing materials, acquisition of the patient's data, work at the microscope, monitoring of smear quality, archiving, quality control measures, using the accounting system, and communicating with requesting physicians.

For economic reasons, larger laboratories have a tendency to allocate staff to separate tasks. For health reasons, however, it should be borne in mind that working exclusively at the microscope may lead to lack of concentration and musculoskeletal strain. In order to vary the daily routine, the sedentary screening work should therefore be interrupted by other activities that are preferably carried out while standing. We also know from experience that screening is most effective when carried out in the morning.

Professional associations of cytologists and medical associations provide recommendations for the **maximal number of specimens to be screened** by a single person (174). The recommendations refer either to the unit of time per specimen or to the number of specimens per hour, day, or year. An indication of the individual daily workload, which usually lies between 60 and 80 specimens in an efficient laboratory, seems to make most sense (59). Unfortunately, these numbers are often surpassed since many laboratories suffer from a shortage of cytotechnologists, and this probably has a negative effect on the quality of the screening.

Performance control. Two types of performance control are recommended: either the **10 % control**, in which 10 % of all specimens are screened a second time, or **rapid rescreening**, in which all specimens are quickly screened once more (145). In part, automation is also used today for quality control (318).

Continuous monitoring of the performance standard is achieved by holding daily internal meetings to **discuss problem cases** at the discussion microscope. The number of detected positive and doubtful findings should not fall below a certain minimum.

Depending on the quality of the smears, the detection rate should lie between 2 % and 5 % of the specimens received (368).

Communication with the submitting physician. Quick and reliable reporting of the results will be appreciated by every physician requesting the Pap test. The results may be sent by regular post, courier, fax, or—more recently—by electronic data transmission or email. However, nothing can replace a personal phone call when **positive findings** still require further examination.

Another important service to the submitting physician are regular **notifications when control smears are overdue**. This is facilitated by setting up a data-processing system.

A good service includes the **reporting of statistical evaluations**. By means of electronic data processing, statistics for any given period of time can be quickly and easily generated for every submitting physician. Comparisons between the specimens received from individual practices and those received in total may actually be very informative. These statistics are usually compiled at the end of every 3-month period and then again at the end of the year. They essentially reflect the frequencies of specific findings.

Overall statistics of the laboratory. Detailed annual laboratory statistics used to be stipulated by Germany's federal association of panel doctors. Their purpose is to compare all positive and doubtful results of Pap tests with the corresponding histological findings and—in the absence of confirmation—with the cytological findings of repeat smears (Table 7.**4**). Even with the help of computer software, providing such detailed statistics requires a considerable amount of organization. It is usually possible only for the last year but one, because of the requirement for time-consuming feedback. Nevertheless, these statistics represent an important contribution to the internal quality control of the laboratory. In the United States, laboratory data must be retrievable for quality assurance purposes and to generate statistical reports required by regulatory agencies and accreditation organizations within the retention period prescribed by CLIA 1988 or applicable state regulations (5).

Other aspects of organization. These include the regular servicing and maintenance of all

7

Table 7.**4** Overall laboratory statistics. Example of the annual internal laboratory statistics showing the cytological follow-ups and histological examinations (Laboratory of Prof. Nauth, Stuttgart, Germany, 1999).

Number of woman screened		95 031	
Suspicious cases		1 245	
Solved cases:			
▶ PAP III		142	
▶ PAP III D		735	
▶ PAP IV/V		218	
Total number of solved cases		1 095	
Unsolved cases		150	
Cases with cytological follow-up			**794 (72.5 %)**
PAP group	97 × PAP III	691 × PAP III D	6 × PAP IV/V
I/II	82 (84.5 %)	553 (80.0 %)	2 (33.3 %)
III	2 (2.1 %)	3 (0.4 %)	1 (16.7 %)
III D	13 (13.4 %)	135 (19.6 %)	2 (33.3 %)
IV/V	0	0	1 (16.7 %)
Cases with histological examination			**301 (27.5 %)**
Histology	45 × PAP III	44 × PAP III D	212 × PAP IV/V
Negative findings	31 (68.9 %)	9 (20.5 %)	5 (2.4 %)
CIN I/II	1 (2.2 %)	24 (54.5 %)	25 (11.8 %)
CIN III	2 (4.45 %)	10 (22.7 %)	144 (67.9 %)
Cervical carcinoma	2 (4.45 %)	1 (2.3 %)	28 (13.2 %)
Uterine corpus carcinoma	9 (20 %)	0	10 (4.7 %)

technical equipment and instruments, review of staining quality, and monitoring of the material inventory. Holding regular lab meetings, planning participation in continuing education, and rational vacation planning contribute to the smooth operation of the laboratory. Further aspects include regular updating of the laboratory's library and of slide collections, specimens, and teaching material, as well as the thorough training of established staff and new employees.

Performance

The following statements on the performance of cytology apply exclusively to the cervix and not to its adjacent organs. Cytological diagnosis for the endometrium, ovary, and vulva is less effective than for the cervix, not least because cancer screening does not include these organs.

In general, it must be said that an extremely sensitive diagnostic method yields less accurate results than a method with a higher threshold which is more specific. High sensitivity is there-

fore often traded in for an increase in false-positive findings, and high specificity for an increase in false-negative findings.

The efficiency of a method depends on the percentages of true positives and true negatives. Whether the method is sufficiently cost-effective is determined by the ratio of operating expenses to the benefit achieved.

Definition of Terms

The following comparison is used to illustrate how **specificity**, **sensitivity**, **effectiveness**, and **cost-effectiveness** are interconnected:

On a busy interstate highway, the police want to identify all trucks of a certain make because they have safety defects. They use two methods for this purpose. One is relatively simple and consists of posting a patrol car at the roadside. The other one is expensive and consists of flying over the interstate highway in a helicopter at a great height in order to spot as many of the trucks in question as possible within a short period.

The second method allows the police to identify almost all trucks among the many vehicles on the highway, but it is difficult or impossible to detect the make of the trucks. The method is highly **sensitive** but not very effective because it is not very specific; it is also **unprofitable** because of the high costs involved.

In contrast, the patrol car is highly **effective**. Because of the high speed of the traffic, it does not register all vehicles, but those it does are accurately identified so that they can be removed from the traffic. The method is highly **specific**, but not very sensitive. For this reason, the police post two more patrol cars further along the road; this considerably improves the sensitivity of the method, and its effectiveness rises dramatically as a result. Furthermore, this method is very **cost-effective**.

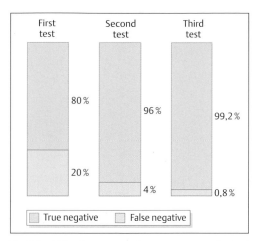

Fig. 7.**1 Effectiveness of repeat cytological screening.** The sensitivity of cervical screening increases from 80% to 99.2% after the third consecutive test.

Specificity

Cervical cytology is characterized by its **high specificity**. After histological examination, 99% of the smears with positive findings turn out to have been correctly assessed as positive (197). There is no other diagnostic method for the early detection of cancer that has such a high positive predictive value as cytology (323).

The **few false-positive findings** are usually due to diagnostic misinterpretation of inflammatory, degenerative, or technical cell damage. However, it occasionally happens that an existing dysplastic lesion has not been included in the surgical sampling or histopathological processing and is therefore not diagnosed by the pathologist. Possible causes of such cytological–histological discrepancy include improper surgical sampling technique, mix-up of specimens, or incomplete histological processing of the tissue. In such cases, the cytological finding is only **apparently false positive**.

If there is **no agreement** between cytological and histological findings, the cytologist should always subject the positive smear to a thorough retrospective assessment and carefully consider whether there has been any misinterpretation of the smear, or whether the finding is still deemed positive in spite of the negative histology (5). If the latter is the case, the cytologist should contact the pathologist to discuss the case and to ask for re-evaluation of the histological specimen. If the discrepancy continues to exist, the patient's physician must be informed. For safety reasons, at least two cytological follow-ups at short intervals (3 months) are required.

Sensitivity

Unfortunately, cervical cytology is **not very sensitive** when based on a single smear. It is quite possible for a Pap test to turn out negative in spite of an existing dysplastic lesion or even carcinoma. The frequency of such false-negative results is difficult to determine. The reported data range from 5% to 50%, and the average frequency is probably around 20% (144). Such a high rate of false-negative findings corresponds to a sensitivity of only 80%.

However, the **sensitivity increases** sharply when the smear is repeated, thus reaching almost 99% after three successive smears (Fig. 7.**1**). Even if the sensitivity of the first smear were only 50%, it would increase to about 97% after the fifth smear (323).

In about two thirds of all cases, the **false-negative findings** result from sampling errors by the submitting physician and, in one third of the cases, from misinterpretations by the cytology laboratory (144). To get an idea of the frequency of false-negative findings, thorough re-evaluation of a negative specimen is advised whenever a positive cytological finding is obtained later. This review may lead to the retrospective detection of any abnormal cells that may have been overlooked (5). It must be taken into consideration, however, that about 6% of all severe cervical precancerous lesions (CIN III) are of the small-cell type and do not develop via a mild precancerous state (CIN I–II) but presumably directly from basal cell hyperplasia (see also p. 171).

Fig. 7.**2** **Incidence and mortality of cervical carcinoma in Germany between 1970 and 2000.** Age-adjusted incidence and mortality per 100 000 women screened each year. (Robert Koch Institute, Berlin, Germany.)

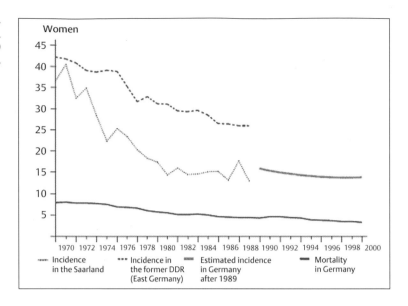

Effectiveness

The gynecological smear is viewed as the most successful cancer test of all time and of all organs. We owe it to this test that the incidence of cervical carcinoma, which affected almost 0.3–0.4‰ of all women in the 1960s, has been reduced to 0.1–0.2‰ today. Nevertheless, in Germany, about 7000 women still die of cervical carcinoma every year, since neither the test nor the screening system are absolutely perfect.

The effectiveness of cytological screening for cancer depends essentially on three factors:

▶ Regular participation of women in the cancer screening program
▶ Correct cell sampling technique employed by the patient's physician
▶ Performance of the cytology laboratory

In cervical cytology, the price for the high proportion of true-positive findings is a certain number of false-negative findings. Nevertheless, the method is **highly effective** because it is repeated at regular intervals, thus dramatically reducing the number of false-negative findings with every repeat smear. In a 10-year study carried out by the Bavarian Cancer Society, 1.65 % of positive cases were detected with the first Pap test, 0.88 % with the second one, and 0.7 % with the third one (366). Regular participation in the cancer screening program is therefore a prerequisite for the high performance of cervical cytology. The effectiveness of the method is reflected by the dramatic fall in the incidence of cervical carcinoma, which was 0.3–0.4‰ in the 1960s and is now only 0.1–0.2‰ (Fig. 7.**2**) (367).

The fact that cervical carcinoma has not yet been completely eliminated is due to several factors. In particular, **regular participation in the cancer screening program** is important; failure to participate is associated with a high risk of developing cervical carcinoma. The number of cytological smears performed increases steadily, but this statement must be qualified because these numbers concern mostly the wrong group of women; for opportunistic reasons, smears are performed too often and on the same women. Other reasons why a certain portion of the population escapes the cytology screening include inadequate information and ignorance.

Improper **sampling or fixation techniques** are additional reasons why the effectiveness of the method is limited. If the necessary skills have not been acquired during medical training, it is essential to make up for this.

The lack of progress in the effectiveness of cytology screening may also be due to a decline in quality regarding the **cytological findings**. During the last 20 years, countless small laboratories have been set up that do not meet the performance standards because of the low numbers of smears that they screen (222).

Numerous attempts have been made to further increase the effectiveness of cytology. So far, most **other methods** have failed because their specificity or cost-effectiveness is low, thus preventing long-term studies that would make it possible to measure their effectiveness. It remains to be seen whether the performance of gynecological cytology will be surpassed by another test in the future.

Recently, it has been reported that specific biomarkers in abnormal epithelial cells are detectable by certain methods. Additional screening for such markers might improve the sensitivity of the conventional Pap test (279a).

Cost-effectiveness

Regular population screening by means of **modern high-resolution techniques**, such as CT or MRI, is presumably far more efficient in detecting diseases than are conventional methods. However, use of these methods in cancer screening programs would be far too uneconomic, and the huge expense would soon lead to the collapse of the health care system.

By contrast, the **screening of cytological smears** has a good cost–benefit ratio and is therefore highly cost-effective. It is easy to do and inexpensive. It covers a large portion of the population in which cervical carcinoma may be expected to develop, over a long period of time, via several precursors for which successful treatment is available.

It is possible to further improve the cost-effectiveness of cervical smears by doing repeat tests at **3-year intervals** after three annual follow-ups have been negative. There are indications that this approach will not lower the effectiveness of the screening (5); however, intervals that are longer than 1 year should meet very high quality requirements. The steep rise in the number of Pap tests constitutes a financial burden on the health care system; nevertheless, for reasons of professional ethics, it is difficult to implement more rational checks and regulations.

So far, **other methods** that are more specific and sensitive than conventional cytology have not gained acceptance because they are far more expensive, require more time and personnel, and are not well accepted by patients (43).

Supplementary Diagnostic Procedures

With respect to the early detection of cervical carcinoma, essentially five procedures have been established that attempt to improve the effectiveness of the conventional smear. They focus on:

- ▶ Detection of the main carcinogen responsible for this type of cancer (HPV analysis)
- ▶ Effects of cancer development on the set of chromosomes (DNA cytometry)
- ▶ Technical prerequisites for automated microscopic screening (liquid-based monolayer cytology)
- ▶ Sensitivity of visual analysis (automation)
- ▶ Morphological signs of cancer development (colposcopy)

The above methods are either highly specific or highly sensitive, but not both at the same time. Because of lack of cost-effectiveness or lack of acceptance, the effectiveness of these methods has not yet been investigated as they have not been in use for long enough.

HPV Analysis

The aim of HPV analysis is to detect infections with low-risk or high-risk types of the virus, in order to calculate the probability of cancer development within the cervical epithelium.

Significance for the Cancer Screening Program

The use of HPV analysis is based on the assumption that **cervical carcinoma** is caused almost exclusively by high-risk types of HPV. Numerous well-founded studies support this assumption, and it is therefore no longer seriously challenged (428).

Nevertheless, HVP analysis is not recommended as a **method of early detection of cervical cancer** as it does not contribute to the diagnosis of morphological changes during carcinogenesis. Even if high-risk viruses are detected, this only means that cancer will develop in 1% of these cases (429). A comparison between the sensitivity and specificity of HPV testing and the sensitivity and specificity of cytology is not feasible, since the two methods have completely different objectives. HVP analysis only detects the carcinogen, whereas cytology detects the disease itself, namely, carcinogenesis (see also pp. 102, 169 ff.). However, HPV analysis may be helpful in cases in which the dysplastic lesions persist for a long time and, in particular, when repeat smears continue to be doubtful (see also pp. 113, 172).

7

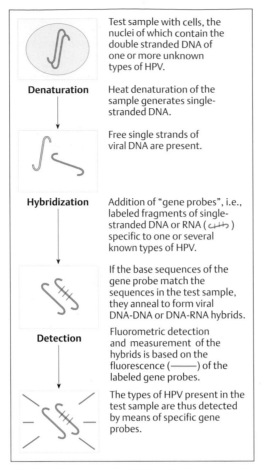

	Test sample with cells, the nuclei of which contain the double stranded DNA of one or more unknown types of HPV.
Denaturation	Heat denaturation of the sample generates single-stranded DNA.
	Free single strands of viral DNA are present.
Hybridization	Addition of "gene probes", i.e., labeled fragments of single-stranded DNA or RNA (⊂╌╍⊃) specific to one or several known types of HPV.
	If the base sequences of the gene probe match the sequences in the test sample, they anneal to form viral DNA-DNA or DNA-RNA hybrids.
Detection	Fluorometric detection and measurement of the hybrids is based on the fluorescence (———) of the labeled gene probes.
	The types of HPV present in the test sample are thus detected by means of specific gene probes.

Fig. 7.**3** **Detection of HPV by hybridization.** Detection of the virus is based on the experimental testing of cervical cells by means of known sequences of HPV DNA.

Detection Methods

Common features. Several different test procedures are available for detecting the many different types of HVP (46a, 101a). The methods differ with respect to the time and costs involved and also with respect to their accuracy. Although most assays aim at detecting viral DNA, tests have also become available that focus on the viral coat protein (capsid) or on mRNA from the transforming genes *E6* and *E7*.

In principle, we distinguish between **direct methods** used for routine diagnosis and **amplification methods** used primarily for research purposes. Furthermore, we differentiate between **suspension methods**, in which the test sample is immersed in a liquid, and **in situ methods**, in which the virus is detected directly in the nuclei of histological sections or cytological smears.

All methods searching for viral DNA are based on the following principle (Fig. 7.**3**). The double-stranded viral DNA to be detected in the test sample is denatured by thermal or chemical means, thus separating the double strands into single strands. Short fragments of single-stranded DNA or RNA derived from one or several known types of the virus are then added. These fragments are labeled with tracer molecules and serve as a "gene probe." They anneal at sites of base homology with the single-stranded viral DNA in the test sample to form DNA–DNA or DNA–RNA hybrids. Depending on the label carried by the gene probe, these hybrids are then detected by photometric or enzymatic means.

Hybrid capture assay. This test uses highly specific RNA probes. It has been developed for **routine diagnosis** as a quick test for detecting an entire group of viruses rather than a single virus (336). The RNA probes represent a selection of the most widespread HPV types, including a specific number of low-risk and high-risk types. If the test sample contains HVP types represented in this "gene cocktail," HPV DNA–RNA hybrids are formed. These hybrids are then captured by hybrid-specific immobilized antibodies and subsequently visualized by enzyme-labeled antibodies and a chemical reaction. The test is also used as a semi-quantitative measure of the viral load.

The base sequences of the gene probes used for specific viral types are archived in data bases, such as the one at the European Molecular Biology Laboratories (EMBL), from where they may be obtained as the need arises. They may be used for the artificial production of short single-stranded DNA or RNA fragments that can then be employed as gene probes.

Polymerase chain reaction. The PCR method is extremely sensitive and specific. It is routinely used for detecting the DNA of one or several HPV types in cervical specimens (66). In order to obtain sufficient amounts of test material for the many known viral types, the viral DNA present in the sample collected from the patient is denatured and then artificially increased (amplified). The amplification process requires primers which are short fragments of single-stranded DNA. Once the primer has annealed to the single-stranded viral DNA, adequate amounts of substrate (bases, cofactors) are added. This triggers a biochemical chain reaction catalyzed by Taq polymerase and, ultimately, leads to amplification of the viral DNA. After another round of denaturation, the newly synthesized single strands are hybridized with one or more of the HPV type-specific gene probes. Depending on the label of the gene probe, these hybrids are then detected by means of photometry, fluorometry, or—as was usually done in the past—gel electrophoresis combined with autoradiography (Southern blotting) (369).

The most advanced development in the application of this qualitative method is the **quantitative determination of viral types**, since it has been demonstrated that the risk of developing cervical cancer increases with the viral load—the amount of high-risk HPV present in the tissue (422).

In-situ hybridization. In this method, labeled gene probes are applied directly to cells on a microscope slide. Like the hybrid capture assay, this method detects low-risk and high-risk HPV types, although not in a liquid test sample but in situ, i.e., directly **in the cytological smear or histological section** (396).

The nucleic acid hybrids are made visible with a special stain. A positive reaction results in the formation of brown granules within the cell nuclei. Unlike the hybrid capture assay, in-situ hybridization is a morphological (histochemical) technique. It is primarily of interest to cytologists and pathologists, and less so to clinicians, because it not only detects the viral infection but also the exact localization of viral particles. The disadvantage of this method is that it is relatively insensitive and labor-intensive.

Tumor markers. An alternative approach to DNA detection is the **in-situ localization of viral mRNA**. Since expression of the transforming genes E6 and E7 is required for initiating and maintaining the malignant phenotype, detection of E6/E7 mRNA in the smears indicates a risk of developing severe cervical dysplasia or even carcinoma. (46a, 101a).

By means of immunocytochemical methods, it is now possible to detect the **viral coat protein (capsid)** in situ (331). The monoclonal antibodies used are directed against the major capsid protein (L1), which is only produced during the late phase of HPV multiplication. A **negative** test result is thought to indicate progression of an existing CIN lesion.

An indirect method for detecting HPV infection is the immunocytochemical detection of the **tumor suppressor gene p16**. Since overexpression of p16 is induced by HPV and associated with cervical carcinogenesis (173b), a **positive** test result is thought to indicate progression of an existing CIN lesion.

At the moment, no data are available on the reliability of these methods. There are therefore only very few appropriate clinical follow-up measures available for women with cervical dysplasia.

DNA Cytometry

DNA cytometry makes use of the fact that integration of DNA from a high-risk type HPV into the genome of cervical cells causes irregular multiplication of chromosomes. This results in **aneuploidy** of nuclei, which can be detected by photometry.

As a measure for cervical cancer screening, DNA cytometry is too time-consuming and costly. However, in cases of **dysplastic** and **doubtful cytological findings**, it provides additional diagnostic information that enables the pathologist to make relatively reliable predictions on the possible progression or regression of a precancerous lesion (398).

The method focuses on the fact that abnormal nuclei show an increased and irregular number of chromosomes, i.e., they have **aneuploid** sets of chromosomes. By contrast, the normal chromosome set is **euploid**; it is usually diploid but may be polyploid in benign lesions (330).

The test requires a special staining technique, the **Feulgen method**, which makes it possible to measure the optical density of the nuclei. For this purpose, the Pap smears need to be destained and restained with Feulgen stain, after marking suspicious cells or groups of cells.

Using **interactive image analysis**, the absorption of light by abnormal nuclei is measured and compared with that of reference nuclei, such as the nuclei of normal intermediate cells or leukocytes. The results are presented in a histogram (Fig. 7.**4a–c**). If the measurements yield euploid (diploid and polyploid) sets of chromosomes—which they do in most mild cases of dysplasia—the probability of **regression** is 85%. If, however, the measurements reveal aneuploid sets of chromosomes—which is the case in the majority of moderately and severely dysplastic lesions—the probability of **progression** is 91% (35). These data basically confirm the findings obtained by conventional cytology.

7

Fig. 7.**4 a–c DNA cytometry.** Typical DNA contents of nuclei in cervical squamous cells.

a Normal finding (diploid chromosome sets with DNA peaks at 2c and 4c).

b Benign lesion (polyploid chromosome sets with DNA peaks at 2c, 4c, 8c,16c).

c Dysplastic lesion (aneuploid chromosome sets with an irregular distribution of DNA peaks, up to 21c).

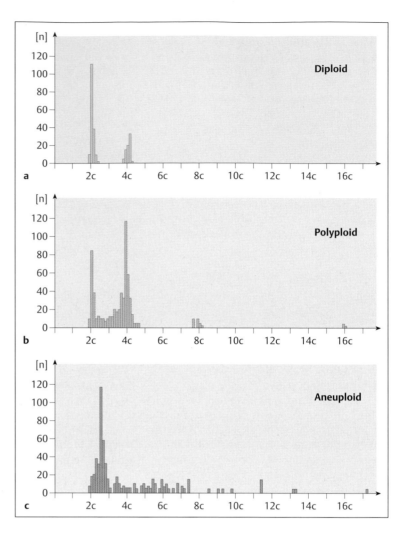

Liquid-based Monolayer Cytology

Liquid-based cytology is used to prepare a monolayer of single cells on the microscope slide. This makes it much easier to analyze the smears microscopically because cellular details are no longer obscured by overlying structures. The method is a technical prerequisite for operating automated screening devices.

Although liquid-based cytology is **very specific and sensitive**, it is doubtful whether this method will gain broad application because of the relatively high operating costs and the sophisticated technical equipment involved.

The aim of any monolayer technique is to optimize cervical smears in such a way that they become more accessible to microscopic analysis than conventional smears are. This is achieved by **dispersion** of the cells in a liquid preservative and their subsequent transfer onto a microscope slide (12).

Currently, the U.S. Food and Drug Administration (FDA) has approved two liquid-based monolayer cytology methods as a replacement for the conventional Pap smear: a density gradient method (SurePath, formerly called AutoCytePrep) and a filter transfer method (ThinPrep). The technical aspects of the underlying principles are similar (see description of methods below).

Preparing a monolayer of single cells is a prerequisite for many automated screening devices because superimposed or overlapping cells are difficult for these devices to analyze (153).

The disadvantages of conventional Pap smears obtained under poor technical conditions are well known. The smear quality depends on the sampling device as well as the sampling method and the method of cell transfer onto the microscope slide. The result is often an unsuitable smear preparation characterized by overlying cells and mucus as well as admixtures of blood and leukocytes. By contrast, liquid-based monolayer cytology yields a clean preparation of single cells with distinct chromatin structures, thus making it easier to identify the cells (253a).

Originally developed for automated screening devices, the liquid-based preparation of a monolayer of cells has many advantages (384a). The technique was first thought to be superior to the conventional smear method, because it supposedly yields more representative cell samples. Numerous studies seem to document the increased diagnostic accuracy of this method (31a, 154a). However, most of these studies were supported by equipment manufacturers, so the results need to be confirmed by independent surveys.

So far, there is no evidence that liquid-based cytology is superior (69a, 229a, 233a). It is quite possible that its undeniable advantages over conventional cytology are counterbalanced by disadvantages. It is therefore doubtful whether this method is really able to provide an improvement in the diagnostic accuracy of cytological cancer screening.

Technical Aspects

Common features of established methods. There are basically three competing commercially available procedures: the ThinPrep method, the SurePath method, and simple centrifugation techniques, such as the Gyno-Prep method. All these techniques have in common that the liquid preservative yields a high-quality fixation of the cell sample and that traces of mucus and blood are almost completely eliminated.

The three methods also have in common that only part of the collected cell suspension is used, and the remaining portion may be stored for further tests. The complete dispersion of cells is achieved by thorough shaking of the sample on a vortex mixer.

The available volume of liquid is relatively large in relation to the number of dispersed cells. Both the SurePath method and the simple centrifugation methods attempt to collect as many cells as possible from the liquid. A cell-counting device, called CellCheck, is used to determine when the right amount of cells for monolayer preparation is obtained.

All three procedures largely remove blood, mucus, and cell debris, and also the microbial flora. The purified cells are then transferred

onto a microscope slide coated with an adhesive. The result is a monolayer of single cells, which is supposed to facilitate a quicker screening. However, this goal is highly questionable because the visual attention is directed to single cells, which often calls for a higher optical magnification, thus lengthening the screening time.

The methods used in routine practice partly differ considerably in terms of time and technical expense. This raises the question of whether the higher costs are justified by increased efficiency. So far, it has not been possible to prove that one or the other of these methods is more successful than its competitors (34a).

At least morphologically, all methods yield nearly equivalent results. However, independent long-term studies regarding their efficiency are still pending (132a).

ThinPrep method. This procedure involves aspiration of the cell suspension onto a micropore filter to remove any accompanying elements, until a maximum of 60 000 cells has been collected on the filter (Fig. 7.**5 a**). The rest of the suspension is stored unchanged.

SurePath method. This procedure involves slow centrifugation of the cell suspension through a separation medium to absorb any accompanying elements and concentration of the cells once a maximum of 60 000 cells has been obtained (Fig. 7.**5 b**).

Simple centrifugation methods. These techniques involve the transfer of dispersed cells to a special centrifugation funnel and sedimentation directly onto the microscope slide, without an intermediate purification step (Fig. 7.**5 c**).

Morphological Aspects

Smear technique. Cell sampling for liquid-based cytology is standardized through the use of a lamellar broom (in the future, possibly a spatula). This technique is much more effective than the conventional method—where sometimes, unfortunately, a cotton swab is still in use. Presumably, this standardization will improve both the condition of cellular preservation and the sampling of endocervical cells.

Nuclear structure. Because of the quick and complete fixation of cells, the liquid-based monolayer technique features a number of morphological criteria for the high quality of

7

Fig. 7.5 a–c Liquid-based monolayer cytology.

a ThinPrep method. The dispersed cells are collected on a micropore filter by aspiration and then transferred to a microscope slide.

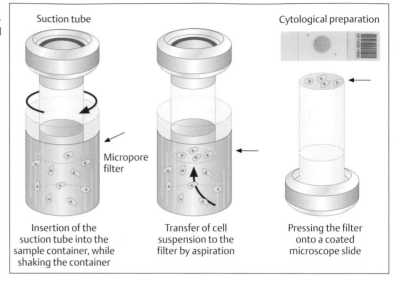

Suction tube

Cytological preparation

Micropore filter

Insertion of the suction tube into the sample container, while shaking the container

Transfer of cell suspension to the filter by aspiration

Pressing the filter onto a coated microscope slide

b SurePath method. The dispersed cells are purified by centrifugation through a density gradient, concentrated by further centrifugation, and collected on a microscope slide through sedimentation.

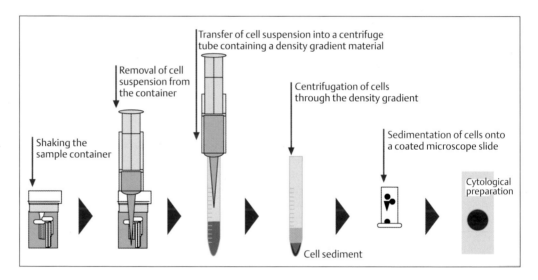

Transfer of cell suspension into a centrifuge tube containing a density gradient material

Removal of cell suspension from the container

Centrifugation of cells through the density gradient

Sedimentation of cells onto a coated microscope slide

Shaking the sample container

Cytological preparation

Cell sediment

c GynoPrep method, one of many centrifugation methods. The tip of the sampling device (a broom-like brush) is placed directly into the transport medium. The cells are dispersed by shaking of the container, transferred to a special centrifuge tube (ECOfunnel), and centrifuged directly onto the microscope slide.

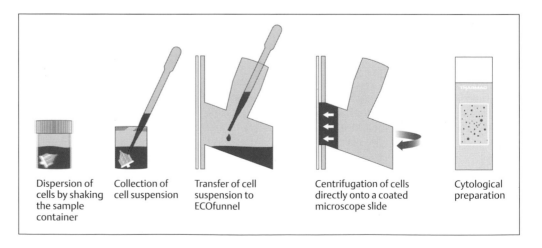

Dispersion of cells by shaking the sample container

Collection of cell suspension

Transfer of cell suspension to ECOfunnel

Centrifugation of cells directly onto a coated microscope slide

Cytological preparation

364

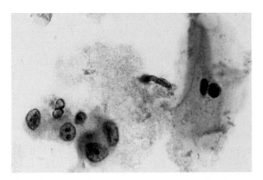

Fig. 7.**6** **Liquid-based monolayer method: reserve cells.** The regular chromatin structure and the delicate chromocenters are easier to recognize than in conventional smears. ×1000.

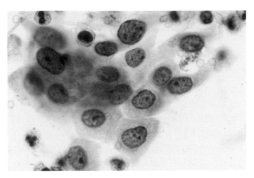

Fig. 7.**7** **Liquid-based monolayer method: CIN III lesion.** The abnormal chromatin structure and the prominent chromocenters are easier to recognize than in conventional smears. ×1000.

dispersed single cells. Because the fixation acts on every cell from all sides (three-dimensional fixation) and the fixative completely engulfs the cell without admixture of air, the preservation of details is probably more effective than in a conventional smear preparation. In particular, chromatin structures, nucleoli, and nuclear membranes, as well as cytoplasmic and intercellular structures like desmosomes, are usually more visible—and therefore easier to interpret—than in conventional smears (Figs. 7.**6**–7.**8**). This might lead to an increase in the number of true-positive and true-negative findings.

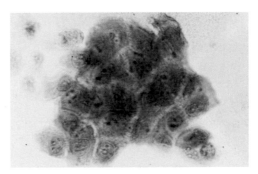

Fig. 7.**8** **Liquid-based monolayer method: desmosomes.** The intercellular spaces are clearly visible between adjacent cells of the endocervical epithelium. ×1000.

Representative cell sample. It is very doubtful whether cell samples obtained by liquid-based methods are always more representative than those obtained using the conventional method. As long as plenty of dysplastic cells have been collected with the sampling device, this may well be true, but it does not hold when only a few cells have been collected. If only a few dysplastic cells are present in a sample derived from a tiny circumscribed lesion, possibly they would still all be detected with the conventional smear technique because they have been collected and smeared while still adhering to one another. By contrast, the suspension of cells used in liquid-based cytology dilutes the sampled cells to such an extent that a few dysplastic cells are difficult to detect because of their low number per volume, thus yielding false-negative findings.

The following examples will serve to illustrate the situation. Let grains of salt in a container represent the normal cervical cells, and a dozen grains of rice lying on top of the salt represent the abnormal cells. In the first example (representing the conventional smear method), salt is removed from the container with a teaspoon without looking. The likelihood that rice kernels are removed together with the salt is relatively high. In the second example (representing the liquid-based method), the container content is stirred so as to distribute the rice kernels evenly before removing the salt. The probability that the sample still contains rice kernels is now extremely low.

Cellular arrangement and topography. Certain cell types tend to be arranged in groups or clusters because of the intensity of intercellular adhesion. This is a specific diagnostic criterion that helps in reaching the correct diagnosis, especially in the case of reactive or degenerative nuclear enlargement that is difficult to interpret. Whereas intercellular adhesion is largely preserved by the conventional smear technique, the liquid-based monolayer

7

methods cause extensive disintegration of cell group populations.

On the other hand, artificial cell aggregations may be generated during aspiration or centrifugation of the single-cell suspension, as a result of clogging of the filter pores or uneven centrifugation. These cell aggregations make diagnosis difficult as they do not share a common origin.

Cells with an increased tendency to adhere to one another include reserve cells, endocervical cells, metaplastic cells, and dysplastic cells. Basal cells, cells of normal and inflamed squamous epithelia, and cells of certain types of carcinoma are less adherent.

The distinction between ectocervical and endocervical localization of a lesion has important clinical and therapeutic consequences (e. g., choosing the size of a cone biopsy). This kind of differentiation is possible in a conventional smear, but not in the liquid-based method because endocervical and ectocervical cells become mixed in the suspension fluid.

Size, shrinkage, and preservation of cells. The conventional smear method ensures that all of the cells sampled are transferred onto the microscope slide. In addition to the epithelial cells that need to be analyzed, nonepithelial cells—such as leukocytes and erythrocytes—are also present in the smear. By contrast, the liquid-based method removes almost all contaminating cells or cell fragments together with the microbial flora, by means of a porous filter or by centrifugation.

Small epithelial cells—such as reserve cells, immature metaplastic cells, basal cells, and endometrial cells—are very similar in size to leukocytes and may be of considerable diagnostic value. Cells that are suspended in a liquid preservative usually undergo some shrinkage that is, so far, little understood. Hence, it is possible that some of the small cells pass through the porous filter, or are eliminated by centrifugation due to low density gradients, and thus escape diagnostic evaluation.

Furthermore, it is worth remembering that centrifugation may destroy the cytoplasm of sensitive cells, such as endocervical cells, endometrial cells, atrophic basal cells, or reserve cells. As a result, only cell debris and naked nuclei may be present in the preparation.

The role of the microbial flora. Finding abnormal microorganisms in the smear often provides important diagnostic clues, thus pointing to a lesion of either reactive or dysplastic nature. These clues, which are easily recognized in a conventional smear, are absent from liquid-based preparations because the microbial flora has largely been eliminated. The few bacteria that directly adhere to the squamous cells are insufficient for the final assessment of the situation (46).

The role of mucus and cell debris. Elimination of mucus and cell debris facilitates the microscopic examination—a feature that seemingly gives the liquid-based cytology an advantage over the conventional method. However, the presence of these accompanying elements may also provide clues for inflammatory diseases or even invasive carcinoma—an advantage that is not offered by the liquid-based method.

Automation

An automated microscopic screening process does not involve the human eye. Rather, it uses a computer-controlled image analysis system. Some systems even take over the subsequent diagnostic evaluation, although most devices transfer this part of the process to the cytologist in an interactive process.

Automation caught on some time ago in areas such as data acquisition, archiving, and statistical evaluation, and also in the processing of specimens, such as staining and mounting. However, it did not meet the high expectations regarding diagnostic evaluation of the smears. Automation has a relatively low degree of specificity, although it has **very high sensitivity** with a rate of only about 3–5 % of false-negative findings (180).

The huge technical and operating costs, as well as the lengthy approval process, have so far prevented this method from being widely used, although it could be extremely effective. There have been cases where smears had to be sent to diagnostic centers for diagnostic re-evaluation and then sent back to the laboratory for final assessment. Since the re-evaluation of the test is not reimbursed by the health insurance system, the costs involved have to be covered either by the laboratory or by the patients themselves. In the future it might be promising to manufacture more reasonably priced devices in larger numbers. This might also help to compensate for the current shortage of cytotechnologists.

Screening process. Devices for the screening of smear specimens almost exclusively operate according to the flowchart shown in Fig. 7.**9**.

▶ **Step 1:** A computer-controlled microscope with integrated technical processing of the cell samples (handling of specimens) screens each specimen and transmits any pictures taken at the correct focus and illumination to the digital image analysis system.

▶ **Step 2:** The evaluation program determines whether the visual field is representative and whether artifacts are present, and then selects cells or cell groups in specific areas (visual segmentation).

▶ **Step 3:** The cell image is analyzed. A number of features (called measured values) are measured to provide information on light distribution, color, shape, and texture. These features are compared with those that have been programmed into the device during the set-up phase and are available from a data bank. This comparison may either occur by means of statistical classifiers or by means of neural networks (129). The automated machine classifies the individual cells into "suspicious" or "unsuspicious" by means of the **cell image classifier**.

▶ **Step 4:** All evaluations of cells or cell groups of a specimen are collected and, in a population analysis, interpreted for the entire specimen using a **specimen classifier**. The entire smear is then classified as "suspicious" or "unsuspicious."

▶ **Step 5:** In this last step, a selection of cells or cell groups are illustrated in a gallery of images, and the final diagnostic evaluation is then made by the cytologist using an interactive system. Because of the relatively low specificity of the method, this step involves about 60% of all specimens. It is hoped that technical improvements may further reduce the rate of false-positive findings in the future.

An important question about future automated machines is whether they will eventually be able to accept the conventional Pap smears with their numerous imperfections, such as blood, mucus, and cell clumps, or whether they will continue to rely on the new liquid-based monolayer methods. Monolayer specimens are far easier to evaluate by automated methods, but they require an additional, time-consuming, automated preparation step. This makes the technique much more expensive and currently prevents it from being widely used.

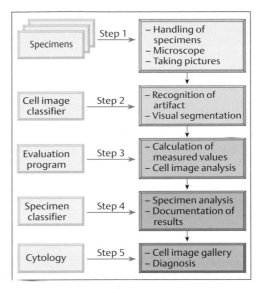

Fig. 7.**9 Automated cytodiagnostic procedure.** Flowchart. After photographic recording of the cells of the smear, segments of the specimen are selected and compared with known data, first individually, then the entire smear. The most suspicious cells are documented in a cell image gallery in order to be evaluated by the cytologist.

Colposcopy

Colposcopy makes use of the fact that the development of cervical cancer is accompanied by morphological changes in the epithelium of the portio vaginalis. These changes can be assessed and, up to a certain point, even classified at low magnification.

Unlike cytology, colposcopy does not allow for an accurate **distinction between the changes involved in cancer development**. It only permits assessment of the epithelial surface and the capillary image. It should be borne in mind that about one third of all precancerous stages and carcinomas are localized inside the cervix and therefore escape visual inspection (49). Colposcopy is slightly less specific than cytology, but far more sensitive; the rate of false-positive findings is therefore relatively high (23).

Although colposcopy was introduced as early as 1925 (148), it has never played an **important role in early cancer detection**. In Germany, it has not been widely practiced because it is not reimbursed by the national health insurance system. So far, it has been difficult to prove that colposcopy is highly effective, although the test is easy, quick, and inexpensive to do. It does not

7

Fig. 7.**10** **Colposcopy.** A scolposcope is a magnifying lens fixed to the gynecological chair. It is used for inspection of the vulva, vagina, and cervic.

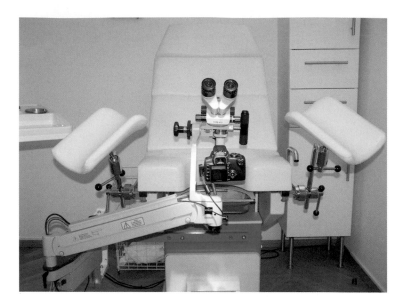

require a huge technical investment and would thus be rather cost-effective.

Colposcopy has been overtaken by highly specific cytodiagnostic procedures and has had no chance to establish itself as a screening method. However, it has become indispensable as a **supplementary method** and helps to uncover important pathological findings that sometimes escape cytology. The combined use of both methods may achieve a sensitivity as high as 98% (250).

The **magnifying technique** employed in colposcopy is easily integrated into the routine gynecological examination by attaching a swivel-mounted optical tube with a magnification of 4× to 40× to the examination chair (Fig. 7.**10**). The visual examination of native epithelium is supplemented by the obligatory use of the acetic acid test, the facultative use of a green filter, and the occasional use of the iodine test (see also p. 29 ff.). The principles of colposcopy are described on p. 28. Colposcopic images and diagrams of benign cervical lesions are shown on pp. 10, 12, 28 ff., 93, 100, 102, 107, 120, 131 ff., 146, and those of malignant cervical lesions on pp. 174 ff. and 213 ff., 216.

References

1. Abdulla, M., S. Hombal, A. Kanbour, M. Becich, D. Stankovich, A. Ries, A. Kanbour-Shakir: Characterizing "blue blobs," immunohisto-chemical staining and ultrastructural study. Acta Cytol. 44, 547 (2000).
2. Achen, M.G., S.A. Stacker: The vascular endo-thelial growth factor family; proteins which guide the development of the vasculature. Int. J. Exp. Pathol. 79, 255 (1998).
3. Affandi, M.Z., V. Doctor, S.S. Rao: Detection of ovulation by single endometrial aspiration smear. Acta Cytol. 25, 157 (1981).
4. Altmann, H.-W., H.A. Müller: Grundlagen der Karyologie. Verh. Dtsch. Ges. Pathol. 57, 12 (1973).
5. American Society of Cytopathology: Cervical cytology practice guidelines. Acta Cytol. 45, 201 (2001).
6. Ammann, M., W. Hänggi, U. Baumann, P.J. Keller: Differentialdiagnostik zystischer Ovarialtumoren durch Bestimmung von Östradiol im Zystenpunktat. Geburtsh. Frauen-heilk. 51, 481 (1991).
7. Anestad, G., O. Lunde, M. Moen, K. Dalaker: In-fertility and chlamydial infection. Fertil. Steril. 48, 787 (1987).
7a. Arbeitsgemeinschaft bevölkerungsbezogener Krebsregister in Deutschland. In: Krebs in Deutschland, Häufigkeiten und Trends. Robert Koch Institute, Berlin, 2004.
8. Arends, M.J., C.H. Buckley, M. Wells: Aetiology, pathogenesis, and pathology of cervical neo-plasia. J. Clin. Pathol. 51, 96 (1998).
9. van Aspert van Erp, A.J., A.B. van't Hof-Grotten-boer, G. Brugal, G.P. Vooijs: Endocervical columnar cell intraepithelial neoplasia. I. Dis-criminating cytomorphologic criteria. Acta Cytol. 39, 1199 (1995a).
10. van Aspert van Erp, A.J., A.B. van't Hof-Grootenboer, G. Brugal, G.P. Vooijs: Endocervi-cal columnar cell intraepithelial neoplasia. II. Grades of expression of cytomorphologic cri-teria. Acta Cytol. 39, 1216 (1995b).
11. Auerbach, R.: Patterns of tumor metastasis: organ selectivity in the spread of cancer cells. Lab. Invest. 58, 361 (1988).
12. Austin, R.M., J. Ramzy: Increased detection of epithelial cell abnormalities by liquid-based gynecologic cytology preparations. A review of accumulated data. Acta Cytol. 42, 178 (1998).
13. Ayer, B., F. Pacey, M. Greenberg, L. Bonsfield: The cytologic diagnosis of adenocarcinoma in situ of the cervix uteri and related lesions. I. Adenocarcinoma in situ. Acta Cytol. 31, 397 (1987). II. Microinvasive adenocarcinoma. Acta Cytol. 32, 318 (1988).
14. Ayer, B., F. Pacey, M. Greenberg: The cytological features of invasive adenocarcinoma of the cer-vix uteri. Cytopathology 2, 181 (1991).
15. Ayre, J.E.: The vaginal smear; precancer cell studies using a modified technique. Am. J. Ob-stet. Gynecol. 58, 1205 (1949).
16. Ayre, J.E.: Role of the halo cell in cervical cancerogenesis: a virus manifestation in pre-malignancy? Obstet. Gynecol. 17, 175 (1960).
17. Babes, A.: Diagnostic du cancer du col uterin par les frottis. Presse med. 36, 451 (1928).
18. Bachmann, R.: Untersuchungen über den Ovu-lationstermin nebst Bemerkungen zur Histo-logie des Corpus luteum. Z. Mikr. Anat. Forsch. 40, 57 (1936).
18a. Badaracco, G., A. Venuti, A. Sedati, M.L. Mar-cante: HPV16 and HPV18 in genital tumors: significantly different levels of viral integration and correlation to tumor invasiveness. J. Med. Virol. 67, 574 (2002).
19. Badawy, S.Z.A., L.J. Elliott, A. Elbadawi: Plasma levels of estrone and estradiol-17 in post-menopausal women. Brit. J. Obstet. Gynecol. 86, 56 (1979).
20. von Baer, K.: De ovi mammalium et hominis genesi epistolam ad Academiam Imperialem Scientiarum Petropolitanam, L. Voss, Leipzig (1927). German translation by B. Ottow: Über die Bildung des Eies der Säugetiere und des Menschen. Mit einer Einführung. Faksimile-druck. L. Voss, Leipzig (1927).
21. Bajardi, F.: Histomorphology of carcinoma in situ. Acta Cytol. 5, 271 (1961).
22. Bajardi, F.: Histomorphology of reserve cell hy-perplasia, basal cell hyperplasia and dysplasia. Acta Cytol. 5, 133 (1961).
23. Bajardi, F.: Kolposkopische Befunde und ihre histologischen Korrelate. Geburtsh. Frauen-heilk. 44, 84 (1984).
24. Baltzer, J., W. Köpcke, K.J. Lohe, C. Kaufmann, K.G. Ober, J. Zander: Die operative Behandlung des Zervixkarzinoms. Geburtsh. Frauenheilk. 5, 279 (1984).
25. Beale, L.S.: Results of the chemical and micro-scopical examination of solid organs and secre-tions. Examination of sputum from a case of cancer of the pharynx and the adjacent parts. Arch. Med., London 2, 44 (1860).
26. Beams, H.W., S. Müller: Effects of ultracentrifu-gation on the interphase nucleus of somatic cells with special reference to the nuclear en-velope-chromatin relationship. Z. Zellforsch. 108, 297 (1970).
27. Bender, H.G.: Stadienangepasste Behandlung des Vulvakarzinoms. Gynäkologe 14, 159 (1981).
28. Bender, H.G., L. Illig, P. Otto, B.P. Robra, R. Schrage, L. Weißbach: Das deutsche Krebsfrüh-erkennungsprogramm, Deutscher Ärzte-Verlag, Cologne (1987).
29. Benson, P.A.: Psammoma bodies found in cer-vicovaginal smears. Acta Cytol. 17, 64 (1973).

30. Benson, R.C., H.F. Traut: The vaginal smear as a diagnostic and prognostic aid in abortion. J. Clin. Endocrinol. 10, 675 (1950).

31. Berg, J.W., G.R. Durfee: The cytological presentation of endometrial carcinoma. Cancer 11, 158 (1958).

31a. Bernstein, S.J., L. Sanchez-Ramos, B., Ndubisi: Liquid-based cervical cytologic smear study and conventional Papanicolaou smears: A metaanalysis of prospective studies comparing cytologic diagnosis and sample adequacy. Am. J. Obstet. Gynecol. 185, 308 (2001).

32. Beutner, K.R.: Patient applied podofilox in the treatment of genital warts: a review. In: Groß, G., S. Jablonska, H.Pfister, H.E. Stegner (eds.), Genital Papillomavirus Infections. Springer, Berlin, Heidelberg, New York (1990).

33. Bibbo, M.L., M.J. Harris, G.L. Wied: Microbiology of the female genital tract. In: Wied, G.L., L. G. Koss, S.W. Reagan (eds.), Compendium on Diagnostic Cytology. International Academy of Cytology, Chicago 108 (1973).

34. Bickenbach, W., H.J. Soost: Berichte über die 1. Tagung der Deutschen Gesellschaft für Angewandte Zytologie, Dr. E.A. Müller, Munich (1964).

34a. Black-Schaffer, W.S.: Choosing between competing technologies in cytology laboratory. Clin. Lab. Med. 23, 681 (2003).

35. Böcking, A.: DNA-Zytometrie und Automatisation in der klinischen Diagnostik. In: J.H. Holzner u. W. Queißer (eds.), Beiträge zur Onkologie, vol. 38, p. 298. Karger, Basel (1990).

36. Boschann, H.W.: Are spindle-shaped squamous cells suggestive of a distinct type of carcinoma or of distinct degree of cellular maturity? Acta Cytol. 2, 272 (1958).

37. Boschann, H.W.: Cytometry on normal and abnormal endometrial cells. Acta Cytol. 2, 520 (1958).

38. Boschann, H.W.: Radiation changes in benign cells. In: Wied, G.L., L.G. Koss, J.W. Reagan (eds.), Compendium on Diagnostic Cytology, 3rd ed. Tutorials of Cytology, Chicago (1974).

39. Bousfield, L., F. Pacey, Q. Young, I. Krumins, R. Osborn: Expanded cytologic criteria for the diagnosis of adenocarcinoma in situ of the cervix and related lesions. Acta Cytol 24, 283 (1980).

40. Breinl, H.: Die Phasenkontastmikroskopie als morphologische Untersuchungsmethode in Biologie und Medizin. In: Witte, S., F. Ruch (eds.), Moderne Untersuchungsmethoden in der Zytologie. Witzstrock, Baden-Baden, p. 11 (1976).

41. Broders, A.C.: The grading of carcinoma. Minn. Med. 8, 726 (1925).

42. Broders, A.C.: Carcinoma-grading and practical application. Arch. Pathol. 2, 376 (1926).

43. Brown, A.D., A.M. Garber: Cost effectiveness of three methods to enhance the sensitivity of Papanicolaou testing. JAMA, 281, 347 (1999).

44. Brown, S.W.: Heterochromatin. Science 151, 417 (1966).

45. Bruch, C.: Die Diagnose der bösartigen Geschwülste. Nach eigenen Untersuchungen. Von Zabern, Mainz (1847).

46. Bur, M., K. Knowles, P. Pekow, O. Corral, J. Donovan: Comparison of ThinPrep preparations with conventional cervicovaginal smears. Practical considerations. Acta Cytol. 39, 631 (1995).

46a. Burd, E.: Human papillomavirus and cervical cancer. Clin. Microbiol. Rev. 16, 1 (2003).

47. Burghardt, E.: Die diagnostische Konisation der Portio vaginalis uteri. Geburtsh. Frauenheilk. 23, 1 (1963).

48. Burghardt, E., E. Holzer: Die Lokalisation des pathologischen Cervixepithels. I. Carcinoma in situ, Dysplasien, und abnormes Plattenepithel. Arch. Gynäkol. 73, 558 (1951).

49. Burghardt, E., E. Holzer: Die Lokalisation des pathologischen Cervixepithels. Arch. Gynäkol. 209, 305 (1970).

50. Cabanne, F., R. Michiels, C. Mottot, H. Bastien: Ciliated bodies in gynecologic cytopathology: Parasite or cellular debris? Acta Cytol. 19, 407, (1975).

51. Calabresi, P., N.V. Arnold, W.D. Stowall: Cytological screening for uterine cancer through physicians offices. J. Am. Med. Ass. 168, 243 (1958).

52. Carmichael, R., B.L. Jeaffreson: Squamous metaplasia of the columnar epithelium in the human cervix. J. Pathol. Bacteriol. 52, 173 (1941).

53. Carter, J.J.: Nuclear morphology and mitotic activity in the human endometrium observed in squash preparations. Am. J. Obstet. Gynecol. 85, 397 (1963).

54. Cartier, R.: Colposcopie Pratique. Karger, Basel (1977).

55. Carvalho, G., W.M. Kramer, S. Kay: The presence of Leptothrix in vaginal smears. Acta Cytol. 9, 244 (1965).

56. Castaño-Almendral, A., M. Beato: Die Frühstadien des Plattenepithelcarcinoms des Collum uteri. II. Histometrische Untersuchungen. Arch. Gynäkol. 205, 428 (1968).

57. Christophers, E.: Epidermopoese und Keratinisation. In: Korting, G.W. (ed.), Dermatologie in Praxis und Klinik, Allgemeine Pathologie, vol. I. Thieme, Stuttgart (1980).

58. Chung, C.K., J.A. Stryker, S.P. Ward, W.A. Nahhas, R. Mortel: Histologic grade and prognosis of carcinoma of the cervix. Obstet. Gynecol. 57, 636 (1981).

59. Clinical Laboratory Improvement Act. Public Law 100–578, 100th Congress of the USA. Government Printing Office, Washington D.C.

60. Couture, M.L., M. Freund, A. Sedlis: The normal exfoliative cytology of menstrual blood. Acta Cytol. 23, 85 (1979).

61. Crabbe, J.G.S.: Exfoliative cytological control in occupational cancer of the bladder. Brit. Med. J. 2, 1072 (1952).

62. Cramer, D.W.: The role of cervical cytology in declining morbidity and mortality of cervical cancer. Cancer 34, 2018 (1974).

63. Cramer, D.W.: Epidemiology of the gynecologic cancers. Compr. Ther. 4, 9 (1978)

64. Cremer, C., C. Munkel, M. Granzow, A. Jauch, S. Dietzel, R. Eils, X.Y. Guan, P.S. Meltzer, J.M. Trent, J. Langowski, T. Cremer: Nuclear architecture and the induction of chromosomal aberrations. Mutat. Res. 366, 97 (1996).

65. Czernobilsky, B.: Common Epithelial Tumors. In: Kurman, R.J. (ed.), Blaustein's Pathology of the Female Genital Tract, 3rd ed. Springer, Heidelberg (1987).

66. Dallas, PB., J.L. Flanagan, B.N. Nightingale, B.J. Morris: Polymerase chain reaction for fast, nonradioactive detection of high- and low-risk papillomavirus types in routine cervical speci-

mens and in biopsies. J. Med. Virol. 27, 105 (1989).

67. Dallenbach-Hellweg, G.: Endometrium. Springer, Berlin (1969).

68. Daly, J.J., K. Balogh: Hemorrhagic necrosis of the senile endometrium ("Apoplexia uteri"). New Engl. J. Med. 278, 709 (1968).

69. Dance, E.F., C.D. Fullmer: Extrauterine carcinoma cells observed in cervico-vaginal smears. Acta Cytol. 14, 187 (1970).

69a. Davey, E., A. Barratt, L. Irwig, et al.: Effect of study design and quality on unsatisfactory rates, cytology classifications and accuracy in liquid-based versus conventional cervical cytology: a symptomatic review. Lancet, 367,122 (2006).

70. David, C., P. Figge, M. Otto, H.K. Tamimi, B.E. Greer: Treatment variables in the management of endometrial cancer. Am. J. Obstet. Gynecol.146, 495 (1983).

71. Dennerstein, G.J.: The cytology of the vulva. J. Obstet. Gynaecol. Brit. Cwlth. 75, 603 (1968).

72. Derenzini, M., D. Ploton: Interphase nucleolar organizer regions in cancer cells. Int. Rev. Exp. Pathol. 32, 149 (1991).

73. Donné, A.: Animalcules observés dans les matières purulentes et le produit des sécrétions des organes genitaux de l'homme et de la femme. Compt. Rend. Acad. Sci. 3, 385 (1836).

73a. Duensing, S., K. Münger: Mechanisms of genomic instability in human cancer: insights from studies with human papillomavirus oncoproteins. Int. J. Cancer 109, 157 (2004).

74. Dukes, C.D., H.L. Gardner: Identification of Haemophilus vaginalis. J. Bacteriol. 81, 227 (1961).

75. Duperroy, G.: Morphological study of the endocervical mucosa in relation to the menstrual cycle and to leucorrhea. Gynaecologia (Basel) 131, 73 (1951).

76. Dyroff, R., K. Michalzik: Untersuchungen über die Biologie von Trichomonas vaginalis. Geburtsh. Frauenheilk. 14, 36 (1954).

77. Eisenstein, R., H. Battifora: Lymph follicles in cervical smears. Acta Cytol. 9, 344 (1965).

78. Engineer, A.D., J.S. Misra, P. Tandon: Long-term cytologic studies of copper-IUD users. Acta Cytol. 25, 550 (1981).

79. Eschenbach, D.A.: Vaginal infection. Clin. Obstet. Gynecol. 26, 186 (1983).

80. Eschenbach, D.A., M.G. Gravett, U.B. Hoyme, K.K. Holmes: Possible complications related to nonspecific vaginosis. Abstract, 5th International Meeting, International Society for STD Research, Seattle, WA, 58 (1983).

81. Feichter, G., P. Dalquen: Zytopathologie. In: W. Remmele (ed.), Pathologie, vol. 8. Springer, Berlin (2000).

82. Feichter, G., S. Heinzl, U. Uehlinger, J. Torhorst, B. Vrabec, P. Dalquen: Vergleichende DNS-Hybridisierung, Zytologie und Histologie an kondylomatösen und präkanzerösen Läsionen der Cervix uteri. Geburtsh. Frauenheilk. 52, 758 (1992).

83. Feichter, G., P.F. Tauber, E. Rosenthal, H. Ludwig: Zur bakteriellen Besiedelung kupferhaltiger Intrauterinpessare. Geburtsh. Frauenheilk. 38, 904 (1978).

84. Figge, D.C., R. de Alvarez: Diagnosis of ovarian carcinoma by vaginal cytology. Obstet. Gynecol. 8, 655 (1956).

85. Fluhmann, C. F.: The Cervix Uteri and its Diseases. Saunders, Philadelphia (1966).

86. Francke, E.: Cytologischer Nachweis von Tumorzellen in Markpunktaten. Z. Klin. Med. 140, 622 (1942).

87. Friedberg, V., R.E. Herzog: Die Therapie der Zervixkarzinome. In: Friedberg, V., Käser, O., Ober K.G. (eds.), Spezielle Gynäkologie, vol. III/ 2. Thieme, Stuttgart, New York (1988).

88. Friedlander, S.: A new method for detecting changes in the surface of human exfoliated cervical cells with the scanning electron microscope. Acta Cytol. 13, 288 (1969).

89. Friedrich, E.G.: Vulvar disease. In: Friedmann, E.A. (ed.), Major Problems in Obstetrics and Gynecology. Saunders, Philadelphia (1976).

90. Frischbier, H.J.: Die Strahlentherapie des Collumcarcinoms. Gynäkologe 8, 93 (1975).

91. Frischbier, H.J., K. Thomsen: Die Strahlenbehandlung der Vulva- und Vaginaltumoren. In: Käser, O., Gynäkologie, vol. 3., Thieme, Stuttgart (1972).

92. Frischkorn, R.: Die Nachbestrahlung des operierten Zervix- und Endometriumkarzinoms—Ergebnisse einer Expertenberatung. Arch. Gynaecol. 232, 175 (1981).

93. Frost, J.K.: The Cell in Health and Disease. Monographs in Clinical Cytology. Karger, Basel (1969).

94. Fu, Y.S., P.J. Parks, J.W. Reagan: The ultrastructure and factors relating to survival of endometrial cancer. Am. J. Obstet. Gynecol. 1, 55 (1979).

95. Fu, Y.S., J.W. Reagan, J.G. Hsiu, J.P. Storaasli, W.B. Wentz: Adenocarcinoma and mixed carcinoma of the uterine cervix. I. A clinicopathologic study. Cancer 49, 2560 (1982a).

96. Fu, Y.S., J.W. Reagan, A.S. Fu, K.E. Janiga: Adenocarcinoma and mixed carcinoma of the uterine cervix. II. Prognostic value of nuclear DNA analysis. Cancer 49, 2571 (1982b).

97. Fuller, A.F., N. Elliot, C. Kosloff, J.L. Lewis: Lymph node metastasis from carcinoma of the cervix stages IB and IIA: implications for prognosis and treatment. Gynecol. Oncol. 13, 165 (1982).

98. Gardner, H.L., R.H. Kaufmann: Benign Diseases of the Vulva and Vagina. Mosby, St. Louis (1969).

99. Geerling, S., J.A. Nettum, L.E. Lindner, S.L. Miller, L. Dutton, S. Wechter: Sensitivity and specificity of the Papanicolaou-stained cervical smear in the diagnosis of Chlamydia trachomatis infection. Acta Cytol. 29, 671 (1985).

100. Geirssom, G., F.E. Woodworth, S.F. Pattern, Jr., T.A. Bonfiglio: Epithelial repair and regeneration in the uterine cervix: an analysis of the cells. Acta Cytol 21, 371 (1977).

101. Gill, G.: Der Goldstaubeffekt (Cornflaking), Cyto Info 2, 44 (2001).

101a. Ginocchio, C.C.: Assays for the detection of human papillomavirus. Pan Am. Soc. Clin. Virol. 31, 1 (2004).

102. Gissmann, L., L. Wolnik, H. Ikenberg, U. Koldovsky, H.G. Schnürch, H. zur Hausen: Human papillomavirus types 6 and 11 DNA sequences in genital and laryngeal papillomas and in some cervical cancers. Proc. Nat. Acad. Sci. 80, 560 (1983).

103. Gissmann, L., M. Boshart, M. Dürst, H. Ikenberg, H. zur Hausen, D. Wagner: Presence of

human papillomavirus (HPV) DNA in genital tumors. J. Invest. Dermatol. 83, 265 (1984).

104. Gloor, E.: Morphology of adenocarcinoma in situ of the uterine cervix. Cancer 49, 294 (1982).

105. Gloor, E., J. Ruzicka: Morphology of adenocarcinoma in situ of the uterine cervix: A study of 14 cases. Cancer 49, 294 (1982).

106. Cherry, C.P., A. Glücksmann: Incidence, histology, and response to radiation of mixed carcinomas of the uterine cervix. Cancer 9, 971 (1956).

107. Goldschmidt, H., A.M. Kligman: Desquamation of the human horny layer. Arch. Derm. 95, 583 (1967).

108. Goldschmidt, H., A.M. Kligman: Exfoliative cytology of human horny layer. Methods of cell removal and microscopic techniques. Arch. Derm. 96, 572 (1967).

109. Gordon, A.N., J. Bornstein, R.H. Kaufman, R.G. Estrada, E. Adams, K. Adler-Storthz: Human papillomavirus associated with adenocarcinoma and adenosquamous carcinoma of the cervix: Analysis by in situ hybridization. Gynecol. Oncol. 35, 345 (1989).

110. Göretzlehner, G., C.Lauritzen: Praktische Hormontherapie in der Gynäkologie. Walter de Gruyter, Berlin, New York (1995).

111. Graham, R.M.: The small histiocyte: Its morphology and significance. Acta Cytol. 77 (1961).

112. Graham, R.M., W.A. van Niekerk: Vaginal cytology in cancer of the ovary. Acta Cytol. 6, 496 (1962).

113. Granberg, S., M. Wikland, B. Karlsson, A. Norström, L.G. Friberg: Endometrial thickness as measured by endovaginal ultrasonography for identifying endometrial abnormality. Am. J. Obstet. Gynecol. 164, 47 (1991).

114. Gross, G., M. Hagedorn, H. Ikenberg, T. Rufli, C. Dahlet, E. Grosshans, L. Gissmann: Bowenoid papulosis. Presence of human papillomavirus (HPV) structural antigens and of HPV 16-related DNA sequences. Arch. Dermatol. 121, 858 (1985).

115. Gross, G., S. Jablonska, H. Pfister, H.E. Stegner: Genital Papillomavirus Infections. Springer, Berlin, Heidelberg, New York (1990).

116. Grundmann, E., F. Stein: Untersuchungen über die Kernstrukturen in normalen Geweben und im Carcinoma. Beitr. Pathol. Anat. 125, 54 (1961).

117. Gupta, P.K., D.H. Hollander, J.K. Frost: Actinomycetes in cervico-vaginal smears: An association with IUD usage. Acta Cytol. 20, 295 (1976).

118. Gupta, P.K., E.F. Lee, Y.S. Erozan, J.K. Frost, S.T. Geddes, P.A. Donovan: Cytologic investigations in Chlamydia infection. Acta Cytol. 23, 315 (1979).

118a. Gustafsson, L., J. Ponten, R. Bergsström, H.-O. Adami: International incidence rates of invasive cervical cancer before cytological screening. Int. J. Cancer 71, 159 (1997).

119. van Haam, E.: Some observations in the field of cytology. Am. J. Clin. Pathol. 24, 652 (1954).

120. van Haam, E.: Symposium of techniques for endometrial cytological examination. Acta Cytol. 2, 580 (1958).

121. van Haam, E.: Radiation cell changes. In: Wied, G.L., L.G. Koss, J.W. Reagan (eds.), Compendium on Diagnostic Cytology, 3rd ed. Tutorials of Cytology, Chicago (1974).

122. van Haam, E., J.W. Old: Reserve cell hyperplasia, squamous metaplasia and epidermization. In: Gray, L.A. (ed.), Dysplasia, Carcinoma in Situ and Micro-Invasive Carcinoma of the Cervix Uteri. Thomas, Springfield, IL, 41 (1974).

123. Hackelöer, B.J., F. Fleming, H.P. Robinson, A.H. Adam, J.R.T. Couts: Correlation of ultrasonic and endocrinologic assessment of human follicular development. Am. J. Obstet. Gynecol. 135, 122 (1979).

124. Haefner, H.K., M.B.S. Kamlapurker: Literature-filing system for vulvar conditions. J. Reprod. Med. 42, 473 (1997).

125. Hajdu, St., E.O. Hajdu: Cytopathology of Sarcomas and Other Nonepithelial Malignant Tumors. Saunders, Philadelphia (1976).

126. Hambert, H., A. Thuilliez, G. Tserrenis, B. Fousier: Le prélèvment a l'aiguille comme moyen diagnostic des tumeurs intrathoraciques. J. Franc. Med. et Chri. Thor. 3, 262 (1950).

127. Hameed, K., D.A. Morgan: Papillary adenocarcinoma of endometrium with psammoma bodies. Cancer 29, 1326 (1972).

128. Hammerstein, J.: Neuere Verfahren der hormonalen Contrazeption. Gynäkologe 5, 120 (1972).

129. Hammerstrom, D.: Neural networks at work. JEEE Spectrum 30, 26 (1993).

130. Hamperl, H., C. Kaufmann, K.G. Ober, P. Schneppenheim: Die Erosion der Portio (Die Entstehung der Pseudoerosion, das Ektropion und die Plattenepithelüberhäutung der Cervixdrüsen auf der Portiooberfläche). Virchows Arch. 331, 51 (1958).

131. Hando, T., D.M. Ikado, L. Zamboni: Atypical cilia in human endometrium. J. Cell. Biol. 39, 475 (1968).

131a. Harper, D.M., E.L. Franco, C. Wheeler, et al.: Efficacy of a bivalent L1 virus-like particle vaccine in prevention of infection with human papillomavirus types 16 and 18 in young women: a randomised controlled trial. Lancet 364, 1757 (2004).

131b. Harro, C.D., Y.Y. Pang, R.B. Roden, et al.: Safety and immunogenicity trial in adult volunteers of a human papillomavirus 16 L1 virus-like particle vaccine. J. Natl. Cancer Inst. 93, 284 (2001).

132. Hart, W.R., I. Zaharvi, B.S. Kaplan, D.E. Townsend, S.O. Aldrich, B.E. Henderson, M. Roy, B. Benton: Cytologic findings in stilbestrol exposed females with emphasis on detection of vaginal adenosis. Acta Cytol. 16, 336 (1972).

132a. Hartmann, K.E., K. Nanda, S. Hall, E. Meyers: Technologic advances for evaluation of cervical cytology: Is newer better? Obstet. Gynecol. Survey 56, 765 (2001).

133. Heber, K.R.: The effect of progesterons on vaginal cytology. Acta Cytol. 19, 103 (1975).

134. Hecht, E.L.: Cytological approach to uterine carcinoma; detection, diagnosis, and therapy. Am. J. Obstet. Gynecol. 64, 81 (1952).

135. Heinen, G., P. Siegel: Die Wirkung von konjugierten Östrogenen auf die Vaginal- und Uterusschleimhaut. Dtsch. Med. Wschr. 91, 1553 (1966).

136. Heinzl, S.: Anwendung des Lasers bei Erkrankungen der Vulva. In: Zander, J., J. Baltzer (eds.), Erkrankungen der Vulva. Urban & Schwarzenberg, Munich, Vienna, Baltimore (1986).

137. Heitz, E.: Heterochromatin, Chromozentren, Chromomeren. Ber. Dtsch. Bot. Ges. 47, 274 (1929).

138. Heitz, E.: Die Herkunft der Chromozentren. Dritter Beitrag zur Kenntnis der Beziehung zwischen Kernstruktur und qualitativer Verschiedenheit der Chromosomen in der Längsrichtung. Planta (Berlin) 18, 571 (1933).

139. Heller, C., V. Hoyt: Squamous cell changes with the presence of Candida sp. in cervical-vaginal Papanicolaou smears. Acta Cytol. 15, 379 (1971).

140. Hemesly, H.D., R.C. Boronow, J.L. Lewis, Jr.: Stage II endometrial adenocarcinoma. Memorial Hospital for Cancer, 1949–1965. Obstet. Gynecol. 49, 604 (1977).

141. Henning, N., S. Witte: Atlas der gastroenterologischen Cytodiagnostik, 1st and 2nd eds., Thieme, Stuttgart (1957, 1968).

142. Herrington, C.S.: Human papillomaviruses, and cervical neoplasia. Classification, virology, pathology, and epidemiology. J. Clin. Pathol. 47, 1066 (1994).

143. Hiersche, H.D., R.Wagner: Die Struktur des cervikalen Drüsenfeldes im menschlichen Uterus (eine rasterelektronenmikroskopische Studie). Arch. Gynäkol. 216, 23 (1974).

144. Hilgarth, M., R. Schultz: Ursachen und Ausmaß falsch negativer Befunde in der gynäkologischen Krebsvorsorge. Frauenarzt 5, 324 (1981).

145. Hilgarth, M.: Aspekte der Qualitätssicherung in der gynäkologischen Zytologie. Gynäkologe 23, 312 (1990).

146. Hillemanns, H.G., K. Rha: Quantitative Untersuchungen über den Beginn bösartigen Wachstums an der Portio uteri. Z. Krebsforsch. 64, 245 (1961).

147. Hills, E., C.R. Laverty: Electron microscopic detection of papillomavirus particles in selected koilocytotic cells in a routine cervical smear. Acta Cytol. 23, 53 (1979).

148. Hinselmann, H.: Verbesserung der Inspektionsmöglichkeiten von Vulva, Vagina und Portio. Münch. Med. Wschr. 72, 1733 (1925).

149. Hinselmann, H.: Die Essigsäureprobe ein Bestandteil der erweiterten Kolposkopie. Dtsch. Med. Wschr. 40 (1938).

150. Hoffmann, P.: Histologie und Prognose des Vulvakarzinoms. Geburtsh. Frauenheilk. 30, 452 (1970).

151. Hollander, D.H.: Curschmann's spirals in cervicovaginal smears (Letter). Acta Cytol. 28, 518 (1984).

152. Horava, A., J.C. de Neef, J.C. Boutselis, E. v. Haam: The exfoliative cytologic characteristics of the endometrium in health and disease. Clin. Obstet. Gynecol. 4, 1128 (1961).

153. Howell, L.P., R.L. Davis, T.I. Belk, R. Agdigos, J. Lowe: The AutoCyte preparation system for gynecologic cytology. Acta Cytol. 42, 171 (1998).

154. Hurt, W.G., S.G. Silverberg, W.J. Frable, R. Belgrad, L.D. Crooks: Adenocarcinoma of the cervix: histopathologic and clinical features. Am J Obstet Gynecol 129, 304 (1977)

154a. Hutchinson, M.L., L.M. Isenstein, A.Goodman, et al.: Homogenous sampling accounts for the increased diagnostic accuracy using the ThinPrep processor. Am. J. Clin. Pathol. 101, 215 (1994).

155. Igel, H.: Die Diagnose des Uteruskarzinoms durch Vaginalabstrich. Zbl. Gynäkol. 69, 1369 (1947).

156. Jäger, J.: Zytodiagnostik während und am Ende der Schwangerschaft. Report of the 1st meeting of the Dtsch. Ges. Angew. Zytolog. Müller, Munich, 105 (1963).

157. Janovski, N.A.: Dysplastische und prämaligne Veränderungen der Vulva. Gynäkol. Rdsch. 9, 161 (1970).

158. Janssen, H., Z. Janssen: Dutch lens grinders and microscope builders from Middelburg, around 1600. Inventors of compound microscope. See: Harting, vol. 3 (1866).

159. Jee, C., I. Krishnan-Hewlett, C.C. Baker, R. Schlegel, P. Howley: Presence and expression of human papillomavirus sequences in human cervical carcinoma lines. Am. J. Pathol. 117, 361 (1985).

160. Jenny, J.: Der entzündliche und der atrophische Abstrich. Huber, Bern (1973).

161. Jenny, J.: Die Phasenkontrastmikroskopie in der täglichen Praxis. Jenny und Artusi, Schaffhausen (1977).

162. Jensen, E.V., E.R. de Sombre, P.W. Jungblut, W.E. Stumpf, L.J. Roth: Biochemical and autoradiographic studies of H3-estradiol localization. In: Roth, L.J., W.E. Stumpf (eds.), Autoradiography of diffusible substances. Academic Press, New York, p. 81 (1969).

163. Jensen, E.V., T. Suzuki, M. Numata, S. Smith, E.R. de Sombre: Estrogen-binding substances of target tissues. Steroids 13, 417 (1969).

164. Johnson, L.D. R.J. Nickersen, C.L. Easterday, R.S. Stuart, A.T. Hertig: Epidemiologic evidence for the spectrum of change from dysplasia through carcinoma in situ to invasive cancer. Cancer 22, 901(1968).

165. Jordan, S.W., E. Erangel, N.L. Smith: Ethnic distribution of cytologically diagnosed herpes simplex genital infections in a cervical cancer screening program. Acta Cytol. 16, 363 (1972).

166. Kassenärztliche Bundesvereinigung und Spitzenverbände der Krankenkassen: Gesetzliche Krankheits-Früherkennungsmaßnahmen. Dokumentation der Untersuchungsergebnisse, Deutscher Ärzte-Verlag, Cologne (1976–1986).

167. Kassenärztliche Bundesvereinigung: Richtlinien der Kassenärztlichen Bundesvereinigung über Voraussetzungen zur Durchführung von zytologischen Untersuchungen im Rahmen der Krebsfrüherkennungsmaßnahmen bei Frauen (Zytologie-Richtlinien). Dtsch. Ärztebl. 77, 154 (1980).

168. Kaufman, R.H.: Hyperplastic dystrophy. J. Reprod. Med. 17, 137 (1976)

169. Kaufmann, C., K.G. Ober, F.O. Huhn: Das beginnende Karzinom der Cervix uteri (sog. Mikrokarzinom). Ein Erfahrungsbericht zur Prognose und Therapie an Hand von 130 Beobachtungen. Geburtsh. Frauenheilk. 25, 112 (1965).

170. Kearns, P.R., J.F. Gray: Mycotic vulvovaginitis. Obstet. Gynecol. 22, 621 (1963).

171. Kepp, R., H.J. Staemmler: Lehrbuch der Gynäkologie. Thieme, Stuttgart (1971).

172. King, L.A., T. Tase, L.B. Twiggs, T. Okagaki, J. E. Savage, L.L. Adcock, K. A. Prem, L. F. Carson: Prognostic significance of the presence of human papillomavirus. DNA in patients with invasive carcinoma of the cervix. Cancer 63, 897 (1989).

173. Kistner, R.W.: Endometriosis. In: Sciarra, J. (ed.), Gynecology and Obstetrics, vol. 1/38. Harper & Row, Hagerstown (1980).

173a. von Knobel Doeberitz, M.: Aspekte der molekularen Pathogenese des Zervixkarzinoms für neue Marker in der Krebsfrüherkennung und Diagnostik. Zbl. Gynäkol. 123, 186 (2001).

173b. Klaes, R., T. Friedrich, D. Spitkovsky, R. Ridder, W. Rudy, U. Petry, G. Dallenbach-Hellweg, F. Schmidt, M. von Knebel Doeberitz: Overexpression of p16(INK4A) as a specific marker for dysplastic and neoplastic epithelial cells of the cervix uteri. Int. J. Cancer 92, 276 (2001).

174. Kolkmann, F.W., A. Kielwein: Richtlinie zur Qualitätssicherung zytologischer Untersuchungen im Rahmen der Früherkennung des Zervixkarzinoms. In: Satzung der Landesärztekammer Baden-Württemberg zur Einführung von Maßnahmen der Qualitätssicherung. ÄBW, 12, 463 (1994).

175. Kolstad, P.: Carcinoma of the cervix, stage 0. Am. J. Obstet. Gynecol. 96, 1098 (1966).

176. Korting, G.W.: Dermatologie in Praxis und Klinik. In: Korting, G.W. (ed.), Spezielle Dermatologie, vol. 4. Thieme, Stuttgart (1981).

177. Koss, L.G.: The role of the vaginal pool smear in the diagnosis of endometrial carcinoma. In: Wied, G.L., L.G. Koss, J.W. Reagan (eds.), Compendium on Diagnostic Cytology, 3rd ed. Tutorials of Cytology, Chicago (1974).

178. Koss, L.G.: Diagnostic Cytology and its Histopathologic Bases, 3rd ed. Lippincott, Philadelphia (1979).

179. Koss, L.G.: Cytologic and histologic manifestations of human papillomavirus infection of the female genital tract and their clinical significance. Cancer 60, 1942 (1987).

180. Koss, L.G., E. Lin, K. Schreiber, P. Elgert, L. Mango: Evaluation of the PapNet cytologic screening system for quality control of cervical smears. Am. J. Clin. Pathol. 101, 220 (1994).

181. Koss, L.G., C.R. Durfee: Unusual patterns of squamous epithelium of the uterine cervix; cytologic and pathologic study of koilocytotic atypia. Ann. N.Y. Acad. Sci. 63, 1245 (1956).

182. Koss, L.G., K. Schreiber, S.G. Oberlander, H. Moussouris, M. Lesser: Detection of endometrial carcinoma and hyperplasia in asymptomatic women. Obstet. Gynecol. 64, 1 (1984).

183. Koss, L.G., M.R. Melamed, W.W. Daniel: In situ epidermoid carcinoma of the cervix and vagina following radiotherapy for cervical cancer. Cancer 14, 353 (1961).

184. Koutsky, L.A., K.K. Holmes, C.W. Critchlow, C.E. Stevens, J. Paavonen, A.M. Beckmann, T.A. DeRouen, D.A. Galloway, D. Vernon, N.B. Kiviat: A cohort study of the risk of cervical intraepithelial neoplasia grade 2 or 3 in relation to papillomavirus infection. New Engl. J. Med. 327, 1272 (1992).

185. Krebs, D., W. Schallenberg: Bakteriologische Untersuchungen unter besonderer Berücksichtigung der Anaerobier. Arch. Gynäkol. 206, 209 (1971).

186. Krebs, H.B.: Prophylactic topical 5-fluorouracil following treatment of human papillomavirus-associated lesions of the vulva and vagina. Obstet. Gynecol. 68, 837 (1986).

187. Kreider, J.W., M.K. Howett, S.A. Wolfe, G.L. Bartlett, R.J. Zaino, T.V. Sedlacek, R. Mortel: Morphological transformation in vivo of human uterine cervix with papillomavirus from condylomata acuminata. Nature 317, 639 (1985).

188. Kreienberg, R., D. Grab: Moderne Krebsfrüherkennung in der gynäkologischen Praxis, Möglichkeiten und Grenzen. Forum DGK, 13, 552 (1998).

189. Krimmenau, R.: Adenocarcinoma in situ, beginnende adeno-carcinomatöse Invasion und Microcarcinoma adenomatosum. Geburtsh. Frauenheilk. 26, 1297 (1966).

190. Kurman, R.J., P.F. Kaminski, H.J. Norris: The behavior of endometrial hyperplasia. A long-term study of "untreated" hyperplasia in 170 patients. Cancer 56, 403 (1985).

191. Kurtz, S.M.: The fine structure of the lamina densa. Lab. Invest. 10, 1189 (1961).

192. Lämmermann, U.: Morbus Paget der Vulva. Doctoral thesis, LMU Munich (1985).

193. Langman, J.: Medizinische Embryologie. Thieme, Stuttgart (1977).

194. Lauritzen, C.: Die Therapie des Klimakteriums. Therapiewoche 45, 5198 (1974).

195. Lee, K.R., T.D. Trainer: Adenocarcinoma of the uterine cervix of small intestinal type containing numerous Paneth cells. Arch. Pathol. Lab. Med. 114, 731 (1990).

196. van Leeuwenhoek, A.: Letters to the Royal Society. Philos. Transact. Roy. Soc. London, 9, 121 (1674), 12, 1040 (1679), 22, 552 (1702).

197. Lehmacher, W., H.J. Lange, B. Ruffing-Kullmann, H.J. Soost: Aussage zur Ermittlung von Prädiktiven Werten, Sensitivitäten und Spezifitäten zytologischer Krebsvorsorgeuntersuchungen in der Gynäkologie. 32. Jahrestagung der GMDS Tübingen (1987). In: Selbmann, H.K., K. Dietz (eds.), Medizinische Informationsverarbeitung und Epidemiologie im Dienste der Gesundheit. Springer, Berlin (1988).

198. Lewis, J.F., S. O'Brien: Diagnosis of Haemophilus vaginalis by Papanicolaou smears. Techn. Bull. Reg. Med. Technol. 39, 34 (1969).

199. Leyendecker, G., Sh. Wardlaw, B.A. Barry, B. Leffekt, E. Jost, W. Nocke: Untersuchungen zur Aktivierung des cyclischen Sexualzentrums durch exogene Steroide ("positiver Feedback-Mechanismus"). Acta Endocrinol. 152, 4 (1971).

200. Leyendecker, G., W. Nocke. Die endokrine Regulation des menstruellen Zyklus. Fortschr. Med. 94, 1910 (1976).

201. Limburg, H.: Die Tumoren der Vulva. In: Uehlinger, E. (ed.), Handbuch der speziellen pathologischen Anatomie und Histologie, vol. VII/4. Springer, Berlin (1972).

202. Loch, E.G., D. Tenhaeff, H. Lohmeyer: Die zytohormonale Diagnostik bei verschiedenen oralen Kontrazeptiva. Geburtsh. Gynäkol. 172, 261 (1970).

203. Lohe, K.J.: Vergleichende Untersuchungen zu Sitz und Ausdehnung von Dysplasien und Carcinomata in situ der Zervix. Arch. Gynäkol. 207, 470 (1969).

204. Lohe, K.J.: Das beginnende Plattenepithelkarzinom der Cervix uteri. Postdoctoral Thesis, Munich (1974).

205. Lohe, K.J., J. Baltzer: Ausbreitung, klinische Stadieneinteilung und Symptome. In: Zander, J. (ed.), Ovarialkarzinom. Urban & Schwarzenberg, Munich, Vienna, Baltimore (1982).

206. Long, M.E., H.C. Taylor, Jr.: Nucleolar variability in human neoplastic cells. Ann. N. Y. Acad. Sci. 63, 1095 (1956).

207. Longcope, E., R. Hunter, C. Franz: Steroid secretion by the postmenopausal ovary. Am. J. Obstet. Gynecol. 138, 564 (1980).

208. Maass, H.: Epidemiologie gynäkologischer Tumoren. In: Käser, O., V. Friedberg, K.G. Ober, K. Thomsen, J. Zander (eds.), Gynäkologie und Geburtshilfe. Thieme, Stuttgart, New York (1988).

209. Martin, H.E., E.B. Ellis: Biopsy by needle puncture and aspiration. Ann. Surg. 92, 169 (1930).

210. Martin, P.L., S.S. Yen, A.M. Burnier, H. Hermann: Systemic absorption and sustained effect of vaginal estrogen creams. J. Am. Med. Ass. 242, 2699 (1979).

211. Martius, H.: Lehrbuch der Geburtshilfe. Thieme, Stuttgart, 7th ed. (1971).

212. de May, R.M.: Practical Principles of Cytopathology, ASCP Press, Chicago (1999).

213. McIndoe, W.A., G.H. Green: Vaginal carcinoma in situ following hysterectomy. Acta Cytol. 13, 158 (1969).

214. Meisels, A.: The menopause—a cytohormonal study. Acta Cytol. 10, 49 (1966)

215. Meisels, A.: The maturation value. Acta Cytol. 11, 249 (1967).

216. Meisels, A.: Developing carcinoma of the uterine cervix. In: Wied, G.L., L.G. Koss, J.W. Reagan (eds.), Compendium on Diagnostic Cytology, 3rd ed. Tutorials of Cytology, Chicago (1974)

217. Meisels, A., R. Fortin: Condylomatous lesions of the cervix and vagina. Acta Cytol. 20, 505 (1976).

218. Meisels, A., R. Fortin, M. Roy: Condylomatous lesions of the cervix II. Cytologic, colposcopic and histopathologic study. Acta Cytol. 21, 379 (1977).

219. Meisels, A., C. Jolicoeur: Criteria for the cytologic assessment of hyperplasias in endometrial samples obtained by the Endopap Endometrial Sampler. Acta Cytol. 29, 297 (1985).

220. Meisels, A., M. Roy, M. Fortier, C. Morin, M. Casas-Cordero, K.V. Shah, H. Turgeon: Human papillomavirus infection of the cervix. The atypical condyloma. Acta Cytol. 25, 7 (1981).

221. Meisels, A., C. Morin: Cytopathology of the Uterine Cervix. ASCP Press, Chicago (1991).

221a. Melsheimer, P., S. Kaul, S. Dobeck, G. Bastert: Immuncytochemical detection of human papillomavirus high risk type L1 capsid proteins in LSIL and HSIL as compared with detection of HPV L1 DNA. Acta Cytol. 47, 124 (2003).

222. Menton, M., D. Wallwiener, M. Hilgarth: Klinische Wertigkeit der kolposkopischen Diagnostik in der Früherkennung und Therapie von Zervixkarzinomvorstufen. Geburtsh. Frauenheilkd. 58, 159 (1998).

223. Mestwerdt, G., H.J. Wespi: Atlas der Kolposkopie, Georg Fischer, Stuttgart (1974).

224. Michalzik, L., K.G. Ober: Positive Zytologie der Cervix, ihre histologische Abklärung und die anschließende Wahl der Behandlung. Geburtsh. Frauenheilk. 26, 202 (1966).

225. Miescher, G.: Über Klinik und Therapie der Melanome. Arch. Derm. Syph. 200, 215 (1955).

226. Miller, R.L., J.F. Gerster, M.L. Owens, H.B. Slade, M.A. Tomai: Imiquimod applied topically: a novel immune response modifier and new class of drug. Int. J. Immunopharm. 21, 4 (1999).

227. Miravete, A.P.: Estudios sobre flora vaginal. IX. Classification de lactobacilli de origen vaginal. Rev. Lat. Amer. Microbiol. Parasitol. 9, 11 (1967).

228. Moghissi, K.S., F.N. Syner, L.C. McBride: Contraceptive mechanism of microdose norethindrone. Obstet. Gynecol. 41, 585 (1973).

229. Morrow, C.P., P. DiSaia: Malignant melanoma of the female genitalia: a clinical analysis. Obstet. Gynecol. Surv. 31, 233 (1976).

229a. Moseley, R.P., S. Paget: Liquid-based cytology: Is this the way forward for cervical screening? Cytopathology 13, 71 (2002).

230. Moulder, J.W.: The relation of basic biology to pathogenic potential in the genus Chlamydia. Infection 10, Suppl. 1, 10 (1982).

231. Müller, J.: Über den feineren Bau und die Formen der krankhaften Geschwülste. G. Reimer, Berlin (1838).

232. Müller Kobold-Walterbeck, A.C., M.E. Beyer-Boon: Ciliocytophthoria in cervical cytology. Acta Cytol. 19, 89 (1975).

233. Murad, T.M., K. Terhart, A. Flint: Atypical cells in pregnancy and postpartum smears. Acta Cytol. 25, 623 (1981).

233a. Muth, C., M. Velasco-Garrido, V. Schneider: Dünnschichtzytologie: Rechtfertigt die Evidenzlage einen breiten Einsatz? Frauenarzt 44, 421 (2003).

234. Nagele, F., H. O'Connor, A. Davies, A. Badawy, H. Mohamed, A. Magos: 2500 outpatient diagnostic hysteroscopies. Obstet. Gynecol. 88, 87 (1996).

235. van Nagell, J.R.: Ovarian cancer screening. Cancer 68, 679 (1991).

236. Naib, Z.M.: Exfoliative cytology of viral cervicovaginitis. Acta Cytol. 10, 126 (1966).

237. Naib, Z.M.: Exfoliative Cytopathology, 2nd ed. Little Brown, Boston (1976).

238. Naib, Z.M., A.J Nahmias, W.E. Josey: Cytology and histopathology of cervical herpes simplex infections. Cancer 17, 1026 (1966).

239. Nasiell, K., G. Auer, M. Nasiell, A. Zetterberg: Retrospective DNA analyses in cervical dysplasia as related to neoplastic progression or regression. Analyt. Quant. Cytol. 1, 103 (1979).

240. Nasiell, M.: Histiocytes and histologic reactions in vaginal cytology. Cancer 14, 1223 (1961).

241. National Cancer Institute: The 1988 Bethesda system for reporting cervical/vaginal cytological diagnoses. Acta Cytol. 33, 567 (1989).

242. Nauth, A., H.F. Nauth, M. Menton, S. Menton: Die Häufigkeit positiver zytologischer Befunde nach kompletter Hysterektomie. Poster presentation 166, 16th Annual Meeting of the Arbeitsgemeinschaft für Zervixpathologie und Kolposkopie, Tübingen (2001).

243. Nauth, H.F.: Zur Definition der Begriffe "Parakeratose" und "Dyskeratose" aus gynäkologischer Sicht. Z. Hautkr. 56, 1552 (1981).

244. Nauth, H.F.: Vulva-Diagnostik. Fortschr. Med. 10, 396, and 11, 478 (1982).

245. Nauth, H.F.: Vulva-Zytologie, Thieme, Stuttgart, New York (1986).

246. Nauth, H.F.: Der schwer beurteilbare gynäkologische Abstrich, Thieme, Stuttgart, New York (1996).

246a. Nauth, H.F.: Häufigkeit und Bedeutung der verschiedenen Wachstumsformen des CIN III und des kleinzelligen invasiven Zervixkarzinoms. Verh. Dtsch. Ges. Zyt. 23, 158 (2003).

247. Nauth, H.F., E. Schilke: Die zytologische Krebsvorsorge am äußeren weiblichen Genitale. Morphologische Studie an einem großen Untersuchungsgut. Geburtsh. Frauenheilk. 10, 739 (1982).

248. Nauth, H.F., M. Haas: Zytologische und histologische Untersuchungen zur Hormonabhängigkeit der Haut des äußeren weiblichen Genitales. Geburtsh. Frauenheilk. 44, 451 (1984).

249. Navab, A., L.G. Koss, J.S. La Due: Estrogen-like activity of digitalis. Its effect on this squamous epithelium of the female genital tract. J. Am. Med. Ass. 194, 30 (1965).

250. Navratil, E.: Frühdiagnose des Uteruskarzinoms. In: Biologie und Pathologie des Weibes, vol. 4. Urban & Schwarzenberg, Munich (1955).

251. Naylor, B.: The century for cytopathology. Acta Cytol. 44, 709 (1999).

252. Nettesheim, P., A.J.P. Klein-Szanto, A.C. Marchok, V.E. Steele, M. Terzaghi, D.C. Topping: Studies of neoplastic development in respiratory tract epithelium. Arch. Pathol. Lab. Med. 105, 1 (1981).

253. Neue Aspekte in der gesetzlichen Mutterschaftsvorsorge. Dtsch. Ärztebl. 92, 30 (1995).

253a. Neumann, H.H., J.P.A de Jonge, G. Kleijer, et al.: Prinzipien flüssigkeitsgestützter Präparationsverfahren für die Krebsvorsorgezytologie und eigene Erfahrungen mit dem AutocytePREP™-System in der Routinediagnostik. [Principles of liquid-based cell preparation for gynaecological samples and report on our own experience with AutoCytePREP™ in a routing setting.] Verh. Dtsch. Ges. Zyt. 22, 57–66 (2001).

254. Ng, A.B.P., J.W. Reagan, E. Lindner: Cellular manifestations of primary and recurrent herpes genitalis. Acta Cytol. 14, 124 (1970).

255. Ng, A.B.P., J.W. Reagan, E. Lindner: The cellular manifestations of microinvasive squamous cell carcinoma of the uterine cervix. Acta Cytol 16, 5 (1972).

256. Ng, A.B.P, J.W. Reagan, R.L. Cechner: The precursors of endometrial cancer: A study of their cellular manifestations. Acta Cytol.17, 439 (1973).

257. Ng, A.B.P, J.W. Reagan: Precursors of endometrial carcinoma. In: Wied, G.L., L.G. Koss, J.W. Reagan (eds.), Compendium on Diagnostic Cytology, 3rd ed. Tutorials of Cytology, Chicago (1974).

258. van Niekerk, W.A.: Cervical cytological abnormalities caused by folic acid deficiency. Acta Cytol. 10, 67 (1966).

259. Nocke, W., G. Leyendecker: Die endokrinen Funktionen von Hypothalamus, Hypophysenvorderlappen und Ovar während der Prä- und Postmenopause. Gynäkologe 2, 113 (1970).

260. Nocke, W., G. Leyendecker: Neue Erkenntnisse über die endokrine Physiologie des menstruellen Cyclus. Gynäkologe 5, 39 (1972).

261. Nogales, F., J. Botella: Untersuchungen über den Cyclus der Cervikalschleimhaut. Arch. Gynäcol. 189, 272 (1957).

262. Novotny, D.B., S.J. Maygarden, D.E. Johnson, W.J. Frable: Tubal metaplasia. Acta Cytol. 36, 1 (1992).

263. Nunez, C.: Cytopathology and fine-needle aspiration in ovarian tumours: its utility in diagnosis and management. Curr. Top. Pathol. 78, 69 (1989).

264. Ober, K.G., P. Schneppenheim, H. Hamperl, C. Kaufmann: Die Epithelgrenzen im Bereich des Isthmus uteri. Arch. Gynäkol. 190, 346 (1958).

265. Ober, K.G., K. Thomsen: Spezielle Gynäkologie, vol. III/1, Thieme, Stuttgart, New York (1985).

266. Oehlert, W.: Regeneration, Hyperplasie und Cancerisierung am Beispiel der epithelialen Wechselgewebe. In: Altmann, H.-W. (ed.), Handbuch der allgemeinen Pathologie, vol. IV/2. Springer, Berlin, p. 244 (1969).

267. Osmers, R., M. Völksen, B. Hinney, H. Kühnle, W. Rath, A. Teichmann, W. Wuttke, W. Kuhn: Klinisches Management von zystischen Ovarialtumoren. Geburtsh. Frauenheilk. 50, 20 (1990).

267a. Östör, A.G.: National history of cervical intraepithelial neoplasia: a critical review. Int. J. Gynecol. Pathol. 12, 186 (1993).

268. Pacey, F., B. Ayer, M. Greenberg: The cytologic diagnosis of adenocarcinoma in situ of the cervix uteri and related lesions. III. Pitfalls and diagnosis. Acta Cytol. 32, 325 (1988).

269. di Paola, G.R., M.G. Belardi: Squamous vulvar intraepithelial neoplasia. In: Knapstein, P.G., F. diRe, P. di Saja, U. Haller, B.U. Sevin (eds.), Malignancies of the Vulva. Thieme, Stuttgart, New York (1991).

270. di Paola, G.R., N. Gòmez Rueda, M.G. Belardi, S. Vighi: Vulvar carcinoma in situ: a report of 28 cases. Gynecol. Oncol. 14, 236 (1982).

271. Papanicolaou, G.N.: Diagnosis of early human pregnancy by the vaginal smear method. Proc. Soc. Exp. Biol. (N.Y) 22, 436 (1925).

272. Papanicolaou, G.N.: New cancer diagnosis. In: Proc. 3rd Race Betterment Conference. Battle Creek: Race Betterment Foundation, 528 (1928).

273. Papanicolaou, G.N.: The sexual cycle in the human female as revealed by vaginal smears. Am. J. Anat. 52, 519 (1933).

274. Papanicolaou, G.N: General survey of a vaginal smear and its use in research and diagnosis. Am. J. Obstet. Gynecol. 51, 316 (1946).

275. Papanicolaou, G.N.: Observations on the origin and specific function of the histiocytes in the female genital tract. Fertil. Steril. 4, 472 (1953).

276. Papanicolaou, G.N.: Atlas of Exfoliative Cytology. Commonwealth Fund, Cambridge, MA (1963).

277. Papanicolaou, G.N., F.V. Maddi: Observations on behavior of human endometrial cells in tissue culture. Am. J. Obstet. Gynecol. 76, 601 (1958).

278. Papanicolaou, G.N., H.F. Traut: Diagnosis of Uterine Cancer by the Vaginal Smear. Commonwealth Fund, New York (1943).

279. Parkinson, R.W.: Evaluation of a new sequential constraceptive agent. Obstet. Gynecol. 28, 239 (1966).

279a. Patterson, B., R. Domanik, P. Wernke, M. Gombrich: Molecular biomarker-based screening for early detection of cervical cancer. Acta Cytol 45, 36 (2001).

280. Pincus, G., J.Rode, G.-R. Garcia, E. Rice-Luchay, M. Panigna, J. Rodriguez: Fertility control with oral medication. Am. J. Obstet. Gynecol. 75, 1333 (1958).

281. Patten, S.F., Jr.: Monographs in Clinical Cytology. Diagnostic Cytopathology of the Uterine Cervix. Karger, Basel (1978).

282. Patten S.F., Jr.: Benign Proliferative Reactions of the Uterine Cervix. In: G.L. Wied, C.M. Keebler, L.G. Koss, J.W. Reagan (eds.), Compendium on Diagnostic Cytology, 6th ed. Tutorials of Cytology, Chicago (1990).

283. Petry, G.: Die Konstruktion des Eierstockbindegewebes und dessen Bedeutung für den ovariellen Zyklus. Z. Zellforsch. 35, 1 (1950).

284. Pierce, G.B., Jr., T.F. Beal, S.J. Ram, A.R. Midgley: Basement membranes. IV. Epithelial origin and

immunologic cross section. Am. J. Pathol, 45, 929 (1964).

285. Pignatelli, M., M.E.F. Smith, W.F. Bodmer: Low expression of collagen receptors in moderate and poorly differentiated colorectal adeno-carcinomas. Brit. J. Cancer, 61, 636 (1990).

286. Piver, M.S., S. Leile, J.J. Barlow: Preoperative and intraoperative evaluation in ovarian malignancy. Obstet. Gynecol. 48, 312 (1976).

287. Plewig, G.: Größe, Form und Anordnung verhornender Schleimhautzellen. Arch. Derm. Forsch. 242, 30 (1971).

288. Potenza, L., O. Pulido, G., Rojas: Zur Wirkung von D-Norgestrel auf die Struktur des menschlichen Endometriums. Med. Mitt. Schering 1, 20 (1972).

289. Pouchet, F.A.: Théorie positive de l'ovulation spontanée et de la fécondation des mammifères et de l'espece humaine. Basée sur l'observation de toute la série animale. J.B. Bailliière, Paris (1847).

290. Poulsen, H.E., C.W. Taylor, L.H. Sobin: Histological Typing of Female Genital Tract Tumours. World Health Organization, Geneva (1975).

291. Pundel, J.P.: Du rapport entre l'indice acidophilique et karyopycnotiques des frottis vaginaux en fonction de l'activité oestrogenique. Acta Clin. Belg. 5, 66 (1950).

292. Pundel, J.P., F. v. Meensel: Gestation et cytologie vaginale, 2nd ed. Masson, Paris 1966.

293. Pundel, J.P.: Les frottis endocriniens. Masson, Paris (1967).

294. Purola, E., E. Savia: Cytology of gynecologic condyloma acuminatum. Acta Cytol. 21, 26 (1977).

295. Quensel, U.: Zur Frage der Zytodiagnostik der Ergüsse seröser Höhlen. Methodologische und pathologisch-anatomische Bemerkungen. Acta med. Scand. 68, 427 (1928).

296. Ranney, B., J.T. Chung: Endometriosis of the cervix uteri. Am. J. Obstet. Gynecol. 64, 1333 (1952).

297. Rauscher, H.: Vergleichende Untersuchungen über das Verhalten des Vaginalabstriches, der Zervixfunktion und der Basaltemperatur in zweiphasischen Zyklen. Geburts. Frauenheilk. 14, 327 (1954).

298. Rauscher, H.: Ovulationszeit und Konzeptionsoptimum im Lichte vergleichender Untersuchungen von Basaltemperatur, Vaginalabstrich, Cervix, Endometrium und Ovar. Arch. Gynäkol. 189, 268 (1957).

299. Reagan, J.W.: The definition of a dyskaryotic cell. Acta Cytol. 1, 27 (1957).

300. Reagan, J.W.: Dysplasia of the uterine cervix. Acta Cytol. 1, 41 (1957).

301. Reagan, J.W.: The nature of the cells originating in so-called precancerous lesions of the uterine cervix. Obstet. Gynecol. Surv. 13157 (1958).

302. Reagan, J.W.: Can screening for endometrial cancer be justified? Acta Cytol. 24, 87 (1980).

303. Reagan, J.W., Y.S. Fu: Pathology of endometrial carcinoma. In: Coppleson, M. (eds.), Gynecologic Cytology, Fundamental Principles and Clinical Practice. Churchill Livingstone, London (1981).

304. Reagan, J.W., M.J. Hamonic: The cellular pathology in carcinoma in situ; a histopathological correlation. Cancer 9, 385 (1956).

305. Reagan, J.W., M.J. Hamonic: Dysplasia of the uterine cervix. Ann. N. Y. Acad. Sci. 63, 1235 (1956).

306. Reagan, J.W., M.J. Hamonic, W.B. Wentz: Analytical study of the cells in cervical squamous cell cancer. Lab. Invest. 6, 241 (1957).

307. Reagan, J.W., R.D. Moore: Morphology of the malignant squamous cell. Am. J. Pathol. 28, 105 (1952).

308. Reagan, J.W., A.B.P. Ng: The Cells of Uterine Adenocarcinoma, 2nd ed. Karger, Basel (1973).

309. Reagan, J.W., A.B.P. Ng: The cellular detection of cervical cancer. In: Wied, G.L., L.G. Koss, S.W. Reagan (eds.), Compendium on Diagnostic Cytology, 3rd ed. Tutorials of Cytology, Chicago (1974).

310. Reagan, J.W., I.L. Seidemann, Y. Saracusa: The cellular morphology of carcinoma in situ and dysplasia or atypical hyperplasia of the uterine cervix. Cancer 6, 224 (1953).

311. Richart, R.M.: The natural history of cervical intraepithelial neoplasia. Am. J. Obstet. Gynecol. 10, 748 (1967).

312. Richart, R.M., B. A. Barron: Follow-up study of patients with cervical dysplasia. Am. J. Obstet. Gynecol. 205, 386 (1969).

313. Riotton, G., W.M. Christopherson, R. Lunt: Cytology of the Female Genital Tract. World Health Organisation, Geneva (1973).

314. Ritzerfeld, W.: Bakeriologie, Virologie und Parasitologie. In: Käser, O., V. Friedberg, K.G. Ober, K. Thomsen, J. Zander (eds.), Gynäkologie und Geburtshilfe, vol. 3. Thieme, Stuttgart (1971).

315. Roberts, T.H., A.B.P. Ng: Chronic lymphocytic cervicitis: Cytologic and histopathologic manifestations. Acta Cytol. 19, 235 (1975).

316. Robra, B.P.: Das "gesetzliche" Krebsfrüherkennungsprogramm aus epidemiologischer Sicht. In: Das deutsche Krebsfrüherkennungsprogramm, Deutscher Ärzte Verlag, Cologne (1987).

317. Rollason, T.P., J. Cullimore, M.G. Bradgate: A suggested columnar cell morphological equivalent of squamous carcinoma in situ with early stromal invasion. Int. J. Gynecol. Pathol. 8, 230 (1989).

318. Rosenthal, D.L., D. Acosta, R.K. Peters: Computer-assisted rescreening of clinically important false negative cervical smears using the PapNet testing system. Acta Cytol. 40, 120 (1996).

319. Roth, O.A.: Über die klinische Brauchbarkeit des Vaginalsmears zur Diagnose und Prognose von Aborten. Gynaecologia 131, 19 (1951).

320. Roth, O.A.: Das Kolpopyknogramm als Kontrollmetholde der Follikelhormonwirkung. Zbl. Gynäkol. 74, 1489 (1952).

321. Rubio, C.A.: The false positive smear. Acta Cytol. 19, 212 (1975).

322. Rubio, C.A., A. Sigurdson, J. Zajicek: Viability tests in exfoliated cells from vaginal and oral epithelia. Acta Cytol. 17, 32 (1973).

323. Ruffing-Kullmann, B., W. Lehmacher, H.J. Soost: Validierung zytologischer Krebsvorsorgeuntersuchungen in der Gynäkologie: Prädiktive Werte, Spezifität, Sensitivität. Verh. Dtsch. Ges. Pathol. 71, 387 (1987).

324. Rupec, M.: Mikroskopische und elektronenmikroskopische Anatomie der Haut. In: Korting, G.W. (ed.), Dermatologie in Praxis und Klinik, Allgemeine Pathologie, vol. 1. Thieme, Stuttgart (1980).

325. Rutledge, F.: Cancer of the vagina. Am. J. Obstet. Gynecol. 97, 635 (1967).

326. Rutledge, F.: The role of radical hysterectomy in adenocarcinoma of the endometrium. Gynecol. Oncol. 2, 331 (1974).

327. Sadeghi, S.B., A. Sadeghi, S.J. Robboy: Prevalence of dysplasia and cancer of the cervix in a nationwide, planned parenthood population. Cancer 61, 2359 (1988).

328. Saigo, P.E., J.M. Cain, W.S. Kim, J.J. Gaynor, K. Johnson, J.L. Lewis: Prognostic factors in adenocarcinoma of the uterine cervix. Cancer 57, 1584 (1986).

329. Salmi, T.: Risk factors in endometrial carcinoma. Acta Obstet. Gynecol. Scand., Suppl. 86 (1979).

330. Sandritter, W., M. Carl, W. Ritter: Cytophotometric measurements of the DNA content of human malignant tumors by means of Feulgen reaction. Acta Cytol. 10, 26 (1966).

331. Sapp, M., U. Kraus, C. Volpers, P.J. Snijders, J.M. Walboomers, R.E. Streeck: Analysis of type-restricted and cross-reactive epitopes on virus-like particles of human papillomavirus type 33 and infected tissues using monoclonal antibodies to the major capsid protein. J. Gen. Virol. 75, 3375 (1994).

332. Sautter, T.: Transvaginalsonographie. Hippokrates, Stuttgart (1990).

333. Schellhas, H.F.: Cell renewal in the human cervix uteri. Am. J. Obstet. Gynecol. 104, 617 (1969).

334. Schenck, U.: Analyse des Screenvorgangs. 10. Fortbildungstagung für Klinische Zytologie, Munich, 3–9 December, vol. 115 (1989).

335. Schieck, R.: Schnellzytologie mittels Phasenkontrast- und Interferenzkontastmikroskopie in der gynäkologischen Sprechstunde. Zbl. Gynäkol. 101, 1208 (1979).

336. Schiffmann, M.H., N.B. Kiviat, R.D. Burk, et al.: Accuracy and interlaboratory reliability of human papillomavirus DNA testing by hybrid capture. J. Clin. Microbiol, 33, 545 (1995).

337. Schiller, W.: Zur klinischen Frühdiagnose des Portiocarcinoms. Zbl. Gynäkol. 1886 (1928).

338. Schirren, C.: Praktische Andrologie. Hartmann, Berlin (1971).

339. Schmidt-Matthiesen, H.: Die operative Behandlung des Vulvakarzinoms. In: Zander, J., J. Baltzer (eds.), Erkrankungen der Vulva. Urban & Schwarzenberg, Munich, Vienna, Baltimore (1986)

340. Schmitt, A.: Eine Gradeinteilung für die funktionelle Zytodiagnostik in der Gynäkologie. Geburtsh. Frauenheilk. 13, 593 (1953).

341. Schneider, A., G. Meinhardt, E.M. de Villiers, L. Gissmann: Sensitivity of the cytologic diagnosis of cervical condyloma in comparison with HPV-DNA hybridization studies. Diagn. Cytopathol. 3, 250 (1987).

342. Schneider, M.L.: Das Ektropium der Zervix, Klinisches Bild, Differentialdiagnosen und Beziehung zur Morphogenese des Zervixkarzinoms. In: Friedberg, V., K. Thomsen (eds.), Spezielle Gynäkologie 2, vol. II/2. Thieme, Stuttgart, New York (1988).

343. Schneider, V.: Cytology in pregnancy. In: Wied, G.L., C.M.Keebler, L.G. Koss, J.W. Reagan (eds.), Compendium on Diagnostic Cytology, 6th ed. Tutorial of Cytology, Chicago (1990).

344. Schöndorf, N.K.: Zytopathologie reaktiver Prozesse. Zytologische Fortbildungstagung, Stuttgart, 511 (1994).

345. Schreiber, H., H. Born: Zum konstruktiven Bau der menschlichen Vaginalwand. Gegenbaurs Morph. 88, 1 (1943).

346. Schulte, E.: Zytologische Kontrolluntersuchungen bei Ovulationshemmereinnahme. Zbl. Gynäkol. 93, 1363 (1971).

347. Scully, R.E.: Definition of endometrial carcinoma precursors. Clin. Obstet. Gynecol. 25, 39 (1982).

348. Scully, R.E., T.A. Bonfiglio, R.J. Kurman, S.G. Silverberg, E.J. Wilkinson: Histological Typing of Female Genital Tract Tumours. World Health Organization. International Histological Classification of Tumours, 2nd ed. Springer, Heidelberg, (1994).

349. Sen, D.K., H. Fox: The lymphoid tissue of the endometrium. Gynecol. (Basel) 163, 371 (1967).

350. Shackney, S.E., T.V. Shankey: Common patterns of genetic evolution in human solid tumors. Cytometry 29, 1 (1997).

351. Shettles, L.B.: Der Zervikalzyklus beim Menschen. Geburtsh. Frauenheilk. 12, 1 (1952).

352. Shiina, Y.: Cytomorphologic and immunocytochemical studies of chlamydial infections in cervical smears. Acta Cytol. 29, 683 (1985).

353. Shrago, S.S.: The Arias-Stella reaction: A case report of a cytologic presentation. Acta Cytol. 21, 310 (1977).

354. Shu, Y.J.: A Histological Study of Reserve Cell Hyperplasia. National Reference (1962).

355. Shu, Y.J., E. Gloor: Comprehensive Cancer Cytopathology of the Cervix Uteri. McGraw-Hill Book Company, New York (1995).

356. Siebert, G., O. Braun-Falco, G. Weber: Isolierung und chemische Charakterisierung von Hyalin aus. Spiegler'schen Tumoren. Naturwissenschaften 10, 300 (1955).

357. Singer, A.: Laser surgery in HPV infections of the female genital tract. In: Groß, G., S. Jablonska, H. Pfister, H.E. Stegner (eds.), Genital Papillomavirus Infections. Springer, Berlin, Heidelberg, New York (1990).

358. Smolka, H.: Über multinukleäre Pseudoriesenzellen in Vaginalabstrichen. Zbl. Gynäkol. 78, 1769 (1956).

359. Smolka, H.: Differentiation of endocervical and endometrial cells. Acta Cytol. 2, 515 (1958).

360. Smolka, H.: Cervicale Zellelemente im Vaginalinhalt und ihre differentialdiagnostische Bedeutung in der cytologischen Abstrichbeurteilung. Arch. Gynäkol. 195, 53 (1961).

361. Smolka, H., L. Kosch: Über zytologische Veränderungen am Vaginalepithel des Neugeborenen. Geburtsh. Frauenheilk. 14, 337 (1954).

362. Soost, H.J.: Statistische Ergebnisse der Krebsvorsorgeuntersuchungen bei Frauen. Fortschr. Med. 92, 360 (1974).

363. Soost, H.J.: Nomenklatur und Befundwiedergabe in der gynäkologischen Zytologie. Rundschr. Dtsch. Ges. Zytol. (1975).

364. Soost, H.J.: Maßnahmen zur Qualitätssicherung in der gynäkologischen Zytologie. In: Stark, G. (ed.), Problematik der Qualitätssicherung in der Gynäkologie. Demeter, Munich (1980).

365. Soost, H.J.: Befundwiedergabe in der gynäkologischen Zytodiagnostik—Münchner Nomenklatur II. Gynäkol. Prax. 14, 433 (1990).

366. Soost, H.J., B. Bockmühl: Effektivität zytologischer Krebsvorsorgeuntersuchungen in der Gynäkologie. Schriftenreihe des Zentralinstituts für die Kassenärztliche Versorgung in der Bundesrepublik Deutschland, vol. 13. Deutscher Ärzte Verlag, Cologne (1979).

367. Soost, H.J., H.J. Lange, W. Lehmacher, B. Ruffing-Kullmann: Ergebnisse zytologischer Krebsfrüherkennungs- und Vorsorgeuntersuchungen bei der Frau—Eine 10-Jahres-Studie. Wissenschaftliche Reihe des Zentralinstituts für die Kassenärztliche Versorgung, vol. 37, Deutscher Ärzte Verlag, Cologne (1987).

368. Soost, H.J., S. Baur: Gynäkologische Zytodiagnostik, Thieme, Stuttgart, New York (1990).

369. Southern E.M.: Detection of specific sequences among DNA fragments separated hy gel electrophoresis. J. Mol. Biol. 98, 503 (1975).

370. Spona, J., W. Schneider, K. Matt: Auswirkungen eines Antikonzeptivums auf die periphere und zentrale Zyklusfunktion. Med. Klin. 68, 505 (1973).

371. Sprenger, E., M. Hilgarth, M. Schaden: Die Zellkern-DNS-Bestimmung als Diagnostikhilfe bei Verlaufsbeobachtung unklarer zytologischer Befunde (a follow-up of doubtful findings in cervical cytology by Feulgen DNA cytophotometry). Beitr. Pathol. 152, 58 (1974).

372. Stafl, A., V. Dohnal, A. Linhartova: Kolposkopische, histologische und Gefäßbefunde an der krankhaft veränderten Portio. Geburtsh. Frauenheilk. 23, 437 (1963).

373. Stegner, H.E.: Zur Klassifikation virusbedingter Dysplasien der Cervix uteri in der zyto-histologischen Routinediagnostik. Gynäkologe 14, 252 (1981).

374. Stegner, H.E.: Geschwülste der Adnexe In: Kaeser O, Friedberg, V.: Spezielle Gynäkologie, vol. III/1. Thieme, Stuttgart, New York (1985).

375. Stegner, H.E., R. Beltermann: Die Elektronenmikroskopie des Cervixdrüsenepithels und der sog. Reservezellen. Arch. Gynäkol. 107, 480 (1969).

376. Stoll, P.: Die zytologische Methode der Karzinomdiagnostik nach Papanicolaou. Med. Klin. 44, 1046 (1949).

377. Stoll, P.: Zelluläre Differenzierungsstufen im Vaginalsekret und ihre Bedeutung für die gynäkologische Zytologie. Z. Geburtsh. Gynäkol. 141, 130 (1954).

378. Stoll, P.: Gynäkologische Vitalzytologie in der Praxis. Springer, Berlin (1969)

379. Stoll, P., J. Jaeger, G. Dallenbach-Hellweg: Gynäkologische Cytologie. Springer, Berlin, Heidelberg, New York (1968).

380. Swammerdam, J.: Biblia naturae sive historia insectorum in classes redacta. Leiden: J. Severinus and B. and P. van der Aa (1737).

381. Syrjänen, K.J., R. Mäntyjärvi, M. Väyrynen, M. Yliskoski, St.M. Syrjänen, S. Saarikoski, T. Nurmi, S. Parkkinen, O. Castrén: Cervical smears in assessment of the natural history of human papillomavirus infections in prospectively followed women. Acta Cytol. 31, 855 (1987).

382. Syrjänen, K.J., M. Hakama, S. Saarikoski: Prevalence, incidence, and estimated life-time risk of cervical human papillomavirus infections in a nonselected Finnish female population. Sex. Trans. Dis. 17, 15 (1990).

383. Szalay, L.: Zervix-Zytologie. Magyar Tavirati Iroda, Budapest (1987).

384. Takahashi, M.: Color Atlas of Cancer Cytology. Thieme, Stuttgart (1971).

384a. Takahashi, M., M., M. Kimura, A. Akagi, M. Naitoh: Autocyte screen interactive automated primary cytology screening system. Acta Cytol. 42, 185 (1998).

385. Teter, J.: The use of selected cytologic indices for evaluation of estrogenicity of synthetic compounds. Acta Cytol. 16, 366 (1972).

386. Timonen, S., O.P. Salo, B. Meyer, H,. Haapoja: Vaginal mycosis. Acta Obstet. Gynecol. Scand. 45, 232 (1966).

387. Tindle, R.W.: Vaccines for Human Papillomavirus Infection and Anogenital Disease, R.G. Landes Comp., Austin, Texas (1999).

388. Tischer, H., E. Schüller: Cytologie der Intravaginalbestrahlung. Zbl. Gynäkol. 75, 409 (1953).

389. Trimbos, J.B., N. F. Hacker: The case against aspirating ovarian cysts. Cancer 72, 828 (1993).

390. Tweeddale, D.N., L.D. Dubilier: Cytopathology of female genital tract neoplasms. Year Book Medical Publishers, Chicago (1972).

391. Ufer, J.: Hormontherapie in der Frauenheilkunde, 5th ed. Walter de Gruyter, Berlin, New York, (1978).

392. Us-Krasovec, M.: Drug-induced changes in malignant cells. 5th Intern. Congress Cytology, Miami Beach, FL, 29.5.–1.6. (1974).

392a. Villa, L.L., R.L. Costa, C.A. Petta et al.: Prophylactic quadrivalent human papillomavirus (types 6, 11, 16, and 18) L1 virus-like particle vaccine in young women: a randomised double-blind placebo-controlled multicentre phase II efficacy trial. Lancet Oncol. 6, 271–278 (2005).

393. de Villiers, E.M., D. Wagner, A. Schneider, H. Miklaw, E. Grussendorf-Conen, H. zur Hausen: A survey on the infection rate in a normal population with genital papillomaviruses. Abstract, Cold Spring Harbor Symposium on Papillomaviruses (1986)

394. Vincent, W.W., O.J. Miller, Jr.: International Symposium on the nucleolus, its structure and function. Nat. Cancer Inst. Monogr. 23, 1–535 (1966).

395. Virchow, R.: Cellularpathologie. Virchows Arch. Pathol. Anat. 8, 3 (1855).

396. Wagner, D., H. Ilkenberg, N. Boehm, L. Gissmann: Identification of human papillomavirus in cervical swabs by deoxyribonucleic acid in situ hybridization. Obstet. Gynecol. 64, 767 (1984).

397. Wagner, D., D. Krieger, N. Freudenberg: Die Geschichte der Deutschen Gesellschaft für Zytologie 1960–1993. Verh. Dtsch. Ges. Zytol. 18. Georg Fischer, Stuttgart (1993).

398. Wagner, D., E. Sprenger, M.H. Blank: DNA-content of dysplastic cells of the uterine cervix. Acta Cytol. 16, 517 (1972).

399. Wagner, D., E.M. de Villiers, L. Gissmann: Der Nachweis verschiedener Papillom-virustypen in zytologischen Abstrichen von Präkanzerosen und Karzinomen der Cervix uteri. Geburtsh. Frauenheilk. 45, 226 (1985).

400. Waschke, G.: Beitrag zum Äthylnortestosteron- und Äthinylnortesteron-Effekt bei verschiedenen Zyklusstörungen. Zbl. Gynäkol. 79, 1199 (1957).

401. de Watteville, H., L. Danon: L'influence des hormones genitales sur la biologie du vagin. J. Genève (1948).

402. Weiss, N.S., D.R. Szekely, D.F. Austin: Increasing incidence of endometrial cancer in the United States. New Engl. J. Med. 294, 1259 (1976).

403. Wells, M., L.J.R. Brown: Glandular lesions of the uterine cervix: The present state of our knowledge. Histopathol. 10, 777 (1986).

404. Wespi, H.J., Entstehung und Früherfassung des Portiocarcinoms. Benno-Schwabe, Basel (1946).

405. Wespi, H.J.: Die Rolle der Kolposkopie bei der Diagnose und beim Ausschluß des Zervixkarzinoms. Minerva Ginec. 22, 1148 (1970).

406. Wied, G.L.: Über die zytologische Karzinomdiagnostik. Zbl. Gynäkol. 75, 1028 (1950).

407. Wied, G.L.: An international agreement on histological terminology of lesions of the uterine cervix. Acta Cytol. 6, 235 (1962).

408. Wied, G.L.: The cytologic indices for hormonal assessment. In: Symposium on Hormonal Cytology. Acta Cytol. 12, 87 (1968).

409. Wied, G.L., W. Christiansen: Der bakterielle Einfluß auf den zytologischen Vaginalabstrich. Zbl. Bakt. I. Abt. Orig. 160, 413 (1953).

410. Wied, G.L., W. Christiansen: Die Zytolyse von Epithelien des Vaginalsekretes. Geburtsh. Frauenheilk. 13, 986 (1953).

411. Wied, G.L., W. Christiansen: Bedeutung und Einfluß der Bakterienflora im zytologischen Vaginalausstrich. Zbl. Bakteriol. 160, 413 (1953).

412. Wied, G.L., C.M. Keebler, L.G. Koss, J.W. Reagan: Compendium on Diagnostic Cytology, 6th ed. Tutorials of Cytology, Chicago (1990).

413. Wied, G.L., G. Legorreta, D. Mohr, A. Raunzy: Cytology of invasive cervical carcinoma and carcinoma in situ. Ann. N. Y. Acad. Sci. 97, 759 (1962).

413a. Winer, R.L., N.B. Kiviat, J.P. Hughes, D.E Adam, S.K. Lee, J.M. Kuypers, L.A. Koutsky: Development and duration of human papillomavirus lesions, after initial infection. J. Infect. Dis. 191, 731 (2005).

414. Wolinska, W.H., M.R. Melamed: Herpes genitalis in women attending planned parenthood of New York City. Acta Cytol. 14, 239 (1970).

415. Woodruff, J.D.: Vulvar atypia and carcinoma in situ. J. Reprod. Med. 17, 155 (1976).

416. Wunder, G.: Funktionszytologische Untersuchungen nach langjähriger Einnahme von oralen Kontrazeptiva. Fortschr. Med. 90, 1223 (1972).

416a. Yahr, L.J., R.L. Kenneth: Cytologic findings in microglandular hyperplasia of the cervix. Diagn. Cytopathol. 7, 248 (1991).

417. Young, R.H., R.E. Scully: Invasive adenocarcinoma and related tumors of the uterine cervix. Sem. Diagn. Pathol. 7, 205 (1990).

418. Zaharopoulos, P., J.Y. Wong, G. Edmonston, N. Keagy: Crystalline bodies in cervico-vaginal smears. A cytochemical and immunochemical study. Acta Cytol. 29, 1035 (1985).

419. Zander, J.: Sources of endogenous progesterone. Acta Cytol. 6, 211 (1962).

420. Zander J.: Ovarialkarzinom-Problemkarzinom. In: Zander, J. (ed.), Ovarialkarzinom, Urban & Schwarzenberg, Munich, Vienna, Baltimore (1982).

421. Zander, J., J. Baltzer (eds.): Erkrankungen der Vulva. Urban & Schwarzenberg, Munich, Vienna, Baltimore (1986).

422. Zerbini, M., S. Venturoli, M. Cricca, G. Gallinella, P. De Simone, S. Costa, D. Santini, M. Musiani: Distribution and viral load of type specific HPVs in different cervical lesions as detected by PCR-ELISA. J. Clin. Pathol. 54, 377 (2001).

423. Zinser, H.K.: Zytologische Karzinomdiagnostik mit dem Phasenkontrastverfahren. Zbl. Gynäkol. 71, 45 (1949).

424. Zinser, H.K.: Die Anwendung der Zytodiagnostik zur Früherkennung des Karzinoms. Zbl. Gynäkol. 72, 1863 (1950).

425. Zinser, H.K.: Kolposkopie und Phasenkontrastmikroskopie im Dienst der Krebsbekämpfung. Röntgen-Bl. 4, 198 (1951).

426. Zollinger, H.U.: Phasenmikroskopische Beobachtungen über Zelltod. Schweiz Z. Pathol. 11, 276 (1948).

427. Zucker, P.K., E.J. Kasdon, M.L. Feldstein: The validity of Pap smear parameters as predictors of endometrial pathology in menopausal women. Cancer 56, 2256 (1985).

428. Zur Hausen, H.: Human papillomaviruses and their possible role in squamous cell carcinomas. Curr. Top. Microbiol. Immunol. 78, 1 (1977).

429. Zur Hausen, H: Papillomaviruses causing cancer: evasion from host-cell control in early events in carcinogenesis. J. Natl. Cancer Inst. 92, 690 (2000).

Index

Page numbers in *italics* refer to illustrations or tables.

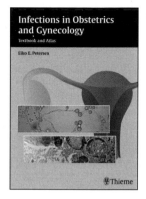